Morphological Methods in Image and Signal Processing

Morphological Methods in Image and Signal Processing

Charles R. Giardina
City University of New York

Edward R. Dougherty
Fairleigh Dickinson University

Prentice Hall, Englewood Cliffs, New Jersey 07632

Library of Congress Cataloging-in-Publication Data

GIARDINA, CHARLES ROBERT.
  Morphological methods in image and signal processing.

  Bibliography
  Includes index.
  1. Image processing. 2. Morphology. I. Dougherty, Edward R. II. Title.
  TA1632.G532   1987   621.36′7   87-2487
  ISBN 0-13-601295-7

Editorial/production supervision and interior design: Ellen B. Greenberg
Cover design: Edsal Enterprises
Manufacturing buyer: Barbara Kittle

©1988 by Prentice-Hall, Inc.
A Division of Simon & Schuster
Englewood Cliffs, New Jersey 07632

All rights reserved. No part of this book may be reproduced, in any form or by any means, without permission in writing from the publisher.

The following figures are being reproduced with the permission of Prentice-Hall, Inc. (Figure numbers refer to the *MATRIX STRUCTURED IMAGE PROCESSING*.) Giardina/Dougherty, ©1987:

Figures 1.1, 1.2, 1.4, 1.6, 1.7, 4.1, 4.3, 4.6, 4.11 through 4.16, 5.8, 5.13 through 5.15.

The following figures are being reproduced with the permission of Prentice-Hall, Inc. (Figure numbers refer to *IMAGE PROCESSING: CONTINUOUS TO DISCRETE—GEOMETRIC, TRANSFORM, AND STATISTICAL, VOL. I.*) Giardina/Dougherty, ©1987:

Figures 3.1, 3.2a, 3.2b, 3.3, 3.4a, 3.5, 3.6, 3.7, 3.8, 3.9, 3.11, 3.12, 3.14, 3.15, 3.16, 3.18, 3.20, 3.21 through 3.32, 3.34 through 3.37, 3.39 through 3.42, 3.49.

Printed in the United States of America

10 9 8 7 6 5 4 3 2 1

ISBN 0-13-601295-7 025

Prentice-Hall International (UK) Limited, *London*
Prentice-Hall of Australia Pty. Limited, *Sydney*
Prentice-Hall Canada Inc., *Toronto*
Prentice-Hall Hispanoamericana, S.A., *Mexico*
Prentice-Hall of India Private Limited, *New Delhi*
Prentice-Hall of Japan, Inc., *Tokyo*
Prentice-Hall of Southeast Asia Pte. Ltd., *Singapore*
Editora Prentice-Hall do Brasil, Ltda., *Rio de Janeiro*

To our parents

Dorothy Dougherty
Guy and Katherine Glandina

and in memory of
Russell and Ann Dougherty

Contents

PREFACE ix

1 MORPHOLOGY IN THE EUCLIDEAN PLANE — 1

1.1. The Morphological Approach 1
1.2. Fundamental Operations 3
1.3. Minkowski Algebra 8
1.4. Opening and Closing 20
1.5. Convex Sets 27

2 DIGITAL MORPHOLOGY — 36

2.1. The Digital Setting 36
2.2. Primitive Digital Morphological Operators 40
2.3. Structural Operators 45
2.4. Constant Images 48
2.5. Digital Minkowski Algebra 53
2.6. Digital Opening and Closing 66
2.7. Digitization 72

3 MORPHOLOGICAL FEATURES — 79

- 3.1. Quantitative Feature Generation — 79
- 3.2. Image Functionals — 81
- 3.3. Euclidean Image Modeling — 88
- 3.4. Integral Geometry and Image Functionals — 89
- 3.5. Granulometries — 97
- 3.6. Digital Size Distributions — 101
- 3.7. A Stochastic Approach — 106

4 TOPOLOGICAL PROCESSING — 111

- 4.1. Topological Preliminaries — 111
- 4.2. Region Processing — 116
- 4.3. Skeleton — 120
- 4.4. Hit and Miss Operator — 125

5 MORPHOLOGICAL FILTERS FOR TWO-VALUED IMAGES — 133

- 5.1. Increasing τ-mappings — 133
- 5.2. Digital Increasing τ-mappings — 137
- 5.3. Basis For the Kernel — 142
- 5.4. Algebraic Openings and Closings of Constant Images — 146
- 5.5. Euclidean Granulometries — 150

6 GRAY-SCALE MORPHOLOGY — 156

- 6.1. Gray-scale Morphological Operators For Euclidean Signals — 156
- 6.2. Umbra Transform — 174
- 6.3. Gray-scale Morphology For Sampled Signals — 181
- 6.4. Umbra Matrix — 191
- 6.5. Algebraic Properties — 198
- 6.6. Gray-scale Morphology For Euclidean Images — 208
- 6.7. Gray-scale Morphology For Digital Images — 216

7 GRAY-SCALE MORPHOLOGICAL FILTERS: THEORY — 225

- 7.1. Extended Signals — 225
- 7.2. Gray-scale Morphological Filters For Signals — 233
- 7.3. Basis For the Kernel — 242
- 7.4. Algebraic Openings Of Signals — 247
- 7.5. Gray-scale Morphological Filters For Images — 252

8 MORPHOLOGICAL FILTERS: APPLICATIONS — 255

- 8.1. Classical Filtering — 255
- 8.2. Morphological Filtering Operations — 265
- 8.3. Order-Statistic Filters — 274
- 8.4. Applications of the Basis Representation — 278

Appendix — 289

Preface

The morphological analysis of black-and-white images was initiated by Georges Matheron in the late 1960's. His early work culminated in the publication in 1975 of his classic treatise, *Random Sets and Integral Geometry*. At the outset, his approach was essentially statistical in nature, synthesizing the geometric probability utilized in stereology with the shape-oriented Minkowski algebra of Hans Hadwiger. Since 1975, the use of the fundamental morphological operations, absent of any significant statistical interpretation, has found an ever-growing field of application in the United States. One need only take account of the number of image-processing software packages and hardware peripherals that include morphological operations such as dilation and erosion.

The plan of the present book is to cover three fundamental areas of morphological analysis: (1) classical two-valued morphological (or Minkowski) algebra, (2) gray-scale morphology, and (3) morphological filtering of both images and signals. In all cases, we introduce the Euclidean (or analog) theory and then proceed to discuss digital implementation. Throughout, digital techniques are presented within the context of overall image algebra, and finite implementation is grounded upon bound matrix image (and signal) representation. The goal is not only to give a geometric theory, but also to provide readily understood algorithms that fit into the universal structure of image and signal processing.

The first two chapters of the book describe the fundamental morphological operations and develop the Minkowski algebra in both the Euclidean and digital settings. This material is the most basic, and it will provide the reader with a working knowledge of the operational capability of most commercially available morphological packages. Though proofs of theorems and properties are not included within the text proper, footnotes are provided which reference a complete theoretical development contained in the appendix.

Morphologically generated feature parameters are discussed in Chapter 3. Of particular importance are the granulometric size distributions of Matheron, and the closely related erosion induced distributions. Included is a detailed discussion of image functional properties and the celebrated Hadwiger theorem concerning such functionals, as well as a brief exposition of the relationship between integral-geometric parameters and the fundamental morphological operations.

Concluding the two-valued operational portion of the book is the topological processing of Chapter 4. Because of its geometric nature, morphological processing is well-suited to topological algorithms. Included in the discussion are morphological algorithms for the boundary, the closure, the pseudoconvex hull, the skeleton, and the filling of a simple closed curve.

The original Matheron theory regarding the morphological filtering of two-valued images is given in Chapter 5. Most notable is the Matheron Representation Theorem for morphological filters; indeed, the more contemporary material of Chapter 7 and most of the applications of Chapter 8 depend directly on this theorem. Also included are Matheron's representation theorems for tau-openings and Euclidean granulometries. Taken together, the three representation theorems of Matheron form the germ of morphological filtering methodology. Included in Chapter 5 is a discussion of the basis for the kernel of a morphological filter. It is the basis representation, not Matheron's original expansion, which makes the mathematical theory workable in the digital setting. Finally, as in Chapter 1, major theorems are footnoted to the rigorous theory given in the appendix.

The lengthiest chapter in the book is Chapter 6, which covers gray-scale morphology, a relatively new subject. Although there have been a number of gray-scale theories presented, we believe most to be fragmentary. In developing the theory as presented herein, we have been guided by four considerations:

(1) The original Hadwiger-Matheron two-valued theory must fall out as a special case of the gray-scale theory.
(2) In the digital setting, the gray-scale morphological algebra must appear as a subalgebra within image algebra as a whole.
(3) The Matheron representation theorems for morphological filters and tau-openings must generalize to the gray-scale setting.
(4) Just as the basis representation provides a digitally implementable version of the basic Matheron Representation Theorem, so too must the gray-scale version of that theorem possess a workable basis-generated form.

The theory of Chapters 6 and 7 satisfies all four requirements.

In Chapter 6, proofs are properly included within the text, and this approach is continued in Chapter 7. The change of style was motivated by the newness of the theory, and our concomitant desire to present the theory in a unified manner.

The theory of gray-scale morphological filters is developed in Chapter 7 within the setting of extended signals. Both the Euclidean and digital theories are included, the major theorems being derived via the original Matheron two-valued theory by use of the umbra transform. The basis representation is studied in detail in the digital setting, and the problem of boundary conditions is addressed.

The last chapter relates morphological filtering to classical engineering-type filtering. It is shown that some classes of customary filters, such as order-statistic filters and some moving-average filters, can be specified exactly within the context of the morphological basis, while others can be approximated morphologically. Due to the preservation of perceptually relevant geometric content within an image or signal when either is acted upon by a morphological filter, it may very well be that the morphological simulation of classical filters for both image and signal processing is the most promising area for future morphological research.

Pedagogically, the text possesses a modular structure so that one might proceed through it in a manner suitable to his or her own interests. Chapters 1, 2, and 6 provide a short course on the fundamental operations of two-valued and gray-scale morphology. Chapters 1, 2, 5, 6, 7, and 8 yield a course on operations and filtering. Chapters 1, 2, 3, 4, and 6 constitute a course on morphological image and signal processing, with the emphasis on processing for enhancement, compression, and signature generation.

The manner of presentation has been designed to make the material available to the widest possible audience in engineering, computer science, mathematics, biology, and the physical sciences. Like most areas of image processing, morphology spans a wide spectrum of application. The text is essentially self-contained and, together with the appendix, is for the most part theoretically complete. Throughout, we have made an effort to include the theory in such a manner as not to interfere with someone interested solely in applications. Moreover, there are a large number of examples and graphical illustrations which provide the nontheoretically-oriented student with a sound, functionally operative grounding in the subject.

It is our hope that the present text will bring to the general scientific audience the most significant aspects of morphological analysis as it exists today. We believe the material herein is both timely and relevant to the needs of current technology; indeed, a good deal of it has already found its way into both industrial and military applications.

Edward R. Dougherty

Charles R. Giardina

Morphological Methods in Image and Signal Processing

1

Morphology in the Euclidean Plane

1.1 THE MORPHOLOGICAL APPROACH

From a general scientific perspective, the word *morphology* refers to the study of form and structure. The term is used in this sense in biology, geography, and linguistics. In image processing, morphology is the name of a specific methodology originated by G. Matheron in his study of porous materials.[1] The term is appropriate because Matheron based his image analysis on the geometric structure inherent within an image.

The morphological approach is generally based upon the analysis of a two-valued image in terms of some predetermined geometric shape known as a *structuring element*. Essentially, the manner in which the structuring element fits into the image is studied. In this chapter, we discuss two-valued Euclidean images, which are subsets of the Euclidean plane.

Consider the image S sketched in Figure 1.1. From the figure, a square of the size shown will fit into S if its center is placed at $(1, 2)$, but not if its center is situated at $(3, 1)$. It should be clear that the manner in which the square fits into S as it is moved about the plane is a reflection of the relationship between the geometric structure of S and that of the square.

Morphological operations can be employed for many purposes, including edge detection, segmentation, and enhancement of images. From the underlying mor-

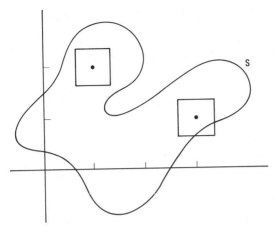

Figure 1.1 Image with structuring elements

phological operations, an entire class of *morphological filters* can be constructed that can often be used in place of the standard linear filters. Whereas linear filters sometimes distort the underlying geometric form of an image, morphological filters leave much of that form intact. Finally, many useful feature parameters can be generated morphologically.

Some of the salient points regarding the morphological approach are as follows:

1. Morphological operations provide for the systematic alteration of the geometric content of an image while maintaining the stability of important geometric characteristics.
2. There exists a well-developed morphological algebra that can be employed for representation and optimization.
3. It is possible to express digital algorithms in terms of a very small class of primitive morphological operations.
4. There exist rigorous representation theorems by means of which one can obtain the expression of morphological filters in terms of the primitive morphological operations.

The import of the preceding properties will become clear as the text unfolds.

The purpose of the present chapter is to develop the basic geometric and algebraic characteristics of the primitive morphological operators in the Euclidean setting. Although the actual implementation of these operators will be in the digital setting, the Euclidean model is essential to the development of an understanding of and intuitive feel for how the operators function in both theory and application. The relationship between Euclidean morphology and digital morphology is akin to that existing, in general, between continuous signal processing and digital signal processing. In the present case, it is the Euclidean model of human visual perception that must support the geometric content of the subject matter. Indeed, one of the most useful areas of morphological analysis is the generation of feature parameters for use in artificial intelligence schemes. The ability of the morphological operations to generate a large class of perceptually intuitive parameters gives the method its power.

1.2 FUNDAMENTAL OPERATIONS

In this section, we introduce the fundamental morphological operations upon which the entire subsequent development depends. As mentioned in the previous section, an image will be a subset of the Euclidean plane R^2. Gray-level morphology, which deals with many-valued images, will be introduced later.

Besides dealing with the usual set-theoretic operations of union and intersection, morphology depends extensively on the translation operation. Given an image (subset) A in R^2, the *translation* of A by the point x in R^2 is defined by

$$A + x = \{a + x : a \in A\}$$

where the plus sign inside the set notation refers to vector addition. Considering x to be a vector in the plane, $A + x$ is A translated along the vector x. This can be seen pictorially in Figure 1.2; the following example illustrates the definition of translation arithmetically.

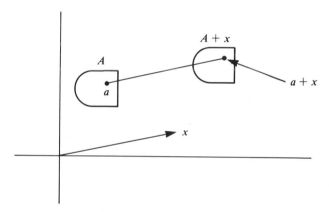

Figure 1.2 Translation of Euclidean image

Example 1.1:

Let $A = \{(0, 0), (1, 0), (0, 1), (1, 1), (2, 2)\}$ and $x = (3, 1)$. Then $A + x = \{(3, 1), (4, 1), (3, 2), (4, 2), (5, 3)\}$. (See Figure 1.3.)

Note that the point z is in the translated set $A + x$ if and only if there exists some point a' in A such that $z = a' + x$. Also, because vector addition is commutative, we can write $x + A$ interchangeably with $A + x$.

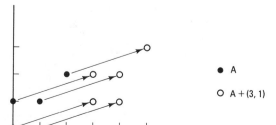

Figure 1.3 Translation of a discrete image

Two fundamental operations are utilized in the morphological analysis of two-valued images. We first consider *Minkowski addition*. Given two images A and B in R^2, we define the Minkowski sum $A \oplus B$ set-theoretically as

$$A \oplus B = \bigcup_{b \in B} A + b$$

$A \oplus B$ is constructed by translating A by each element of B and then taking the union of all the resulting translates.

Example 1.2:

Let A be the unit disk centered at $(2, 2)$ and let $B = \{(4, 1), (5, 1), (5, 2)\}$. Then $A \oplus B$ is the union of the sets $A + (4, 1)$, $A + (5, 1)$, and $A + (5, 2)$. A, B, and $A \oplus B$ are depicted in Figure 1.4.

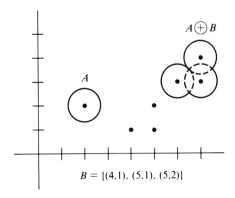

$A \oplus B = [A + (4,1)] \cup [A + (5,1)] \cup [A + (5,2)]$ **Figure 1.4** Minkowski addition

Example 1.3:

This time, let A be the unit disk centered at $a = (\frac{3}{2}, 3)$ and let B be the closed line segment running from $b_1 = (\frac{5}{2}, 1)$ to $b_2 = (\frac{9}{2}, 2)$. Then $A + b_1$ is the unit disk centered at $(4, 4)$ and $A + b_2$ is the unit disk centered at $(6, 5)$. The Minkowski addition $A \oplus B$, depicted in Figure 1.5, consists of the union of all unit disks having centers on the line segment running from $a + b_1$ to $a + b_2$.

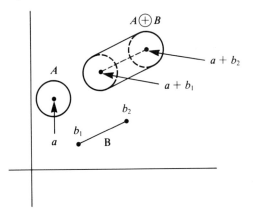

Figure 1.5 Minkowski addition

Sec. 1.2   Fundamental Operations

Some immediate properties of Minkowski addition are:

1. $A + \{(0, 0)\} = A$
2. $A + \{x\} = A + x$ for any point x in R^2.

The important algebraic properties of Minkowski addition will be given in Section 1.3.

The second fundamental morphological operation is *Minkowski subtraction*. Given images A and B in R^2, we define the Minkowski difference

$$A \ominus B = \bigcap_{b \in B} A + b$$

In this operation, A is translated by every element of B and then the intersection is taken.

Example 1.4:

Consider the 3 by 2 rectangle A in Figure 1.6. Let $B = \{(4, 0), (5, 1)\}$. Then $A \ominus B$ is the intersection of the translates $A + (4, 0)$ and $A + (5, 1)$. That is, $A \ominus B$ is the 2 by 1 rectangle depicted in Figure 1.6.

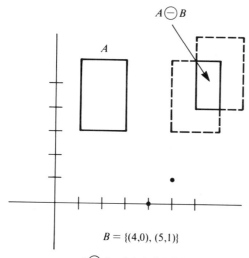

$B = \{(4,0), (5,1)\}$

$A \ominus B = [A + (4,0)] \cap [A + (5,1)]$ **Figure 1.6**   Minkowski subtraction

Example 1.5:

Let A be the rectangle given in Figure 1.7 and let B be the unit segment emanating at the origin and making a 45° angle with the horizontal axis. Then one end point of B is $(0, 0)$ and the other is $(\sqrt{2}/2, \sqrt{2}/2)$. $A \ominus B$ is the intersection of all the translates $A + b$ such that b lies on the segment B. This intersection is the shaded region in the figure.

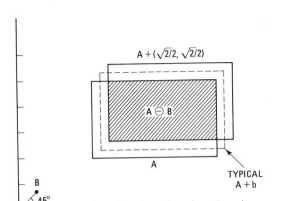

Figure 1.7 Minkowski subtraction of line segment from rectangle

Before proceeding to the general algebraic properties of Minkowski addition and subtraction, an alternative characterization of Minkowski subtraction will be presented. This new characterization, which will prove to be useful throughout the remainder of the text, will be stated as a theorem. Note the use of the notation $-B = \{-b: b \in B\}$, where $-b$ is the scalar mutliple of the vector b by -1. Thus, $-B$ is simply B rotated 180° around the origin. (See Figure 1.8.)

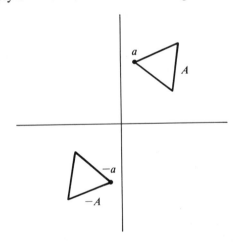

Figure 1.8 Reflection operation

Theorem 1.1.[2] $A \ominus B = \{x: -B + x \subset A\}$

According to Theorem 1.1, $A \ominus B$ can be found by first rotating B 180° around the origin and then finding all points x such that the translate by x of that rotated image is a subimage (subset) of A. Figure 1.9 demonstrates this "fitting" procedure. Note that the output image $A \ominus B$ is *not* necessarily a subimage of the original image A; we are assured that $A \ominus B$ is a subimage of A only if B contains the origin.

Since $-(-B) = B$, Theorem 1.1 can be rewritten in the form

$$A \ominus (-B) = \{x: B + x \subset A\}$$

In other words, the direct fitting of B without rotating it by 180° is the Minkowski

Sec. 1.2 Fundamental Operations

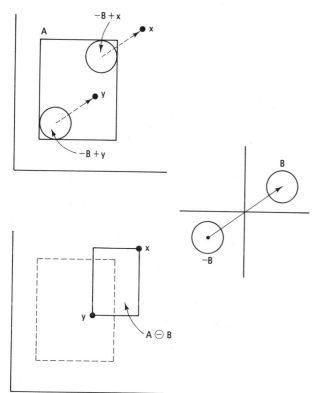

Figure 1.9 Minkowski subtraction by fitting

subtraction of A by $-B$. Using this direct fitting technique, we define the *erosion* of A by B to be $\mathcal{E}(A, B) = A \ominus (-B)$. When A is eroded by B, the latter is called a *structuring element*. If $B = -B$, the erosion is equal to the Minkowski subtraction; otherwise the two are related by the preceding definition.

Eroding an image by a structuring element B has the effect of "shrinking" the image in a manner determined by B. In Figure 1.10, B is a closed disk centered at the origin. As such, it is symmetric, and hence $\mathcal{E}(A, B) = A \ominus B$. In any event, notice the manner in which erosion by the circular surface of the disk has shrunk the original image. This effect can also be seen in Figure 1.7, where $A \ominus B = \mathcal{E}(A, -B)$.

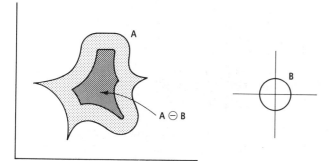

Figure 1.10 Erosion as shrinking

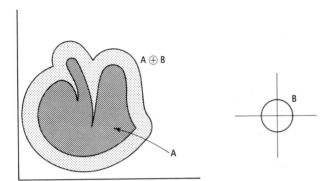

Figure 1.11 Dilation as expansion

Corresponding to the erosion operation is the operation of *dilation*, which is defined simply as Minkowski addition: $\mathcal{D}(A, B) = A \oplus B$. Here again, B is called a structuring element. Dilation has the effect of "expanding" an image. (See Figure 1.11.)

Minkowski addition and subtraction are the basic morphological transformations. Their importance and their extensive capability of representing image processing alogrithms will become evident in the sections that follow. The next section discusses their numerous useful algebraic properties.

1.3 MINKOWSKI ALGEBRA

In this section, the algebraic properties of Minkowski addition and subtraction defined on images in the Euclidean plane will be presented. For the purposes of algebra, $\oplus$ and $\ominus$ are binary (two input) operations with image inputs.

To begin with, note that we have the operational schemes

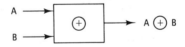

and

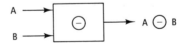

In Chapter 2, such block diagrams are used to discuss the digital implementation of Minkowski algebra. Here, it should simply be recognized that the algebraic properties to be developed have block specifications as operational analogues.

We now proceed to the principal properties of Minkowski algebra.

Property M-1:[3] $A \oplus B = B \oplus A$ (commutativity)

The commutative property states that forming the union of translates of A by elements of B is equivalent to forming the union of translates of B by elements of A. In terms of block specifications, it does not matter which input image is listed first.

Sec. 1.3 Minkowski Algebra

Property M-2:[4] $A \oplus (B \oplus C) = (A \oplus B) \oplus C$ (associativity)

The associative law allows us to form the Minkowski sum of any finite number of input images by applying the binary operation $\oplus$. It allows us to write $A \oplus B \oplus C$ without worrying about which $\oplus$ is performed first. For instance, the Minkowski addition $A \oplus B \oplus C \oplus D$ can be implemented by $A \oplus [B \oplus (C \oplus D)]$ or by $(A \oplus B) \oplus (C \oplus D)$. For those familiar with logic design, the latter formulation is superior since in the digital setting there can be parallel implementation. Indeed, we have the block specification

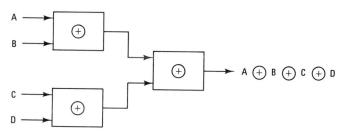

Example 1.6:

Let $A = \{(1, 1), (1, 2), (2, 1)\}$, $B = \{(0, 0), (1, 3)\}$ and $C = \{(2, 1), (2, 2)\}$. Then

$$A \oplus B = \{(1, 1), (1, 2), (2, 1), (2, 4), (2, 5), (3, 4)\}$$

and

$$(A \oplus B) \oplus C = \{(3, 2), (3, 3), (4, 2), (4, 5), (4, 6), (5, 5),$$
$$(3, 4), (4, 3), (4, 7), (5, 6)\}$$

On the other hand,

$$B \oplus C = \{(2, 1), (2, 2), (3, 4), (3, 5)\}$$

and

$$A \oplus (B \oplus C) = \{(3, 2), (3, 3), (4, 5), (4, 6), (3, 4), (4, 7),$$
$$(4, 2), (4, 3), (5, 5), (5, 6)\}$$

By comparison, $(A \oplus B) \oplus C = A \oplus (B \oplus C)$.

The next property gives the translation invariance of Minkowski addition. Put simply, translation followed by Minkowski addition is equivalent to Minkowski addition followed by translation. Figure 1.12 gives an illustration of the property.

Property M-3: $A \oplus (B + x) = (A \oplus B) + x$

In fact, M-3 is a special case of M-2. To see this, recall that $A + x = A + \{x\}$. Hence, by associativity (i.e., M-2),

$$A \oplus (B + x) = A \oplus (B \oplus \{x\}) = (A \oplus B) \oplus \{x\} = (A \oplus B) + x$$

Before considering further properties, we need to define the *scalar multiplication* of a set by a real number. For any real number t and any set A in R^2, we define

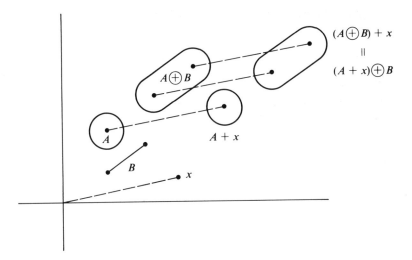

Figure 1.12 Translational invariance of Minkowski addition

$tA = \{ta: a \in A\}$. For instance, if A is the unit disk centered at the origin, then $3A$ is the disk of radius 3 centered at the origin. Similarly, if $B = \{(3, 4), (-1, 0), (2, \frac{1}{3})\}$, then $2B = \{(6, 8), (-2, 0), (4, \frac{2}{3})\}$. Finally, note that $(-1)A = -A$.

Although we have treated Minkowski addition and Minkowski subtraction as equally fundamental, each can actually be obtained from the other by use of the complement operation $A^c = \{x: x \notin A\}$. The following identities are known as the *duality* properties:

Property M-4:[5] $A \oplus B = [A^c \ominus B]^c$

Property M-5:[6] $A \ominus B = [A^c \oplus B]^c$

From M-4, and the fact that successive complementations yield the original image, it follows that

$$[A \oplus B]^c = A^c \ominus B$$

This relation is illustrated in Figure 1.13. Similarly, from M-5, it is immediate that

$$[A \ominus B]^c = A^c \oplus B$$

In terms of dilation and erosion, the last two relations become, respectively,

Property M-6: $[\mathcal{D}(A, B)]^c = \mathcal{E}(A^c, -B)$

Property M-7: $[\mathcal{E}(A, B)]^c = \mathcal{D}(A^c, -B)$

In other words, dilating an image can be accomplished by eroding the complementary image, while eroding an image can be accomplished by dilating the complementary image.

Like Minkowski addition, Minkowski subtraction is invariant with respect to translation. However, unlike property M-3, which could be stated in terms of only one of the inputs because of the commutativity of $\oplus$, the statement of the invariance of

Sec. 1.3 Minkowski Algebra

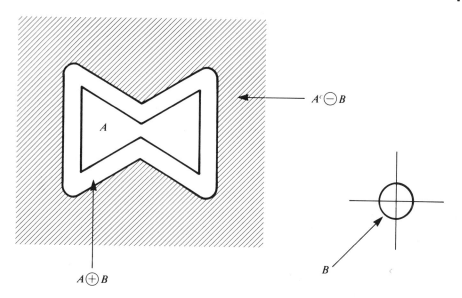

Figure 1.13 Illustration of duality

Minkowski subtraction requires two equalities since $\ominus$ is not commutative:

Property M-8:[7] $A \ominus (B + x) = (A + x) \ominus B = (A \ominus B) + x$

Example 1.7:

Consider the image A of horizontal edge length 4 and the square B of edge length 2 shown in Figure 1.14. Using the fitting characterization of erosion given in Theorem 1.1, we construct the images $A \ominus B$, $A \ominus (B + x)$, and $(A + x) \ominus B$. Clearly, property M-8 holds in this instance.

Careful attention should be paid to M-8 when employing erosion. Since

$$-(B + x) = -B - x$$

we have

$$\mathscr{E}(A, B + x) = A \ominus [-(B + x)] = A \ominus [(-B) + (-x)]$$
$$= [A \ominus (-B)] + (-x) = \mathscr{E}(A, B) - x$$

and

$$\mathscr{E}(A + x, B) = (A + x) \ominus (-B)$$
$$= [A \ominus (-B)] + x = \mathscr{E}(A, B) + x$$

Consequently, the double equality given in M-8 for Minkowski subtraction becomes two different equalities when stated in terms of erosion.

An operation on images in the plane, say $P(A)$, is said to be *increasing* if, whenever A is a subimage of B, then $P(A)$ is a subimage of $P(B)$. P is said to be

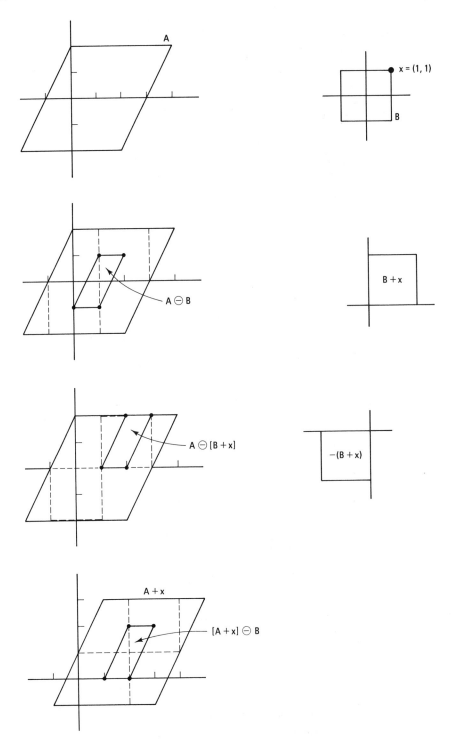

Figure 1.14 Translation invariance for Minkowski subtraction

Sec. 1.3 Minkowski Algebra

decreasing if $A \subset B$ implies $P(A) \supset P(B)$. Increasing set operations play a significant role in morphological analysis, especially as regards morphological filters. The following properties state, respectively, that for a fixed image B, both dilation, $\mathcal{D}(\cdot, B)$, and erosion, $\mathcal{E}(\cdot, B)$, are image-to-image mappings that are increasing in the first variable. That is to say, dilating or eroding a larger image by a fixed image yields a larger output image, in accordance with our intuitions.

Property M-9:[8] If B is fixed and $A_1 \subset A_2$, then

$$\mathcal{D}(A_1, B) \subset \mathcal{D}(A_2, B)$$

Property M-10:[9] If B is fixed and $A_1 \subset A_2$, then

$$\mathcal{E}(A_1, B) \subset \mathcal{E}(A_2, B)$$

The next property states that erosion is decreasing in the second variable. It should be clear that eroding a fixed image by a smaller image will produce a greater output than eroding that same image by a larger image, because the smaller structuring element fits more readily.

Property M-11:[10] If A is fixed and $B_1 \subset B_2$, then

$$\mathcal{E}(A, B_1) \supset \mathcal{E}(A, B_2)$$

Property M-11 is illustrated in Figure 1.15.

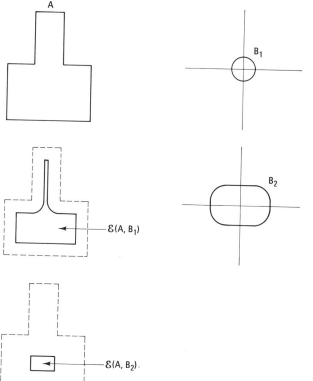

Figure 1.15 Decreasing monotonicity in second variable for erosion

The next two properties concern a type of *distributivity* satisfied by both Minkowski addition and Minkowski subtraction.

Property M-12:[11]   For any real number t, $t[A \oplus B] = tA \oplus tB$

Property M-13:[12]   For any real number t, $t[A \ominus B] = tA \ominus tB$

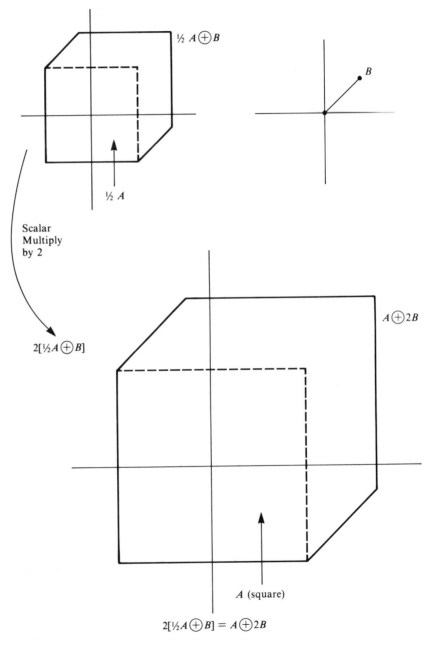

Figure 1.16   Illustration of Property M-12

Sec. 1.3 Minkowski Algebra

Both of these properties deal with the ability to "multiply through" by a scalar. For instance, by M-12, $4[(\frac{1}{4})A \oplus B] = A \oplus 4B$. We might say that scalar multiplication distributes over both Minkowski addition and Minkowski subtraction. Figure 1.16 illustrates the geometry in the case of dilation.

Corresponding to the structuring-element fitting characterization of Minkowski subtraction (Theorem 1.1) is a dual property concerning Minkowski addition: $A \oplus B$ is the set of all points x such that the translate of $-B$ by x intersects A. Formally, we have

Property M-14:[13] $A \oplus B = \{x: (-B + x) \cap A \neq \emptyset\}$

Property M-14 is illustrated in Figure 1.17.

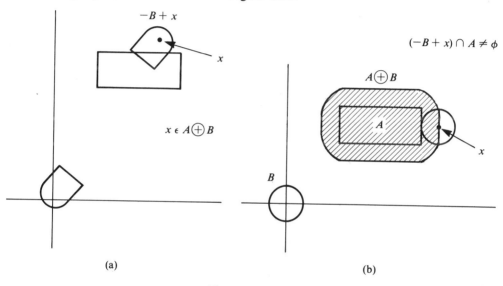

Figure 1.17 Dilation operation

In combination with union and intersection, $\oplus$ and $\ominus$ satisfy many algebraic relations. For instance, $\oplus$ distributes over union from both the right and the left:

Property M-15:[14] (i) $A \oplus (B \cup C) = (A \oplus B) \cup (A \oplus C)$

(ii) $(B \cup C) \oplus A = (B \oplus A) \cup (C \oplus A)$

Schematically, (i) states that the following two diagrams produce the same output:

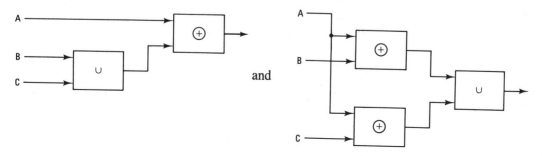

It will be seen in Chapter 2 that M-15 holds in the case of digital images as well. Consequently, one can optimize the implementation of digital algorithms by the use of diagrams such as the preceding. In terms of complexity, the first block specification requires a single dilation and a single union, whereas the second requires two dilations and a single union.

Example 1.8:

Let A be a unit square centered at $(\frac{1}{2}, \frac{1}{2})$, $B = \{(0, 1), (1, 1)\}$, and $C = \{(1, 0), (1, 1)\}$. Then $B \cup C = \{(0, 1), (1, 0), (1, 1)\}$, and $A \oplus (B \cup C)$ is the union of the translates $A + (0, 1)$, $A + (1, 0)$, and $A + (1, 1)$. The distributivity of dilation by A over the union of B with C is illustrated in Figure 1.18.

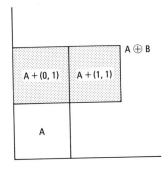

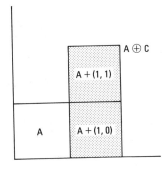

Figure 1.18 Distributivity of Minkowski addition over union

It should be noted that Minkowski addition does not distribute over intersection; indeed, the most that can be said is that

$$A \oplus (B \cap C) \subset (A \oplus B) \cap (A \oplus C)$$

and

$$(B \cap C) \oplus A \subset (B \oplus A) \cap (C \oplus A)$$

Insofar as the dual operation is concerned, Minkowski subtraction satisfies a sort of left *antidistributivity* over union. First we have

Property M-16:[15] $\quad A \ominus (B \cup C) = (A \ominus B) \cap (A \ominus C)$

Then, in terms of erosion, M-16 becomes

$$\mathscr{E}[A, B \cup C] = A \ominus [(-B) \cup (-C)] = [A \ominus (-B)] \cap [A \ominus (-C)]$$
$$= \mathscr{E}(A, B) \cap \mathscr{E}(A, C).$$

Thus, eroding by the union is equivalent to eroding by each structuring element independently and then intersecting the two outputs. On the other hand, from M-15, dilating by the union is equivalent to dilating separately and then taking the union of the resulting outputs. In terms of block diagrams,

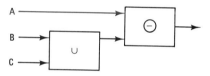

yields the same output as

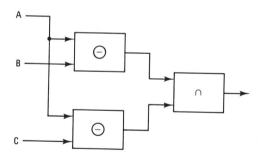

Corresponding to property M-16 is a right distributivity of Minkowski subtraction over intersection:

Property M-17:[16] $\quad (B \cap C) \ominus A = (B \ominus A) \cap (C \ominus A)$

By M-17, it follows that eroding the intersection of two sets by a given structuring element produces the same output as first eroding each set separately and then intersecting the results. (See Figure 1.19.)

The last property to be presented in this section concerns an algebraic relationship between dilation and erosion that has proven to have significant practical value.

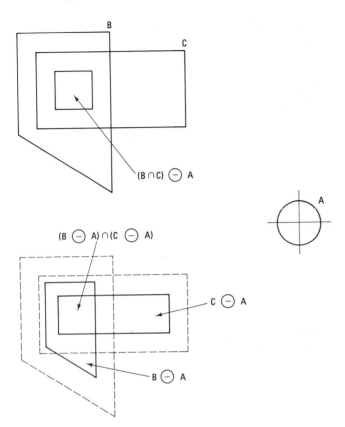

Figure 1.19 Right distributivity of Minkowski subtraction over intersection

Property M-18:[17] $(A \ominus B) \ominus C = A \ominus (B \oplus C)$

Diagrammatically, M-18 says that

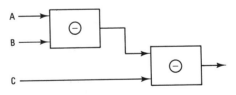

is equivalent to

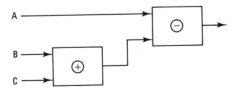

It is easy to see the use to which M-18 can be put. Suppose we wish to erode by a disk of radius 2 centered at the origin. If B is the unit disk centered at the origin, then the scalar multiple rB is a disk centered at the origin having radius r. For any

Sec. 1.3 Minkowski Algebra   19

image A, it is possible to find $A \ominus 2B$ by a succession of erosions by B. Since $2B = B \oplus B$, M-18 yields

$$A \ominus 2B = A \ominus (B \oplus B) = (A \ominus B) \ominus B$$

An example of the technique is given in Figure 1.20. Note that $-B = B$ implies that erosion and Minkowski subtraction are identical.

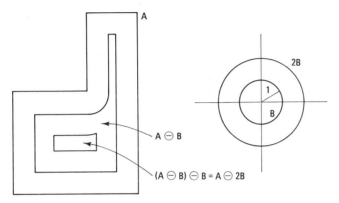

Figure 1.20 Illustration of Property M-18

Two points are noteworthy here. First, the preceding example employed a disk as a structuring element. But we could just as well have used a square centered at the origin; the only requirement is that $2B = B \oplus B$. Second, the scheme is iterative. Thus, we have

$$A \ominus 3B = A \ominus (B \oplus B \oplus B) = (A \ominus B) \ominus (B \oplus B)$$
$$= [(A \ominus B) \ominus B] \ominus B$$

and higher integer scalar multiples of B can be subtracted by longer iterations.

Before leaving this section, let us note the importance of algebra in image analysis. Image processing algorithms consist of algebraic expressions whose terms consist of more primitive expressions. The ability to manipulate algebraic strings of component operations is crucial to both the construction of successful algorithms and the optimal implementation of those algorithms into code. Though we shall not go into details in the present text, Minkowski algebra is an important subalgebra of a more general image algebra that is structurally induced by the primitive mathematical operations upon which general image processing algorithms are founded.[18] Those image processing algorithms which employ morphological techniques form a sub-collection of image processing algorithms in general. Specifically, the morphological subcollection is precisely defined by the primitive operations that constitute the generating terms of the Minkowski subalgebra. A close look at what has been presented in this section reveals the use of unions, intersections, and translations—operations in terms of which the fundamental morphological operations of Minkowski addition and Minkowski subtraction were defined. Further properties were developed by employing set-theoretic complementation and set-theoretic scalar multiplication of subsets in the plane. The manner in which these elemental operations fit into the overall structure of image algebra defines exactly the role of morphology within the context of image processing. Since our main interest is electronic digital processing, we shall be careful to specify a precise *morphological basis* for the digital operators to be presented in Chapter 2.

1.4 OPENING AND CLOSING

Besides the fundamental morphological operations of Minkowski addition and subtraction, two other operations play a central role in image analysis. These two secondary operations, called the *opening* and the *closing*, are respectively defined by

$$O(A, B) = [A \ominus (-B)] \oplus B$$

and

$$C(A, B) = [A \oplus (-B)] \ominus B$$

Employing erosion and dilation terminology, we have

$$O(A, B) = \mathcal{D}[\mathcal{E}(A, B), B]$$

and

$$C(A, B) = \mathcal{E}[\mathcal{D}(A, -B), -B]$$

The opening and the closing are illustrated in Figures 1.21 and 1.22, respectively. In both figures, one should pay particular attention to the composite manner in which the final output is obtained. Moreover, note that in both cases $-B = B$.

A close examination of Figure 1.21 reveals that the original image has been *smoothed*. That is, the output $O(A, B)$ is similar to the input except that the corners

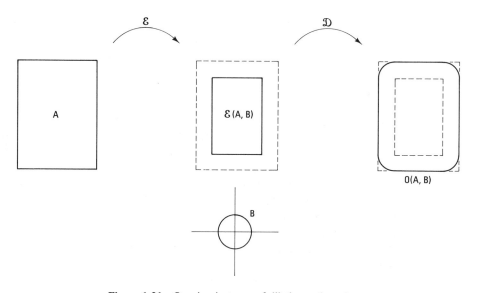

Figure 1.21 Opening in terms of dilation and erosion

Sec. 1.4   Opening and Closing   21

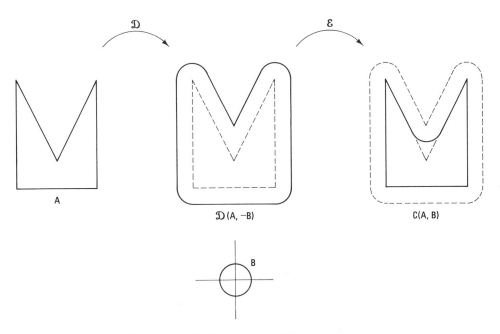

Figure 1.22   Closing in terms of dilation and erosion

have been rounded from the inside. This smoothing effect is a result of the definition of the opening together with the shape of the structuring element. Intuitively, $O(A, B)$ is obtained from A in this instance by "rolling the ball" B about the inside of the image. The next theorem gives a rigorous set-theoretic characterization of this "fitting" property. It states that the opening of A by B is obtained by taking the union of all translates of B which fit into A. Figure 1.23 gives an illustration of the result when the structuring element is not symmetric with respect to the origin.

Theorem 1.2.[19] $O(A, B) = \cup \{B + x : B + x \subset A\}$.

The output $O(A, B)$ in Figure 1.21 appears to be the result of rolling B about the inside of A. On the other hand, the output $C(A, B)$ in Figure 1.22 appears to result from rolling B around the outside of A. This inside–outside *duality* between the opening and the closing is formalized in property M-19, which states that the complement of the closing is equal to the opening of the complement.

Property M-19:[20] $C(A, B)^c = O(A^c, B)$

Insofar as the situation in Figure 1.22 is concerned, taking complements in M-19 gives

$$C(A, B) = O(A^c, B)^c$$

Together with Theorem 1.2, this says that the closing can be obtained (in the case of a disk B) by rolling the ball (really, the surface of B) about the outside of A—in other words, open A^c and then complement.

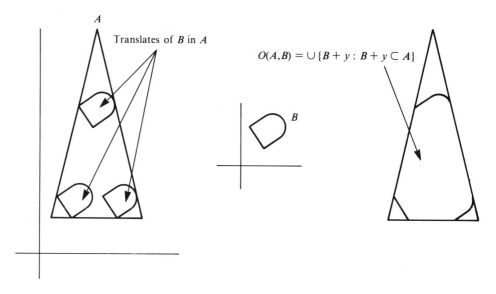

Figure 1.23 Opening by fitting

Other forms of M-19 are

$$O(A, B)^c = C(A^c, B)$$

and, taking complements,

$$O(A, B) = C(A^c, B)^c$$

All of the preceding forms of the duality principle will be used interchangeably.

Applying Theorem 1.2, we can easily obtain a pointwise characterization of the opening, namely, a point z is in $O(A, B)$ if and only if there exists some translate $B + x$ of B such that $z \in B + x \subset A$. (See Figure 1.24.) By duality, we have

Property M-20:[21] z is an element of $C(A, B)$ if and only if $(B + y) \cap A \neq \emptyset$ for any translate $B + y$ containing z.

Geometrically, z is an element of the closing if and only if, whenever the structuring

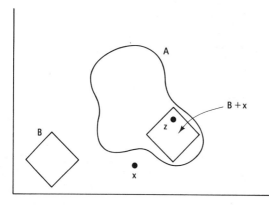

Figure 1.24 Pointwise characterization of opening

Sec. 1.4   Opening and Closing

Translates of B containing z. Each intersects A, and therefore, $z \in C(A,B)$.

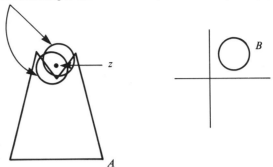

Figure 1.25   Closing using Property M-20

element is situated in such a manner as to contain z, then that placement of the structuring element intersects A. (See Figure 1.25.)

Before proceeding, it is important to note that in employing dilation and erosion, the position of the structuring element in the plane affects the output. By contrast, the relation of the origin to the structuring template (element) is of no importance in employing the opening or the closing.

The next theorem gives the most fundamental algebraic properties of the opening. These properties are essential, for both theoretical and practical reasons, to the morphological methodology. The first states that opening an image produces an output that is a subimage of the original image. The second says that, given a fixed structuring element, the opening is an increasing image-to-image mapping in the first variable. The third states that successive openings by the same structuring element do not alter the image after the primary application.

Theorem 1.3.[22]   The opening satisfies:

(i) $O(A, B)$ is a subimage of A (antiextensivity)
(ii) If A_1 is a subimage of A_2, then $O(A_1, B)$ is a subimage of $O(A_2, B)$ (increasing monotonicity)
(iii) $O[O(A, B), B] = O(A, B)$ (idempotence)

The properties of the opening specified in Theorem 1.3 play a key role in the use of the opening for the construction of morphological filters. As mentioned previously, opening by a disk can have a smoothing effect: the opening acts as a filter, the exact result being dependent upon the shape of the structuring template. According to Theorem 1.3, this filter, $O(\cdot, B)$, produces a subimage, is increasing, and is idempotent. Put succinctly, it filters in a manner which behaves quite well with respect to geometry.

As an example, consider the image A shown in Figure 1.26. It represents a rectangle on a background distorted by noise. Opening A by the disk B has produced a fairly clean copy of the rectangle.

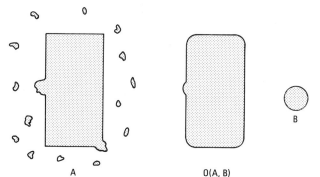

Figure 1.26 Rectangle on noisy background

Theorem 1.4 is the dual of Theorem 1.3; it states the corresponding algebraic properties of the closing.

Theorem 1.4.[23]   The closing satisfies:

(i) A is a subimage of $C(A, B)$ (extensivity)
(ii) If A_1 is a subimage of A_2, then $C(A_1, B)$ is a subimage of $C(A_2, B)$ (increasing monotonicity)
(iii) $C[C(A, B), B] = C(A, B)$ (idempotence)

To illustrate Theorem 1.4, consider image A of Figure 1.27. This time the image is noisy throughout, not just over the background. To clean it up, we apply the filter $C[O(\cdot, B), B]$. We first open by the disk B and then close by the same disk. This composite operation gets rid of the exterior noise, but leaves the rectangle unacceptably altered. To remedy the situation, we filter A by using the smaller disk D in Figure 1.28, again following an opening by a closing. This time, except for the large circular spot on the left, the rectangle is left essentially intact. Indeed, because the circular spot is so large, one might hesitate to call it noise: more than likely, it possesses some significance in the image. Much more will be said about morphological filters in Chapter 5.

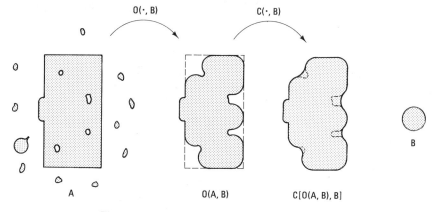

Figure 1.27 Filtering of rectangle in noisy image

Sec. 1.4 Opening and Closing

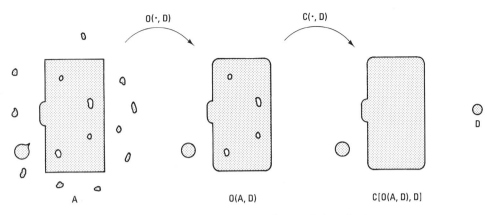

Figure 1.28 Alternative filtering of rectangle in noisy image

In Section 1.3, it was stated that both dilation and erosion are compatible with translation (properties M-3 and M-8, respectively). The same can be said for the opening and the closing, since each is defined as an iteration of a dilation and an erosion:

Property M-21: $O(A + x, B) = O(A, B) + x$

Property M-22: $C(A + x, B) = C(A, B) + x$

If the opening is regarded as a filter, M-21 states that filtering the translate is equivalent to filtering and then translating. M-22 states a similar thing for the closing.

In another direction, suppose we open a disk B_1 of radius 3 by a disk B_2 of radius 1. Then the output is the original disk B_1; i.e., $O(B_1, B_2) = B_1$. When this holds, we say that B_1 is *open with respect to* B_2. More generally, A is open with respect to B if $O(A, B) = A$, and A is closed with respect to B if $C(A, B) = A$. In the first case we say that A is *B-open*, while in the latter we say that A is *B-closed*. In particular, if B is a disk of radius 1, and r and s are positive numbers such that $r \geq s$, then rB is sB-open.

Now, just as following an opening by B with another opening by B yields the output of the original opening (an illustration of the idempotence property), following an opening by a B-open image with an opening by B also yields the output of the original opening. Moreover, following an opening by B with an opening by a B-open image gives the same output as simply opening by the B-open image. These properties are illustrated in Figures 1.29 and 1.30, respectively, and stated formally, together with their respective duals concerning the closing, as follows:

Property M-23:[24] If D is B-open, then

(i) $O[O(A, D), B] = O(A, D)$
(ii) $O[O(A, B), D] = O(A, D)$

Property M-24:[25] If D is B-open, then

(i) $C[C(A, D), B] = C(A, D)$
(ii) $C[C(A, B), D] = C(A, D)$

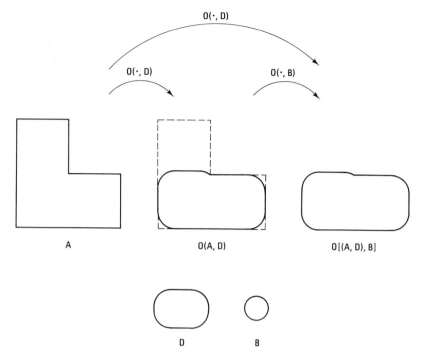

Figure 1.29 Illustration of M-23(i)

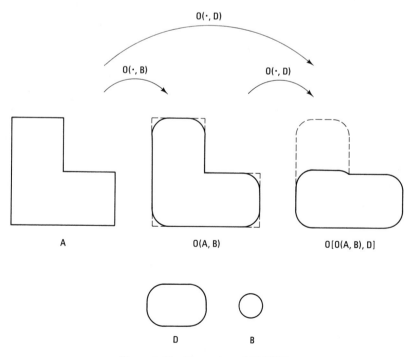

Figure 1.30 Illustration of M-23(ii)

Sec. 1.5 Convex Sets

Properties M-23 and M-24 are generalizations of the idempotence property, which as has been seen, holds for both the opening and the closing.

Two further generalizations of idempotence are given in the next property. Property M-25 (*i*) states that opening by a *B*-open set will produce a subimage of the opening by *B*. Intuitively, if *D* is *B*-open, opening by *B* produces a finer filter than opening by *D*, so that *D* removes more of the original image than *B*. Not only is this property important for morphological filtering, it is fundamental to the development of the granulometric feature-generation methodology. Property M-26 (*ii*) states a dual result for the closing: if *D* is *B*-open, closing by *B* produces a finer filter "from the outside" than closing by *D*.

Property M-25:[26] If *D* is *B*-open, then

(*i*) $O(A, D) \subset O(A, B)$
(*ii*) $C(A, B) \subset C(A, D)$

Note that M-25 is closely related to M-23 and M-24. Indeed, in Figures 1.29 and 1.30, *D* is *B*-open. $O(A, D)$ is computed directly in Figure 1.29, and $O(A, B)$ is similarly computed in Figure 1.30. Comparing the two figures, we see that $O(A, D)$ is a subimage of $O(A, B)$.

1.5 CONVEX SETS

Convex sets play a significant role in Euclidean morphology, from both a granulometric and an image-functional point of view. For that reason, some basic definitions and properties need to be presented.

If *x* and *y* are points in the plane R^2, then the line segment between *x* and *y* is denoted by $\overline{xy}$ and is rigorously defined by

$$\overline{xy} = \{ax + by: a \geq 0, b \geq 0, \text{ and } a + b = 1\}$$

where the plus sign in '*ax* + *by*' denotes vector addition. An image (set) in R^2 is said to be *convex* if, for any points *x* and *y* in *A*, the entire segment $\overline{xy}$ lies in *A*. (See Figure 1.31.) Thus, a disk, a line, a half-plane, an ellipse, and any regular polygon are all convex sets.

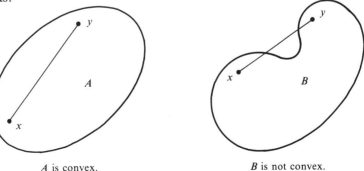

A is convex.   B is not convex.

Figure 1.31 Convex and nonconvex sets

Given a set A, it is often useful to consider the "smallest" or *minimal* convex set containing A. This minimal convex enclosing set is called the *convex hull* of A and is denoted by $H(A)$. In general, any intersection of convex sets is convex, and it turns out that the convex hull of A is the intersection of all convex sets containing A. (See Figure 1.32.) Previously, it was noted that a half-plane—that is, that part of the plane lying on one side of a line—is convex. The convex hull of A is given by the intersection of all half-planes containing A, and the image A is convex if and only if it equals its convex hull.

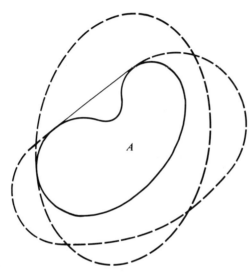

$H(A) = \cap \{B: B \supset A \text{ and } B \text{ is convex}\}$

Figure 1.32 Convex hull

Example 1.9:

Let $A = \{(-1, -1), (0, 0), (1, 1), (1, 0), (-1, 1)\}$. Then the convex hull of A, $H(A)$, is the shaded four-sided polygon depicted in Figure 1.33. Notice that $H(A)$ is the intersection of the lower half-plane determined by the line $y = 1$, the upper half-plane

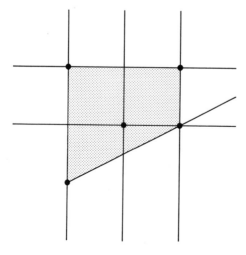

Figure 1.33 A polytope

Sec. 1.5 Convex Sets

determined by the line $y = x/2 - \frac{1}{2}$, the right half-plane determined by the line $x = -1$, and the left half-plane determined by the line $x = 1$. $H(A)$ is an example of a convex set known as a *polytope*. A polytope is the convex hull of a finite collection of points. Triangles, rectangles, and all regular polygons are examples of polytopes.

In morphology, we are often interested in convex sets that are *compact*. Topologically, a compact set is a set that is *closed* and *bounded*. A closed set is a set that contains its *boundary*, while a bounded set is a set for which there exists a number M such that all points in the set have magnitude less than or equal to M. A point x is in the boundary of A if and only if each disk centered at x intersects both A and its complement. (See Figure 1.34.)

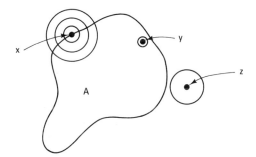

POINT x IS IN THE BOUNDARY OF A.
POINTS y AND z ARE NOT IN THE BOUNDARY OF A.

Figure 1.34 Characterization of a boundary point

Two other topological definitions are of use. First, the *closure* of a set A, denoted $\overline{A}$, is equal to the union of A with its boundary. (A well-known result in topology states that A is closed if and only if A is equal to its closure.) Second, the *interior* of set A is the set-theoretic difference between A and its boundary.

Example 1.10:

Let

$$A = \{(x, y): 0 \leq x \leq 1, 0 \leq y < 1\}$$
$$B = \{(x, y): x^2 + y^2 \leq 1\}$$
$$C = \{(x, y): y = 3x + 2\}$$

Then A is not closed, since $\{(x, y): 0 \leq x \leq 1, y = 1\}$ is part of the boundary of A but is not contained in A. Moreover, the closure of A and the interior of A are respectively given by

$$\{(x, y): 0 \leq x \leq 1, 0 \leq y \leq 1\}$$

and

$$\{(x, y): 0 < x < 1, 0 < y < 1\}$$

The set B is closed and bounded, and hence compact. Set C, which is a line, is closed but not bounded. (See Figure 1.35.)

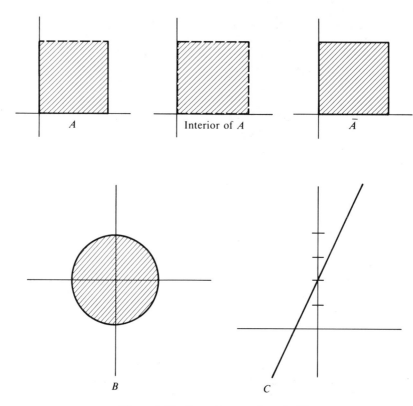

Figure 1.35 Several types of sets in R^2

A line L is said to *support* the set A at x in A if x lies on L and A lies on one side of L. If A is closed and has a nonempty interior, then A is convex if and only if, for any boundary point x in A, there exists at least one line supporting A at x. (See Figure 1.36.)

It is of significance in morphology that the basic operations behave well with respect to convexity. That is, if A and B are convex, then so are $A \oplus B$, $A \ominus B$, $O(A, B)$ and $C(A, B)$.[27] As binary operations, dilation, erosion, opening, and closing

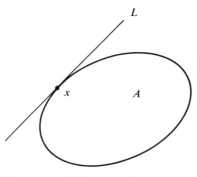

L supports A at x **Figure 1.36** Line that supports set A

Sec. 1.5 Convex Sets

satisfy the *closure* property on the class of convex sets. Moreover, if A is convex and t is a positive real number, then tA is convex.

According to property M-25, if D is B-open, then the opening of an image by D is a subimage of the opening of that same image by B. It is a fundamental property of convex sets that if B is convex and $r \geq s > 0$, then rB is sB-open. This property, which is apparent for disks, actually holds for any convex set. Hence, M-25 leads to the following theorem:

Theorem 1.5.[28] If $r \geq s > 0$ and B is convex, then for any image A, $O(A, rB) \subset O(A, sB)$.

Figure 1.37 illustrates Theorem 1.5 for the case of a disk B.

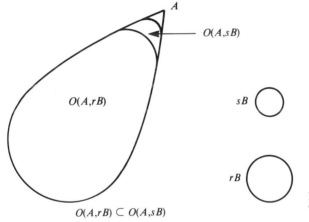

Figure 1.37 Illustration of Theorem 1.5

To say that rB is sB-open whenever $r \geq s > 0$ is equivalent to saying that tB is B-open for any $t \geq 1$. Any set for which the latter property holds is called *expansive*. Convex sets are expansive. Indeed, it is precisely their expansiveness that makes Theorem 1.5 work. But, as has been demonstrated by Matheron, when we restrict our attention to compact images, expansiveness is equivalent to convexity. This is the meaning of the next theorem.

Theorem 1.6.[29] Suppose B is compact. Then B is expansive if and only if B is convex.

Theorem 1.6 is one of the most significant results in Euclidean morphology. Theorem 1.5, which depends upon the expansiveness of convex sets, is crucial to the granulometric methodology. That methodology results from the successive opening of an image by increasing scalar multiples of a given structuring element. As long as the structuring element is expansive, we are assured that the successive outputs will form a *nested* sequence of sets—that is, that $r \geq s$ will result in $O(A, sB)$ containing $O(A, rB)$. According to Theorem 1.6, the only expansive compact structuring elements are those that are convex. Since we desire nesting of the output images when applying granulometries, and since our structuring elements will be compact, Theorem 1.6 tells us that we should employ convex, compact structuring elements.

Earlier, it was shown how an erosion could be performed iteratively by a structuring element formed from a dilation or a sequence of dilations. Given this structuring element, say B, the key to the iteration scheme is to decompose B into a Minkowski sum of more elementary structuring elements. If B happens to be compact and convex, not only do there exist such decompositions, but there exist decompositions for which the summands are identical, each being a scalar multiple of B. Formally, an image B in R^2 is said to be *infinitely divisible* with respect to dilation if, for any positive integer k,

$$B = \frac{1}{k}B \oplus \frac{1}{k}B \oplus \cdots \oplus \frac{1}{k}B$$

where there are k summands of the form $(1/k)B$. This leads to the following theorem:

Theorem 1.7.[30] Suppose B is compact. Then B is infinitely divisible if and only if B is convex.

This theorem is closely related to Theorem 1.6. In fact, putting them together gives, for compact sets, the equivalence of the properties of convexity, expansiveness, and infinite divisibility.

Example 1.11:

Suppose B is a square of edge length 2. Then

$$B = \frac{1}{4}B \oplus \frac{1}{4}B \oplus \frac{1}{4}B \oplus \frac{1}{4}B$$

where each summand is a square of edge length $\frac{1}{2}$. It is the convexity of B that guarantees this decomposition.

As long as a structuring element is convex, Theorem 1.7 can be used to perform the primitive operations iteratively. For instance, when applied to convex structuring elements, the associativity property, M-2, gives

$$A \oplus B = \left[\left(A \oplus \frac{1}{3}B\right) \oplus \frac{1}{3}B\right] \oplus \frac{1}{3}B$$

Similarly, property M-18 yields

$$A \ominus B = \left[\left(A \ominus \frac{1}{3}B\right) \ominus \frac{1}{3}B\right] \ominus \frac{1}{3}B$$

Note that instead of decomposing by thirds, we could have decomposed by any scalar factor $1/k$. Also, since the opening and the closing are iterations of dilation and erosion, given a convex structuring element, they, too, can be formed by iterations using identical smaller structuring elements. This technique has proved to be very useful in the implementation of cellular architecture, where dilations and erosions by large convex templates are performed by iterations employing small, fixed-size tem-

EXERCISES

1.1. Let
$$A = \{(0, 0), (0, 2), (2, 0), (2, 2)\}$$
$$B = \{(0, 0), (1, 0), (2, 0), (3, 0)\}$$
$$C = \{(0, 0), (1, 1), (2, 2)\}$$

Find each of the following sets (in set notation and pictorially):

(a) $(1, 1) + A$
(b) $A \oplus B$
(c) $(A \oplus B) \oplus C$
(d) $A \ominus B$
(e) $B \ominus A$
(f) $\mathscr{E}(A, B)$
(g) $A \oplus (2C)$
(h) $2[\frac{1}{2}A \oplus C]$
(i) $(C \oplus A) \oplus B$
(j) $[A + (-1, 2)] \oplus B$
(k) $A \oplus (B \cup C)$
(l) $A \oplus (B \cap C)$
(m) $(A \oplus B) \cup (A \oplus C)$
(n) $(A \oplus B) \cap (A \oplus C)$
(o) $A \ominus (B \cup C)$
(p) $A \ominus (B \cap C)$
(q) $(A \oplus B) \cap (A \oplus C)$
(r) $(A \ominus B) \ominus C$

1.2. Let
$$A = \{(x, y): 0 \le x \le 2, 0 \le y \le 3\}$$
$$B = \{(x, y): x^2 + y^2 \le 1\}$$
$$C = \{(x, y): y = x \text{ and } 0 \le x \le 1\}$$
$$D = \{(x, y): 0 \le y \le |2x| \text{ and } -1 \le x \le 1\}$$

(See Figure 1.38.) Give graphical representations of the following:

(a) $A \oplus B$
(b) $A \oplus C$
(c) $B \oplus C$
(d) $A \ominus B$

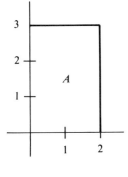
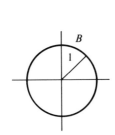
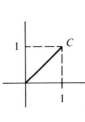
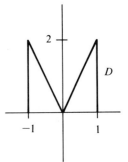

Figure 1.38 Images for Exercise 1.2

(e) $A \ominus C$
(f) $A \ominus \frac{1}{2}B$
(g) $A \ominus \frac{1}{4}C$
(h) $B \oplus \frac{1}{2}B$
(i) $A \oplus \frac{1}{2}B$
(j) $A \oplus \frac{1}{2}C$
(k) $O(A, B)$
(l) $O(A, C)$
(m) $O[B, \frac{1}{2}C]$
(n) $O[A, \frac{1}{2}B]$
(o) $O[A, \frac{1}{4}B]$
(p) $O(D, B)$
(q) $C(D, B)$
(r) $C(D, 2B)$

1.3. Using the properties thus far presented, show that the following identities hold:
(a) $[A^c \oplus B^c]^c = A \ominus B^c$
(b) $[(A + x) \oplus B)] + x = (A \oplus B) + 2x$
(c) $\mathcal{E}(A + x, B + x) = \mathcal{E}(A, B)$
(d) $O(A^c, B^c)^c = C(A, B^c)$
(e) $O(A + x, B)^c = C(A^c, B) + x$
(f) $O(A + x, B + x) = O(A, B) + x$

1.4. Use mathematical induction to show that, for any positive integer n,

$$A \ominus nB = (\cdots((A \ominus B) \ominus B)\cdots) \ominus B$$

where there are n occurrences of the set B on the right-hand side of the equation.

1.5. Derive Theorem 1.4 from Theorem 1.3 by using the duality relation between the opening and the closing.

1.6. Find the convex hull of the image

$$\{(x, y): 0 \le y \le x^2, -1 \le x \le 2\}$$

1.7. The set D of Exercise 1.2 is not convex. Find all boundary points at which there does not exist at least one supporting line.

1.8. Construct an example of a compact set that is not expansive. According to Theorem 1.5, the set cannot be convex.

1.9. Graphically illustrate Example 1.11.

FOOTNOTES FOR CHAPTER 1

1. Georges Matheron, *Random Sets and Integral Geometry* (New York: Wiley, 1975).
2. See Appendix: Proposition A.7.
3. See Appendix: Proposition A.2.
4. See Appendix: Proposition A.2.
5. See Appendix: Proposition A.4.
6. See Appendix: Proposition A.5.
7. See Appendix: Proposition A.6.
8. See Appendix: Proposition A.10.
9. See Appendix: Proposition A.10.
10. See Appendix: Proposition A.11.
11. See Appendix: Proposition A.12.
12. See Appendix: Proposition A.13.
13. See Appendix: Proposition A.9.
14. See Appendix: Proposition A.14.
15. See Appendix: Proposition A.16.
16. See Appendix: Proposition A.17.

17. See Appendix: Proposition A.18.
18. E. R. Doughtery and C. R. Giardina, "A Structurally Induced Image Algebra," Computers and Mathematics, Stanford (1986).
19. See Appendix: Theorem A.2.
20. See Appendix: Proposition A.19.
21. See Appendix: Proposition A.23.
22. See Appendix: Theorem A.3.
23. See Appendix: Theorem A.4.
24. See Appendix: Proposition A.28.
25. See Appendix: Proposition A.29.
26. See Appendix: Proposition A.27.
27. See Appendix: Proposition A.30.
28. See Appendix: Proposition A.32.
29. Matheron, *Random Sets and Integral Geometry*, 196.
30. Ibid., p. 22.

2

Digital Morphology

2.1 THE DIGITAL SETTING

The digital implementation of the Minkowski algebra must begin with a rigorous description of the setting in which the digital form of the algebra is to be placed. Whereas the appropriate model of the two-valued Euclidean form is the Euclidean plane, the digital setting is somewhat different. In order to avoid undue replication at a later stage, the concept of a general gray-valued digital image will be introduced in this section. The *bound matrix*, a suitable vehicle for digital image representation, will also be introduced.

Consider the xy-plane as being partitioned into square regions, much like graph paper, with each square centered at a lattice point (i, j), where i and j are in the set Z of integers. Each square centered at a lattice point is called a *pixel*; in general, the square centered at (i, j) is called the (i, j)th pixel. The set of all lattice points is denoted by $Z \times Z$. A digital image is obtained by assigning a real number, called a *gray value*, to each pixel in some collection of pixels. We define the digital image f as the function $f:D \to R$, where D is a subset of the set of all lattice points $Z \times Z$ and R is the set of real numbers. D is called the *domain* of the image f, and R is called the *codomain*. Sometimes we will employ the notation D_f to denote the domain of f.

Although the domain of an image is rigorously defined to be a subset of $Z \times Z$, we intuitively think of the small squares as being the domain. Consequently, even

Sec. 2.1   The Digital Setting   37

though the small squares are technically the pixels, the lattice points are also referred to as pixels. In effect, each lattice point (i, j) is the address of the small square (pixel) centered at (i, j). We do not differentiate between the square and its center when we speak of the gray value of a pixel; we simply write $f(i, j)$. The precise relationship between the squares and the lattice points will be set out in Section 2.7, where digitization is discussed.

Very often, the domain of a digital image will be rectangular in shape and contain a finite number of elements. In such a case, a digital image will be represented in a manner similar to a matrix or a two-dimensional array. The image will have a gray value for each position (specified by a row number and column number) within the array. However, unlike a conventional matrix, the matrix representation of an image must have a location within the set of lattice points $Z \times Z$. Some of these concepts are illustrated in Example 2.1.

Example 2.1:

Consider the digital image f illustrated in Figure 2.1. The domain consists of the six pixels darkly outlined. Rigorously, the image is $f: D \to R$, where $D = \{(i, j): i = 0, 1$ and $j = -1, 0, 1\}$. The gray value at (i, j) in D [denoted $(i, j) \in D$] is $f((i, j))$, or, more conveniently, $f(i, j)$. Thus, $f(0, 0) = 2, f(0, -1) = 0, f(1, -1) = -3, f(1, 0) = 4$, $f(0, 1) = 3$, and $f(1, 1) = 5$.

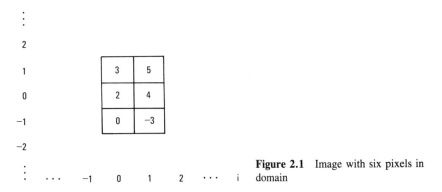

Figure 2.1 Image with six pixels in domain

For the image f of Example 2.1, the matrix representation is

$$f = \begin{pmatrix} 3 & 5 \\ ② & 4 \\ 0 & -3 \end{pmatrix} \quad \text{or equivalently} \quad f = \begin{pmatrix} 3 & 5 \\ 2 & 4 \\ 0 & -3 \end{pmatrix}_{0,1}$$

Both representations of f are examples of bound matrices, a topic to be more formally introduced shortly. In the first representation, the element of the matrix corresponding to the $(0, 0)$ lattice point of the xy coordinate system is circled. This means that the gray level denoted by 2, which occurs in the second row, first column, is the number indicating the gray value assigned to pixel $(0, 0)$, the origin. The values for other pixels in D are determined by positioning the matrix at the specified $(0, 0)$ location and reading off the corresponding matrix values.

In the second matrix representation for f, the two subscripted integers outside the lower right part of the matrix specify the absolute location in the xy lattice of the uppermost, leftmost entry in the matrix. In other words, the first-row, first-column entry of the matrix is positioned at $(0, 1)$, so that $f(0, 1) = 3$. It is as if the matrix were to be "hung" on the $(0, 1)$ "peg" of the lattice, with the "slot" for hanging being through the first-row, first-column entry of the matrix.

A matrix representation can be constructed for any finite digital image with rectangular-type domain D. The matrix involved will be called a *bound matrix* since it has a fixed location on graph paper. The situation is analogous to that of bound vectors in mechanics, where the vectors are not free to translate or slide. If the image f has an $m \times n$ rectangular domain given by $D = \{(i, j): r \leq i \leq r + n - 1$ and $s \leq j \leq s + m - 1 = t\}$, then f can be represented as

$$\begin{pmatrix} a_{11} & a_{12} & \cdots & a_{1n} \\ a_{21} & a_{22} & \cdots & a_{2n} \\ \vdots & \vdots & & \vdots \\ a_{m1} & a_{m2} & \cdots & a_{mn} \end{pmatrix}_{r,t}$$

The values of gray for the image are given as a_{pq} in the bound matrix $(a_{pq})_{rt}$. Here p denotes the matrix row and q the matrix column for the location of the value a_{pq}. The integers r and t specify the leftmost, uppermost pixel (r, t) in D, the domain of the image f. The gray value at this pixel is a_{11}; that is, $f(r, t) = a_{11}$. For any other pixel (i, j) in D, it can be seen that

$$f(i, j) = a_{m+s-j, i+1-r} = a_{t+1-j, i+1-r}$$

or, equivalently,

$$a_{pq} = f(q + r - 1, m + s - p) = f(q + r - 1, t + 1 - p)$$

The preceding relationships can best be seen by overlaying the matrix representation on the pixel grid structure ("hanging" the matrix), as in Figure 2.2.

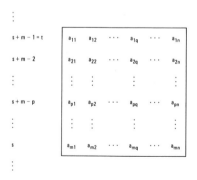

Figure 2.2 Overlay of matrix and grid structures

Example 2.2:

Let f be the image given in Example 2.1. Then, employing the notation of bound matrices, one obtains

Sec. 2.1   The Digital Setting

$$f = \begin{pmatrix} a_{11} & a_{12} \\ a_{21} & a_{22} \\ a_{31} & a_{32} \end{pmatrix}_{0,1} = \begin{pmatrix} 3 & 5 \\ 2 & 4 \\ 0 & -3 \end{pmatrix}_{0,1}$$

This is the gray value at location $(0, 1)$. Thus, by definition, $a_{11} = f(0, 1) = 3$.

This is the gray value for the pixel located one unit to the right and two units down from $(0, 1)$. Thus, $a_{32} = f(1, -1) = -3$.

While it can be seen from the positioning that $a_{32} = f(1, -1)$, this relationship can be directly computed from $f(i, j) = a_{t+1-j, i+1-r}$. The algorithm for obtaining the values of f from the values in the bound matrix, and vice versa, is important for computer implementation and for mechanization purposes. However, in this text it will generally suffice to have a good understanding of the geometric explanation for identifying the appropriate pixel locations.

Now consider the m by n array-type structure

$$\begin{pmatrix} a_{11} & a_{12} & \cdots & a_{1n} \\ a_{21} & a_{22} & \cdots & a_{2n} \\ \vdots & \vdots & & \vdots \\ a_{m1} & a_{m2} & \cdots & a_{mn} \end{pmatrix}_{r,t}$$

where

(i) each a_{pq} is a real number or a star $(*)$
(ii) $1 \leq p \leq m$, $1 \leq q \leq n$
(iii) r and t are integers.

Such a data structure is called a *bound matrix*, or an $m \times n$ bound matrix, and the stars denote values that are not known. The location in $Z \times Z$ of the a_{11} entry, which may be a star, is (r, t). The location of the entry a_{pq} in $Z \times Z$ is $(q + r - 1, t + 1 - p)$. Examples of bound matrices without stars were given previously. The star will be utilized to allow nonrectangular images to be represented by bound matrices. It is helpful to visualize all values *outside* a bound matrix to be stars.

Example 2.3:

Consider the image $f:D \to R$, where $D = \{(1, 1), (1, 2), (2, 2)\}$ and $f(1, 1) = 8$, $f(1, 2) = 4$, and $f(2, 2) = 6$. Then a bound matrix representing f is given by

$$\begin{pmatrix} 4 & 6 \\ 8 & * \end{pmatrix}_{1,2}$$

Two other bound matrices representing f are

$$\begin{pmatrix} * & 4 & 6 \\ * & 8 & * \\ * & * & * \end{pmatrix}_{0,2} \quad \text{and} \quad \begin{pmatrix} * & 4 & 6 \\ * & 8 & * \\ \circledast & * & * \end{pmatrix}$$

Note that in the specification of the origin the circle is always around the gray value for the $(0, 0)$ pixel. In this example that pixel happens not to be part of the image; nevertheless, the notation must be used in a consistent fashion.

When we say that an $m \times n$ bound matrix $(a_{pq})_{rt}$ *represents* a (necessarily finite) image $f:D \to R$, we mean that the gray values of f lie in $(a_{pq})_{rt}$ in the proper position and that all other values (if any) of a_{pq} are $*$. Rigorously, $(a_{pq})_{rt}$ represents the finite image $f:D \to R$ if for every (i, j) in D there corresponds the element $a_{t+1-j, i+1-r}$ in $(a_{pq})_{rt}$, where $a_{t+1-j, i+1-r} = f(i, j)$. Furthermore, all other a_{pq} in $(a_{pq})_{rt}$ must have the entry $*$.

It is important to know how to go from a picture representation of an image to a bound matrix representation, and vice versa. This is most simply done geometrically by overlaying one structure on the other. There is no need to memorize the formula for relating gray values using matrix notation a_{pq} with the function notation $f(i, j)$. The complexity of the relation arises from the somewhat backwardly rotated notation used in representing tuples in matrices relative to the xy coordinate labeling.

For any digital image with finite, nonempty domain, a *minimal bound matrix* can be found to be represent f. The minimal bound matrix for f is simply that representation for which m and n are as small as possible. In Example 2.3, the first bound matrix given is the minimal one. It is simply the representation that contains no extraneous rows or columns of stars.

Using bound matrices to represent images is convenient and often space saving. It is an illustration of a data-compression technique. This is particularly true when there are few stars in the bound matrix.

2.2 PRIMITIVE DIGITAL MORPHOLOGICAL OPERATORS

In Chapter 1, the operators $\oplus$ and $\ominus$ were considered fundamental from the perspective of the Minkowski algebra. Yet $\oplus$ is formed from the more elementary operations of set-theoretic union and translation, and $\ominus$ is formed from intersection and translation. Consequently, from a lower level point of view, Minkowski addition and subtraction are not elemental. They result from the composition of more primitive operations, and with respect to those, $\oplus$ and $\ominus$ can be considered as macro-operators.

The notion of which operators are elemental as opposed to which are macro-operators is fundamental to both image algebra and structured algorithm development. Consequently, even at this early stage one should pay careful attention to the manner in which the morphological operations of Minkowski addition and Minkowski subtraction are built up from more elemental operators which serve as algebraic primitives. Since one of our abiding purposes in the digital setting concerns the algebraically structured implementation of morphological algorithms, close attention will be given to precise algorithm specification in terms of lower level operators. It is these low-level operators that provide the building blocks for the higher level macro-operators. Moreover, the domain of application of the morphological processes is determined from the outset by the collection of elemental operators from which the higher level operators are structured.

A word of caution, however: the choice of precisely which operators are to be considered elemental and which are to be developed as macro-operators is not dictated by mathematical considerations alone, but rather by considerations that pay attention

Sec. 2.2 Primitive Digital Morphological Operators

to both mathematics and practice. Thus, fundamental modeling decisions are involved in the selection of algebraic primitives.

In the present section, we shall introduce a set of elemental operators on the class of digital images which will generate the digital Minkowski (morphological) algebra. We begin with the two binary operations EXTMAX and MIN. The *extended maximum* operator EXTMAX compares two images in a pixelwise manner and outputs the maximum, or highest, value at each pixel at which both input images are defined. If both images are undefined at a specific pixel, then so is the output of EXTMAX. However, if only one image is undefined at a pixel while the other is defined, then the output of EXTMAX is the gray value of the latter image at the given pixel. Rigorously, we have

$$[\text{EXTMAX}(f, g)](i, j) = \begin{cases} \max [f(i, j), g(i, j)], & \text{if both } f \text{ and } g \text{ are defined at } (i, j) \\ f(i, j), & \text{if } f(i, j) \neq * \text{ and } g(i, j) = * \\ g(i, j), & \text{if } g(i, j) \neq * \text{ and } f(i, j) = * \\ *, & \text{if } f(i, j) = g(i, j) = * \end{cases}$$

If an image is finite, then it possesses the value $*$ at every pixel outside its minimal bound matrix. Consequently, for two finite images f and g, the domain of EXTMAX(f, g) is a subset of the union of the frames determined by the minimal bound matrices of f and g. More generally, for f and g finite or not, the domain of EXTMAX(f, g) is the union of the domains of f and g.

In employing EXTMAX in algorithm specification, we shall utilize the block diagram

Example 2.4:

Consider the digital images

$$f = \begin{pmatrix} 2 & 1 & 4 \\ 3 & * & * \end{pmatrix}_{0,1}$$

and

$$g = \begin{pmatrix} 8 & * \\ -7 & 4 \\ 3 & * \end{pmatrix}_{0,2}$$

The domain of f is $D_f = \{(0, 0), (0, 1), (1, 1), (2, 1)\}$, and the domain of g is $D_g = \{(0, 0), (0, 1), (0, 2), (1, 1)\}$. EXTMAX is defined on $D_f \cup D_g = \{(0, 0), (0, 1), (1, 1), (2, 1), (0, 2)\}$ and is given by

$$\text{EXTMAX}(f, g) = \begin{pmatrix} 8 & * & * \\ 2 & 4 & 4 \\ 3 & * & * \end{pmatrix}_{0,2}$$

For instance, $[\text{EXTMAX}(f, g)](0, 1) = \max [2, -7] = 2$, $[\text{EXTMAX}(f, g)](0, 2) = 8$ since $g(0, 2) = 8$ and $f(0, 2) = *$, and $[\text{EXTMAX}(f, g)](0, 5) = *$ since $f(0, 5) = g(0, 5) = *$.

Note that in applying EXTMAX, we treat * as if it were $-\infty$. Indeed, no matter what the value of the input image f at pixel (i, j), if $g(i, j) = *$, then $[\text{EXTMAX}(f, g)](i, j) = f(i, j)$. That is, the star is miminal with respect to all values.

The *minimum* operator MIN is similar to EXTMAX in that a pixelwise comparison is taken; however, the domain of the output image is the intersection of the input domains instead of their union. This is because * is again treated as $-\infty$ when applying MIN: it "absorbs" all other values. Rigorously, we have

$$[\text{MIN}(f, g)](i, j) = \begin{cases} \min[f(i, j), g(i, j)], & \text{if both } f \text{ and } g \text{ are defined at } (i, j) \\ *, & \text{if either } f \text{ or } g \text{ is not defined at } (i, j) \end{cases}$$

Note that if either $f(i, j) = *$ or $g(i, j) = *$, then $[\text{MIN}(f, g)](i, j) = *$.

Example 2.5:

Using the images f and g of Example 2.4, we have, for MIN(f, g),

$$\begin{pmatrix} 2 & 1 & 4 \\ 3 & * & * \end{pmatrix}_{0,1}$$

$$\begin{pmatrix} 8 & * \\ -7 & 4 \\ 3 & * \end{pmatrix}_{0,2}$$

$$\xrightarrow{\text{MIN}} \begin{pmatrix} -7 & 1 \\ 3 & * \end{pmatrix}_{0,1}$$

Whereas, in two-value digital morphology, EXTMAX and MIN play the roles played by union and intersection in the Euclidean morphology of Chapter 1, the *translation* operator takes the part of set translation. A ternary operator, TRAN has the image f and two integers i and j as inputs, and the image that is identical to f but moved over i pixels to the right and j pixels up as output. In terms of bound matrices, if $f = (a_{pq})_{rt}$, then

$$\text{TRAN}(f; i, j) = (a_{pq})_{r+i, t+j}$$

or, in full bound matrix format, if

$$f = \begin{pmatrix} a_{11} & a_{12} & \cdots & a_{1n} \\ \vdots & \vdots & & \vdots \\ a_{m1} & a_{m2} & \cdots & a_{mn} \end{pmatrix}_{r,t}$$

then

$$\text{TRAN}(f; i, j) = \begin{pmatrix} a_{11} & a_{12} & \cdots & a_{1n} \\ \vdots & \vdots & & \vdots \\ a_{m1} & a_{m2} & \cdots & a_{mn} \end{pmatrix}_{r+i, t+j}$$

Note that no values a_{pq} change; only the location of the gray values change by a translation. The block diagram for TRAN is given by

Sec. 2.2 Primitive Digital Morphological Operators

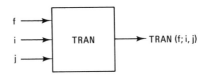

The semicolon is used to indicate that the input f is a different kind of entity than the inputs i and j: whereas f is an image, i and j are integers.

Translation can also be defined pixelwise. In reading the pixelwise description, keep in mind that i and j represent inputs, while u and v correspond to the coordinates of the pixel under consideration. Pixelwise, we have

$$[\text{TRAN}(f; i, j)](u, v) = f(u - i, v - j)$$

Pay particular attention to the minus signs in the argument for f: the somewhat confusing situation of a minus sign on the variable for a right translation is similar to that for real-valued functions in calculus.

While the bound matrix definition of TRAN is certainly easier to remember and use than the functional definition, it is only good for finite digital images. The functional definition is good for all digital images, infinite as well as finite.

Example 2.6:

Let

$$f = \begin{pmatrix} 3 & 4 \\ 1 & * \end{pmatrix}_{1,0} = \begin{pmatrix} \circledast & 3 & 4 \\ * & 1 & * \end{pmatrix}$$

Then

$$\text{TRAN}(f; 4, 1) = \begin{pmatrix} 3 & 4 \\ 1 & * \end{pmatrix}_{5,1} = \begin{pmatrix} * & * & * & * & * & 3 & 4 \\ \circledast & * & * & * & * & 1 & * \end{pmatrix}$$

The input has been shifted 4 pixels to the right and 1 pixel up, with the gray values remaining unchanged in their relative positions.

The *rotation* operator NINETY is similar to TRAN in that it leaves the gray values of an input image intact while altering the domain of the image. NINETY rotates an input image 90° in the counterclockwise direction about the origin and is defined pixelwise by

$$[\text{NINETY}(f)](i, j) = f(j, -i)$$

For instance, if $f(3, 1) = 5$, then $[\text{NINETY}(f)](-1, 3) = 5$.

In general, if f is given by the m by n minimal bound matrix $(a_{pq})_{rt}$, then NINETY(f) will be given by an n by m bound matrix whose first-row, first-column value is located at $(-t, r + n - 1)$.

The block diagram for NINETY is given by

Example 2.7:

$$f = \begin{pmatrix} 3 & 2 \\ 1 & * \\ 5 & 2 \end{pmatrix}_{3,4} = \begin{pmatrix} * & * & * & 3 & 2 \\ * & * & * & 1 & * \\ * & * & * & 5 & 2 \\ * & * & * & * & * \\ \circledast & * & * & * & * \end{pmatrix}$$

ROTATE f 90° WITH ORIGIN AS PIVOT

ORIGIN

THEN

$$f \longrightarrow \boxed{\text{NINETY}} \longrightarrow \begin{pmatrix} 2 & * & 2 \\ 3 & 1 & 5 \end{pmatrix}_{-4,4} = \begin{pmatrix} 2 & * & 2 & * & * \\ 3 & 1 & 5 & * & * \\ * & * & * & * & * \\ * & * & * & * & * \\ * & * & * & * & \circledast \end{pmatrix}$$

This can be seen from a pointwise perspective by letting $h = \text{NINETY}(f)$. Then

$$h(-4, 4) = f(4, 4) = 2$$
$$h(-3, 4) = f(4, 3) = *$$
$$h(-2, 4) = f(4, 2) = 2$$
$$h(-4, 3) = f(3, 4) = 3$$
$$h(-3, 3) = f(3, 3) = 1$$
$$h(-2, 3) = f(3, 2) = 5$$

and

$$h(i, j) = * \quad \text{for all other } (i, j)$$

From a structural standpoint, a 90° rotation keeps the image within the $Z \times Z$ lattice. It is not possible to rotate a digital image 30° while remaining within the square lattice structure. That is not to stay that a 30° rotation cannot be accomplished; it just cannot be accomplished without leaving the $Z \times Z$ structure.

The four elemental operators thus far introduced, EXTMAX, MIN, TRAN, and NINETY, serve as the fundamental set of primitives for morphological image processing. In Section 2.4, one more operator, COMP (the complement of an image), will be added to the collection to make an elemental set of five; however, most morphological processing requires only the first four. We shall refer to the collection of four as the *morphological basis*. The digital processing power of the morphological method is delimited by nature of the four elements within the basis.

Since our primary goals include the explication of digital feature generation and the construction of digital morphological filters, this last point cannot be over-emphasized. Much confusion regarding the morphological method has arisen from a lack of clarity regarding the morphological basis (and a lack of clarity regarding the morphological basis within image algebra in general). The full import of the Matheron representation theorems for morphological filters (Sections 5.1 and 5.4) can be appreciated only insofar as those theorems are seen within the context of the elemental operators that generate the Minkowski algebra.

Before concluding this section, it is appropriate to comment on the relativity of the morphological basis insofar as the choice of a particular set of four primitive operators is concerned. Other choices could be made which would, in the end, lead to exactly the same digital system as that generated by the four operators we have chosen. However, our choice of these operators has been made upon grounds that are both pragmatic and theoretical. An analogy is instructive.

Consider the relational operators AND and OR used in PASCAL. It is the selection of particular low-level operators (from the perspective of the language at hand) that gives a language its character. From the machine-level viewpoint, an AND operation is not low level; nevertheless, from the viewpoint of the PASCAL user, it is.

Different languages employ different low-level operations. The compiler then allows the user to think at a particular level and in a manner appropriate to his or her needs. Thus, while LISP and PROLOG have proved useful for certain artificial intelligence purposes, they do so because the operations they utilize facilitate algorithm development at a certain level of thinking. An image processing language based upon an appropriate set of primitive operators and employing carefully chosen macro-operators facilitates the development of structured image processing algorithms.

2.3 STRUCTURAL OPERATORS

From a mathematical perspective, morphological algorithm development rests upon the morphological basis. However, from both a logical and programmatical perspective, certain structural transformations concerning the makeup of an image must be specified. These transformations are not peculiar to morphology: they exist as a subcollection within image algebra, where they play essentially a linguistic role.

Imagine two stacks called DOMAIN and RANGE. Each stack contains the same number of entries, the first containing ordered pairs (i, j) and the second containing real numbers. (See Figure 2.3.) Together the stacks implicitly contain an image, for if they were popped simultaneously, the corresponding words would form a location together with its gray value. One could logically go so far as to say that a finite image

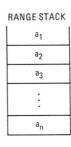

Figure 2.3 Decomposition of an image into stacks

is a *pair of stacks* (DOMAIN, RANGE). It simply depends on one's point of view. Yet when the stacks are separated, each can be treated as an individual data structure, and its contents can be operated on independently of the other. Once such operations are completed, the stacks can once again be considered as a pair and a new image created, or the results of the independent operations can be output.

With the preceding in mind, the *creation* operator is defined. CREATE, as it will be called, takes an array consisting of real numbers or stars and an array of integer-ordered pairs (or, equivalently, a two-dimensional array of integers) as inputs and outputs an image (or bound matrix). For each (i, j) in the array of ordered pairs, pixel (i, j) is given the corresponding value, be it a real number or a star, in the other array. All remaining pixels are given the value $*$. Except for the addressing techniques required to change a three-dimensional array to a bound matrix, the operation CREATE is purely *structural* in that it simply alters structure. Nevertheless, for logical reasons it needs to be materially articulated. Moreover, from a low-level programming point of view or from an architectural perspective, it does involve an actual operation. The block diagram for CREATE is given by

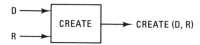

where D is an array of ordered integer pairs, R is an array of real numbers or stars, and each array contains the same number of entries. A schematic is given in Figure 2.4. Keep in mind that the two input stacks are not considered an image until *joined* by CREATE. This point is crucial since the image results not only from the contents of the stacks but also from the ordering within the stacks. As long as the stacks remain independent, each can have its ordering permuted by some nonimage-type operation. The output of CREATE depends on those orderings.

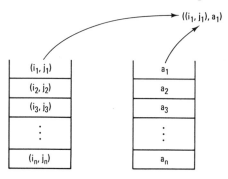

Figure 2.4 Creating image by popping stacks

Example 2.8:

Let arrays A and B be given by

$$A = [(0, 0), (0, 1), (1, 0), (2, 0), (2, 1)]$$

$$B = [1, *, 0, 2, 6]$$

Then

$$\text{CREATE}(A, B) = \begin{pmatrix} * & * & 6 \\ 1 & 0 & 2 \end{pmatrix}_{0, 1}$$

Sec. 2.3 Structural Operators

Just as two arrays of the appropriate types can be joined to form an image, an image can be *disjoined* to form two arrays, each of which is not an image. Two operations, called DOMAIN and RANGE, collectively invert CREATE. DOMAIN takes an image input and yields an array of ordered pairs that make up the domain of the image. RANGE takes an image and yields an array consisting of the gray values of the input image. If f is the input image, their respective block diagrams are given by

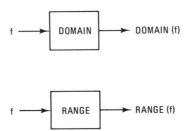

and

It will be assumed that the data are taken from the image f and put into DOMAIN(f) and RANGE(f) in a compatible manner. In other words, the image is *read* in both instances from top down and from left to right. This uniformity of approach allows the exact image to be reconstructed by CREATE as long as the arrays DOMAIN(f) and RANGE(f) have not been transformed in any manner. Thus,

$$\text{CREATE [DOMAIN}(f), \text{RANGE}(f)] = f$$

Example 2.9:

Let

$$f = \begin{pmatrix} 2 & 4 & 0 \\ * & 2 & 1 \end{pmatrix}_{1,1}$$

Then

$$\text{DOMAIN}(f) = [(1, 1), (2, 1), (2, 0), (3, 1), (3, 0)]$$

and

$$\text{RANGE}(f) = [2, 4, 2, 0, 1]$$

Note in Example 2.9 that composing CREATE with the outputs DOMAIN(f) and RANGE(f) returns f. But this is not usually what occurs in practice. First, there might be a permutation of the elements in DOMAIN(f) or there might be an arithmetic operation performed on the gray levels in RANGE(f). Second, a new stack might be popped simultaneously with DOMAIN(f) to form a new image. Whatever the case might be, the intention is to utilize these operators to go from the image world to the number or set world (RANGE and DOMAIN), and to go from the number world and set world back to the image world (CREATE).

Henceforth, the collection {CREATE, DOMAIN, RANGE} will be referred to as the *structural basis* of digital image processing. Morphological image processing is characterized by those algorithms that can be formed through the use of operators from the morphological basis and the structural basis.

2.4 CONSTANT IMAGES

In Chapter 1, it was assumed that images were two-valued and defined throughout the plane. Intuitively, one might think of black-and-white images. However, the black-and-white model leads to problems in dealing with actual images. As has been pointed out in Section 2.1, real digital images always possess an infinite number of pixels at which no gray value is defined. So in effect, a black-and-white, *binary* image actually possesses three values: 0 (white), 1 (black), and ∗ (undefined).

Consider the black-and-white digital image in Figure 2.5, which represents a black square on a white background. Since the image f is represented by a bound matrix, it is undefined outside the frame of the matrix. Moreover, there are two pixels within the frame of the bound matrix at which no gray value is defined. This lack of definition could be due to noise or inoperative sensors. In any case, there is no information to record at either star-valued pixel in the bound matrix or at any pixel outside the minimal bound matrix.

$$f = \begin{pmatrix} 0 & 0 & 0 & 0 & 0 & 0 & 0 & 0 & 0 & 0 & 0 & 0 \\ 0 & 0 & 0 & 0 & 0 & 0 & 0 & 0 & 0 & 0 & 0 & 0 \\ 0 & 0 & 0 & 0 & 0 & 0 & 0 & 0 & 0 & 0 & 0 & 0 \\ 0 & 0 & * & 0 & 1 & 1 & 1 & 1 & 0 & 0 & 0 & 0 \\ 0 & 0 & 0 & 0 & 1 & 1 & 1 & 1 & 0 & 0 & 0 & 0 \\ 0 & 0 & 0 & 0 & 1 & * & 1 & 1 & 0 & 0 & 0 & 0 \\ 0 & 0 & 0 & 0 & 1 & 1 & 1 & 1 & 0 & 0 & 0 & 0 \\ 0 & 0 & 0 & 0 & 0 & 0 & 0 & 0 & 0 & 0 & 0 & 0 \\ 0 & 0 & 0 & 0 & 0 & 0 & 0 & 0 & 0 & 0 & 0 & 0 \\ 0 & 0 & 0 & 0 & 0 & 0 & 0 & 0 & 0 & 0 & 0 & 0 \end{pmatrix}_{8,-4}$$

Figure 2.5 Binary image

The definition of image we employ throughout this text is faithful to the model that it represents, not to a convenient mathematical apparatus. Given that the value of f at (i, j) reflects the intensity registered by a sensor, it is clear that a binary sensor can actually yield three results, one of those results being no reading whatsoever. For such an occurrence, we use the symbol ∗. On the other hand, if a sensor can merely become activated or remain deactivated, this is a true two-valued case, and faithful modeling requires us to use the gray value 1 and the nongray value ∗. Deactivation on the part of the sensor does not yield positive information; indeed, as noted previously, the sensor could simply be inoperative, or, more trivially, the pixel (i, j) might be outside the field of sensors altogether. In accordance with these considerations, we say that pixel (i, j) is *activated* if $f(i, j) \neq *$, and is *deactivated* if $f(i, j) = *$.

Accordingly, a truly two-valued image is a *constant image*—that is, an image possessing only a single gray value. For instance, the image

$$S = \begin{pmatrix} 1 & 1 & * \\ 1 & \circledast & * \end{pmatrix}$$

Sec. 2.4 Constant Images

is a constant image having the single gray value 1. Similarly, the image

$$T = \begin{pmatrix} 1 & 1 & 1 \\ \boxed{1} & 1 & 1 \end{pmatrix}$$

is also a constant image. Note that $T(0, -1) = *$ since $(0, -1)$ lies outside the frame of the minimal bound matrix.

When dealing solely with constant images, we shall usually employ the letters S and T to denote the images, and the constant value will always be 1.

Constant images have a natural set-theoretic interpretation. We can view the grid of pixels as a universal class of locations and say that a set is defined by the collection of activated pixels. Indeed, under this interpretation a constant image S is identified with a set, and a pixel may be either in the set or outside the set.

The identification of a constant image with its domain of activated pixels leads to the comcomitant identification of image operations with set-theoretic operations. For instance, the domain of EXTMAX(S, T) equals the union of the two input domains, and consequently EXTMAX plays the role of union under the identification. Analogously, the domain of MIN(S, T) equals the intersection of the two input domains, and hence MIN plays the role of intersection.

In general, an image f is said to be a *subimage* of an image g if the domain of f is a subset of the domain of g and if $f(i, j) = g(i, j)$ for any pixel (i, j) in the domain of f. Under the identification of a constant image with its domain, the notion of subimage is equivalent to that of subset. Moreover, the constant image S is a subimage (subset) of the constant image T if and only if EXTMAX(S, T) = T (i.e., if and only if, in terms of domains, $D_S \cup D_T = D_T$, which is exactly analogous to the situation in ordinary set theory). For constant images, we write $S \ll T$ to denote that S is a subimage of T.

In sum, the Boolean structure of union and intersection is induced upon the collection of constant images by the operators EXTMAX and MIN, and the subset relation is induced by the subimage relation. A natural extension of the identification of constant images with subsets of $Z \times Z$ and the operations of EXTMAX and MIN with the operations of union and intersection is the notion of a complementary image. Given a constant image S, we define the *complementary* image of S, denoted COMP(S), by

$$[\text{COMP}(S)](i, j) = \begin{cases} 1, & \text{if } S(i, j) = * \\ *, & \text{if } S(i, j) = 1 \end{cases}$$

and we call COMP the *complementation* operator. Its block diagram is given by

We call the collection of image-to-image operations {EXTMAX, MIN, TRAN, NINETY, COMP} the *augmented morphological basis*. For a number of reasons, we do not classify COMP with the four operators of the morphological basis itself. First, although it is primitive with respect to morphology, it is not required for the development of the subject (although, to be sure, its use does allow the extension of certain results from Euclidean Minkowski algebra into digital Minkowski algebra). Second, it is inherently different from the other elemental operators in that it is not defined on general gray-scale images and, consequently, will not play a role in gray-scale morphology. Finally, it is derivable in general image algebra as a macro-operation by using nonmorphological primitives.

The notion of a complementary constant image leads naturally to that of a complementary bound matrix, where a technique analogous to the sign–magnitude representation of negative numbers will prove to be useful. If $T = (a_{pq})_{rt}$ is a bound matrix representation of T, then the complementary image T^c will have the *complementary bound matrix* representation $(b_{pq})_{rt}^c$ defined by

$$b_{pq} = * \quad \text{if} \quad a_{pq} = 1$$

$$b_{pq} = 1 \quad \text{if} \quad a_{pq} = *$$

$$T^c(i, j) = 1 \quad \text{for all } (i, j) \text{ outside the bound matrix frame}$$

Whereas the location of the 1's within the complementary bound matrix must be kept in computer memory, the value 1 for all pixels outside the frame can be represented by a single word. In reality, the situation with T^c, the complementary image, is no different from that with T itself. By convention, we assume the value $T(i, j) = *$ for any pixel (i, j) outside the frame. Yet such a value has no specific location in memory. We simply operate "as though it were," our operators being defined in such a manner as to give us both consistency and a virtually infinite image. An example will be given shortly.

As regards the identification of constant images with subsets of $Z \times Z$, it is immediate that the fundamental laws of set theory possess constant-image analogues. Some of these are as follows:

Idempotence:

$$\text{EXTMAX}(S, S) = S$$

$$\text{MIN}(S, S) = S$$

Commutativity:

$$\text{EXTMAX}(S, T) = \text{EXTMAX}(T, S)$$

$$\text{MIN}(S, T) = \text{MIN}(T, S)$$

Associativity:

$$\text{EXTMAX}[\text{EXTMAX}(S, T), U] = \text{EXTMAX}[S, \text{EXTMAX}(T, U)]$$

$$\text{MIN}[\text{MIN}(S, T), U] = \text{MIN}[S, \text{MIN}(T, U)]$$

De Morgan's Laws:

$$\text{COMP}[\text{EXTMAX}(S, T)] = \text{MIN}[\text{COMP}(S), \text{COMP}(T)]$$

$$\text{COMP}[\text{MIN}(S, T)] = \text{EXTMAX}[\text{COMP}(S), \text{COMP}(T)]$$

Sec. 2.4   Constant Images

It is important to note that the idempotent, commutative, and associative properties hold for general gray-scale images.

Example 2.10:

Consider the images

$$S = \begin{pmatrix} 1 & * & * \\ 1 & ① & 1 \end{pmatrix}$$

$$T = \begin{pmatrix} 1 & 1 \\ * & 1 \\ ① & 1 \end{pmatrix}$$

$$U = \begin{pmatrix} 1 & 1 & 1 \\ ① & 1 & 1 \\ 1 & 1 & 1 \end{pmatrix}$$

We have

$$\text{COMP}(S) = \begin{pmatrix} * & 1 & 1 \\ * & ⓧ & * \end{pmatrix}^c = \begin{pmatrix} 1 & 1 & 1 & 1 \\ * & 1 & 1 & 1 \\ * & ⓧ & * & 1 \end{pmatrix}^c$$

where the first form of the complementary bound matrix is the *minimal complementary bound matrix* for COMP(S). A second example of complementation is given by

$$\text{COMP}(U) = \begin{pmatrix} * & * & * \\ ⓧ & * & * \\ * & * & * \end{pmatrix}^c = \begin{pmatrix} 1 & * & * & * \\ 1 & ⓧ & * & * \\ 1 & * & * & * \end{pmatrix}^c$$

where once again the first form is minimal.

As an illustration of De Morgan's laws, consider

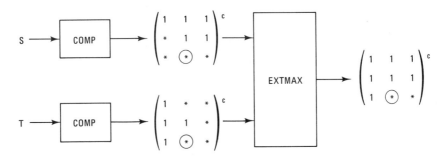

where COMP(S) and COMP(T) have deliberately not been written in minimal form in order to help clarify the EXTMAX operation. Moreover,

$$S \longrightarrow \boxed{\text{MIN}} \longrightarrow (①\ 1) \longrightarrow \boxed{\text{COMP}} \longrightarrow (ⓧ\ *)^c$$
$$T \longrightarrow$$

which is precisely the minimal form of the preceding output.

The following pair of block diagrams illustrate the associativity of MIN.

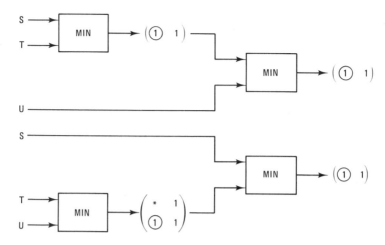

Due to the isomorphic operational relation between EXTMAX and MIN, and their set-theoretic counterparts, union and intersection, we often employ an infix notation for the two image operations that is similar to the set-theoretic symbols $\cup$ and $\cap$. We define $S \vee T = \text{EXTMAX}(S, T)$ and $S \wedge T = \text{MIN}(S, T)$, where $\vee$ and $\wedge$ are the logical symbols called cup and cap, respectively. Although the prefix notation is more standard from the perspective of programmatical procedures, this infix notation helps to relate the set-theoretic interpretation of constant images to ordinary set theory. In terms of the notation $\vee$ and $\wedge$, the idempotent, commutative, and associative properties, as well as De Morgan's laws, are identical to their set-theoretic counterparts. For instance, the associative law for extended maximum takes the form

$$(S \vee T) \vee U = S \vee (T \vee U)$$

In the midst of this new symbolization, a word of warning is in order to those familiar with the cup and cap notation for maximum and minimum as usually employed in analysis: in image algebra the cup denotes *extended* maximum; furthermore, a maximum operator, distinct from EXTMAX, also exists within image algebra.

In order to further the notational analogy between subsets of the $Z \times Z$ lattice and constant images, we shall often employ the notation S^c to denote $\text{COMP}(S)$. While at first glance it might appear that this notational convention is innocuous, a bit of confusion can arise due to the use of the superscript c to denote a complementary bound matrix. If S denotes the bound matrix $(b_{pq})_{rt}^c$, then

$$S^c = \text{COMP}(S) = [(b_{pq})_{rt}^c]^c = (a_{pq})_{rt}$$

where $a_{pq} = 1$ or $*$, depending on whether $b_{pq} = *$ or 1, and where $(a_{pq})_{rt}$ is not a complementary bound matrix. In other words, S^c does not necessarily represent a complementary bound matrix. No confusion will arise if one keeps in mind that a superscript c on an image name denotes the COMP operator, while a superscript c on a bound matrix denotes the corresponding complementary bound matrix. In terms of superscript notation, De Morgan's laws take the usual form

$$(S \vee T)^c = S^c \wedge T^c$$

and
$$(S \wedge T)^c = S^c \vee T^c$$

Example 2.11:

For the image S of Example 2.10, let

$$V = S^c = \text{COMP}(S) = \begin{pmatrix} * & 1 & 1 \\ * & \circledast & * \end{pmatrix}^c$$

Then

$$V^c = \text{COMP}(V) = \text{COMP}(S^c) = \begin{pmatrix} 1 & * & * \\ 1 & \textcircled{1} & 1 \end{pmatrix} = S$$

One might wonder why we do not simply deal with images having values 0 and 1, and make the supposition that 0 denotes a deactivated pixel. After all, this would leave us in an appropriate two-value setting, and it would let us work with the usual maximum and minimum operations. There are two arguments against this approach, however: (1) The value 0 usually denotes a gray value, so that confusion might result. Indeed, the 0–1 approach would lead to inconsistency when morphological procedures are utilized in conjunction with nonmorphological procedures. (2) In the case of gray-scale morphology (Chapter 6), the problem of deactivation would return. For it is in the gray-scale case that the role of the star becomes more crucial, as there the morphological operations must be defined carefully in terms of function domains. Specifically, $*$ is not a gray value and its role must be carefully delineated, especially in terms of the behavior of algorithms at the outer edges of the defining bound matrix.

More generally, the entire question is one of modeling. If one is really interested in images based upon sensor activation and deactivation, and one's desire is to interpret activation as black and deactivation as white, then there is no loss in utilizing the 1–$*$ methodology. When the image is transformed back into the Euclidean mode, one need simply send 1 into black and $*$ into white, in an analogous fashion to what has been done in the graphical illustrations of Chapter 1. Indeed, this methodology will be employed in the graphical representation of two-valued, 1–$*$ digital images. Since no modeling constraints result from the 1–$*$ representation, and since it is more in line with the representation of general gray-scale images, its usage is clearly preferable.

2.5 DIGITAL MINKOWSKI ALGEBRA

In this section we define the digital versions of Minkowski addition and subtraction for constant (1–$*$) images. Using these, we develop the digital Minkowski algebra in a manner paralleling that of the Euclidean Minkowski algebra in Section 1.3.

First, a bit of notation: if $S_1, S_2, \ldots$ are images, then $\bigvee_k S_k$ denotes the image that is 1 on the union of the domains of the S_k and is undefined elsewhere; similarly, $\bigwedge_k S_k$ denotes the image that is 1 on the intersection of the domains of the S_k and is undefined elsewhere. Pixelwise, we have

$$\left[\bigvee_k S_k\right](i, j) = \begin{cases} 1, & \text{if there exists at least one } k' \text{ for which } S_{k'}(i, j) = 1 \\ *, & \text{if } S_k(i, j) = * \text{ for all } k \end{cases}$$

and

$$\left[\bigwedge_k S_k\right](i,j) = \begin{cases} 1, & \text{if } S_k(i,j) = 1 \text{ for all } k \\ *, & \text{if there exists at least one } k' \text{ for which } S_{k'}(i,j) = * \end{cases}$$

In practice, the collection of images $\{S_k\}$ will be finite, and in that case the operations $\bigvee_k$ and $\bigwedge_k$ simply reduce to finite iterations of EXTMAX ($\vee$) and MIN($\wedge$), respectively.

Example 2.12:

Let

$$S_1 = \begin{pmatrix} ① & 1 & * \\ 1 & 1 & 1 \end{pmatrix}$$

$$S_2 = \begin{pmatrix} ① & 1 & 1 \\ 1 & * & * \end{pmatrix}$$

$$S_3 = \begin{pmatrix} 1 & 1 \\ ⊛ & 1 \\ 1 & 1 \end{pmatrix}$$

$$S_4 = \begin{pmatrix} ① & 1 & 1 & 1 \\ 1 & * & * & 1 \end{pmatrix}$$

Then

$$\bigvee_{k=1}^{4} S_k = \begin{pmatrix} 1 & 1 & * & * \\ ① & 1 & 1 & 1 \\ 1 & 1 & 1 & 1 \end{pmatrix}$$

and

$$\bigwedge_{k=1}^{4} S_k = \begin{pmatrix} ⊛ & 1 \\ 1 & * \end{pmatrix}$$

Note that we could write these outputs as $S_1 \vee S_2 \vee S_3 \vee S_4$ and $S_1 \wedge S_2 \wedge S_3 \wedge S_4$, respectively.

Minkowski addition, or *dilation*, is defined by

$$S \boxplus E = \bigvee_{(i,j) \in D_S} \text{TRAN}(E; i, j)$$

where D_S denotes the domain of S. Notice the correspondence between the digital definition and the Euclidean one. The domain of $S \boxplus E$ equals the union of the domains of the translates $\text{TRAN}(E; i, j)$.

As in the Euclidean case, the image E in $S \boxplus E$ plays the role of a template. If E is represented by a bound matrix, as it will be in practice, then the center of the template is the pixel of the bound matrix that is located at the origin. The Minkowski sum is found by placing the center of the template over each of the activated pixels of S and then taking the union of all the resulting copies of E, produced by using the translation operation. As in the Euclidean case, E is referred to as a structuring element. As usual, if the origin is contained in E, then the original image S will be a subimage of $S \boxplus E$.

Sec. 2.5 Digital Minkowski Algebra

Example 2.13:

Consider the two images

$$S = \begin{pmatrix} * & 1 & * & 1 & * \\ * & 1 & 1 & * & 1 \\ \circledast & 1 & 1 & 1 & * \end{pmatrix} \quad \text{and} \quad E = \begin{pmatrix} 1 & * \\ 1 & \textcircled{1} \end{pmatrix} \leftarrow \text{origin or center}$$

The domain of S is

$$D_S = \{(1, 2), (1, 1), (1, 0), (2, 1), (2, 0), (3, 2), (3, 0), (4, 1)\}$$

The translation operation should be used eight times (once for each element in D_S). We first use it to move the center of E to $(1, 0)$. Thus, the translation of E by $(i, j) = (1, 0)$ yields the image

$$\text{TRAN}(E; 1, 0) = \begin{pmatrix} 1 & * \\ \textcircled{1} & 1 \end{pmatrix}$$

This shows that $S \boxplus E$ must have a 1 at $(0, 0)$, $(1, 0)$, and $(0, 1)$. Also,

$$\text{TRAN}(E; 1, 2) = \begin{pmatrix} 1 & * \\ 1 & 1 \\ * & * \\ \circledast & * \end{pmatrix}$$

Thus, $S \boxplus E$ must have a 1 at $(0, 2)$, $(0, 3)$, and $(1, 2)$. When the additional six translations have been formed and the union has been taken, the resulting image is the Minkowski addition

$$S \boxplus E = \begin{pmatrix} 1 & * & 1 & * & * \\ 1 & 1 & 1 & 1 & * \\ 1 & 1 & 1 & 1 & 1 \\ \textcircled{1} & 1 & 1 & 1 & * \end{pmatrix}$$

Note that E has value 1 at the origin and hence S is a subimage of the dilation. (It is recommended that the reader also form the image E on cellophane or some other transparent material and overlay translates of this image on S to visually obtain $S \boxplus E$. See Figure 2.6.)

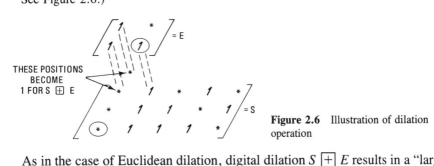

Figure 2.6 Illustration of dilation operation

As in the case of Euclidean dilation, digital dilation $S \boxplus E$ results in a "larger" image than S wherein the "small" holes of S have been filled in a manner depending upon the size and shape of the structuring element E. Figure 2.7 shows an image S and its dilation by the structuring element.

$$E = \begin{pmatrix} 1 & 1 \\ \textcircled{1} & 1 \end{pmatrix}$$

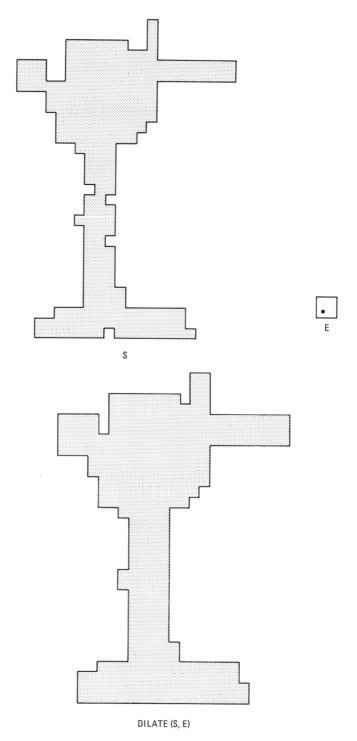

Figure 2.7 Digital dilation

A black-and-white drawing has been employed to depict the 1–∗ digital model. Specifically, if (i, j) is activated, then the pixel square with center (i, j) is colored black. However, if (i, j) is deactivated, the square is colored white. The net result is a black figure on an infinite white background.

The block diagram for dilation is

In accordance with this diagram, we shall often employ the notation DILATE(S, E) in place of $S \boxplus E$.

When inputs S and E are finite images, DILATE is a macro-operator relative to the primitives of the morphological and structural bases. In other words, it can be implemented through the linking of operators in those two bases—specifically, TRAN, EXTMAX, and DOMAIN. The block diagram specification of this linkage is given in Figure 2.8, where each arrow running from the DOMAIN box to a TRAN box is interpreted as a single ordered pair (i, j) from DOMAIN(S), and where the output of the TRAN box in question is TRAN($E; i, j$).

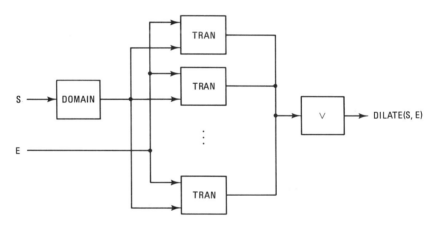

Figure 2.8 Block diagram of DILATE

Since our main interest is with finite images, because only these are amenable to machine processing, the restriction on the specification of Figure 2.8 regarding the finiteness of the inputs is of no practical concern. Consequently, DILATE has been effectively implemented in terms of just three primitives. The representation of higher level operators in terms of more primitive operators is a fundamental part of image processing. Among other things, it allows for a structured approach to algorithm development and a ready format for programming implementation. Moreover, the primitive operators of the morphological and structural bases, when used in conjunction with block diagrams, become a graphical language for the expression of morphological algorithms.

Again as in the Euclidean case, digital Minkowski subtraction and erosion are related by means of a 180° rotation of the structuring element. Since a general gray-level image can be rotated 180° by means of two successive applications of NINETY, we define the macro-operator NINETY² by the block diagram

Due to the identification of a constant (1–*) image with its domain, for such images we will often write $-S$ instead of NINETY²(S). In any event, what is important is that the domain of NINETY²(f) is $\{(-i, -j): (i, j) \in D_f\}$. (See Figure 2.9.)

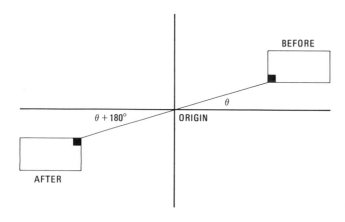

Figure 2.9 180° rotation about origin

We define the *Minkowski subtraction* $S \boxminus E$ by

$$[S \boxminus E](i, j) = \begin{cases} 1, & \text{if TRAN}(-E; i, j) \leq S \\ *, & \text{otherwise} \end{cases}$$

As in the Euclidean case, we are usually concerned with the *erosion*, ERODE$(S, E) = S \boxminus (-E)$. Pixelwise,

$$[\text{ERODE}(S, E)](i, j) = \begin{cases} 1, & \text{if TRAN}(E; i, j) \leq S \\ *, & \text{otherwise} \end{cases}$$

Erosion yields a "smaller" image than the original. As in the Euclidean case, the erosion of an image will be a subimage of the original if the origin is an activated pixel of the structuring element.

Like dilation, the erosion of S by E can be described intuitively by template translation, and it is again advised that a physical model be employed to help see this. (See Figure 2.10.) The template is moved across the minimal bound matrix of S. If, for a given pixel, say (i, j), the translated copy of E, TRAN$(E; i, j)$, is a subimage of S, then (i, j) is activated in the erosion; otherwise, (i, j) is given the value * in the eroded image. From this description it should be clear that, if the origin is activated in E, erosion eliminates those parts of the image that are small in comparison to the structuring element. The manner of the elimination is of course highly dependent on the shape of the element.

Sec. 2.5 Digital Minkowski Algebra

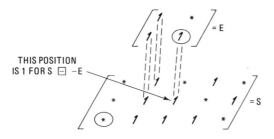

Figure 2.10 Illustration of erosion operation

Example 2.14:

Consider Example 2.13 again. TRAN(E; 1, 2) is certainly not a subimage of S. Therefore, (1, 2) will not be activated in the eroded image. On the other hand,

$$\text{TRAN}(E; 2, 1) = \begin{pmatrix} * & 1 & * \\ * & 1 & 1 \\ \circledast & * & * \end{pmatrix}$$

is a subimage of S. Hence, (2, 1) will be activated in ERODE(S, E) = $S \boxminus (-E)$. When all translations to activated pixels are checked, we obtain

$$\text{ERODE}(S, E) = \begin{pmatrix} * & * & 1 & * \\ \circledast & * & 1 & 1 \end{pmatrix}$$

As in the Euclidean case, Minkowski subtraction can be defined in terms of the intersection operator MIN($\wedge$) as follows:

$$S \boxminus E = \bigwedge_{(i,j) \in \text{DOMAIN}(E)} \text{TRAN}(S; i, j)$$

Since ERODE(S, E) = $S \boxminus (-E)$, a corresponding formulation of erosion is

$$\text{ERODE}(S, E) = \bigwedge_{(i,j) \in \text{DOMAIN}(E)} \text{TRAN}(S; -i, -j)$$

$$= \bigwedge_{(i,j) \in \text{DOMAIN}[\text{NINETY}^2(E)]} \text{TRAN}(S; i, j)$$

Example 2.15:

Let

$$S = \begin{pmatrix} 1 & 1 & 1 & * \\ 1 & 1 & 1 & * \\ * & 1 & * & 1 \\ \circledast & 1 & 1 & 1 \end{pmatrix}$$

and

$$E = \begin{pmatrix} * & 1 \\ \circledone & 1 \end{pmatrix}$$

There are three translations of S by pixels in the domain of E: TRAN(S; 0, 0) = S,

$$\text{TRAN}(S; 1, 0) = \begin{pmatrix} * & 1 & 1 & * \\ * & 1 & 1 & 1 & * \\ * & * & 1 & * & 1 \\ \circledast & * & 1 & 1 & 1 \end{pmatrix}$$

and

$$\text{TRAN}(S; 1, 1) = \begin{pmatrix} * & 1 & 1 & 1 & * \\ * & 1 & 1 & 1 & * \\ * & * & 1 & * & 1 \\ * & * & 1 & 1 & 1 \\ \circledast & * & * & * & * \end{pmatrix}$$

Application of $\wedge$ = MIN to the three translates yields

$$S \;\boxminus\; E = \begin{pmatrix} * & 1 & 1 \\ * & * & 1 \\ * & * & * \\ \circledast & * & * \end{pmatrix}$$

The block diagram for erosion is given by

A block diagram specification of ERODE in terms of primitive operations that employs the preceding MIN formulation is given in Figure 2.11. Figure 2.12 gives the erosion of the image S in Figure 2.7 by the structuring element E of the same figure.

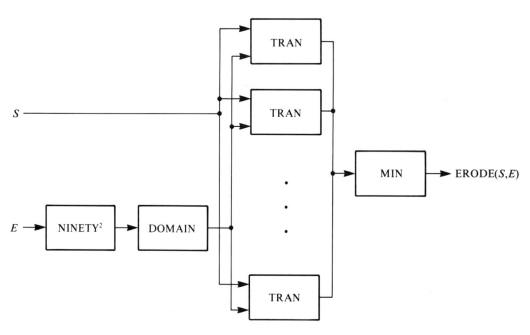

Figure 2.11 Block diagram of ERODE

Sec. 2.5   Digital Minkowski Algebra 61

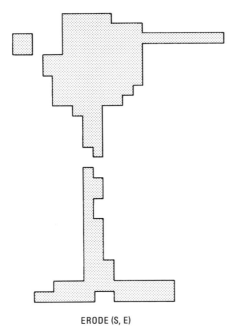

ERODE (S, E)

Figure 2.12   Digital erosion

In Section 1.3, some fundamental properties of $\oplus$ and $\ominus$ were studied. In the current chapter, our thrust has been toward the primitive operators of the morphological basis. From this point of view, $\boxed{+}$ and $\boxed{-}$ are macro-operators. In other words, Minkowski algebra might better be called *morphological algebra* in that it consists of operations derived from the primitives of the morphological basis (or the augmented morphological basis). We stay with the *Minkowski* terminology, however, in order to remain consistent with the terminology of the current literature.

We now proceed to the basic properties of digital dilation and erosion. Since these properties are analogues of Euclidean properties, we need only state them and consider some illustrative computational examples without going into intricate details or explanations. Keep in mind that human visual perception, and consequently understanding, is Euclidean; hence, an intuitive appreciation of the digital properties rests upon a concomitant intuitive grasp of the Euclidean analogues. Table 2.1 lists the basic digital properties of dilation and erosion; the numbers are the same as the numbers of the corresponding properties given in Chapter 1.

Note the absence in Table 2.1 of properties M-12 and M-13. The reason for their omission is that if one tries to define a digital analogue to the set operation $tS = \{(ti, tj): (i, j) \in S\}$, problems arise. In the first place, unless t is restricted to integer values, the points of tS will not lie within the lattice $Z \times Z$. Moreover, even if such a restriction is adopted, "holes" are created where none existed prior to the operation. (See Figure 2.13.) As a result, the Euclidean intuition of magnification as a continuous operation is lost. Consequently, we have chosen not to include a magnification operation within the morphological basis.

TABLE 2.1

M-1: $A \boxplus B = B \boxplus A$
M-2: $A \boxplus (B \boxplus C) = (A \boxplus B) \boxplus C$
M-3: $A \boxplus \text{TRAN}(B; i, j) = \text{TRAN}(A \boxplus B; i, j)$
M-4: $A \boxplus B = (A^c \boxminus B)^c$
M-5: $A \boxminus B = (A^c \boxplus B)^c$
M-6: $[\text{DILATE}(A, B)]^c = \text{ERODE}(A^c, -B)$
M-7: $[\text{ERODE}(A, B)]^c = \text{DILATE}(A^c, -B)$
M-8: $A \boxminus \text{TRAN}(B; i, j) = \text{TRAN}(A; i, j) \boxminus B$
         $= \text{TRAN}(A \boxminus B; i, j)$
M-9: If $A_1 \ll A_2$, then $\text{DILATE}(A_1, E) \ll \text{DILATE}(A_2, E)$
M-10: If $A_1 \ll A_2$, then $\text{ERODE}(A_1, E) \ll \text{ERODE}(A_2, E)$
M-11: If $E_1 \ll E_2$, then $\text{ERODE}(A, E_2) \ll \text{ERODE}(A, E_1)$
M-14: $A \boxplus B = \{(i, j): \text{TRAN}(-B; i, j) \wedge A \neq \emptyset\}$
M-15: $A \boxplus \text{EXTMAX}(B, C) = \text{EXTMAX}(A \boxplus B, A \boxplus C)$
M-16: $A \boxminus \text{EXTMAX}(B, C) = \text{MIN}(A \boxminus B, A \boxminus C)$
M-17: $\text{MIN}(B, C) \boxminus A = \text{MIN}(B \boxminus A, C \boxminus A)$
M-18: $(A \boxminus B) \boxminus C = A \boxminus (B \boxplus C)$

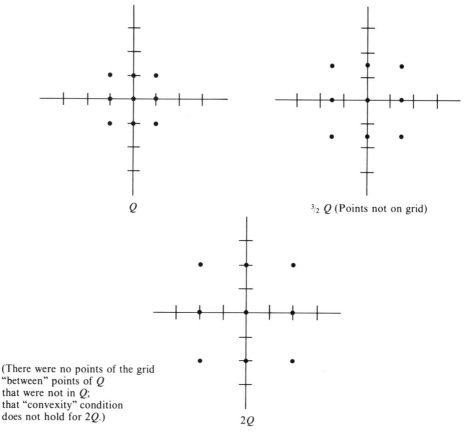

Q

$\tfrac{3}{2} Q$ (Points not on grid)

(There were no points of the grid "between" points of Q that were not in Q; that "convexity" condition does not hold for $2Q$.)

$2Q$

Figure 2.13 Difficulty with digital image scalar multiplication

Sec. 2.5 Digital Minkowski Algebra   63

We now proceed to illustrate some of the properties of Table 2.1.

Example 2.16:

Consider the image S and the structuring element E of Example 2.13. The following block diagram illustrates the application of M-3 to S and E.

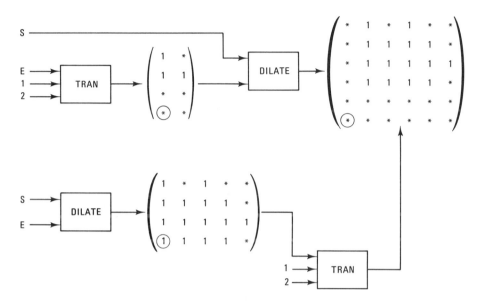

In other words,

$$\text{DILATE}[S, \text{TRAN}(E; 1, 2)] = \text{TRAN}[\text{DILATE}(S, E); 1, 2]$$

Before illustrating duality, it is necessary to explain in what manner the digital implementation of the erosion of a complementary bound matrix is to be accomplished. Let $S = (a_{pq})_{rt}$ be a finite digital image in minimal bound matrix form, $S^c = (b_{pq})_{rt}^c$ be its complement, and E be a structuring element. Note that ERODE(S^c, E) consists of all those pixels (i, j) for which the translated image TRAN($E; i, j$) is a subimage of S^c. Since all pixels outside the minimal complementary bound matrix $(b_{pq})_{rt}^c$ are assumed to be activated, we recognize that all pixels for which the translated image TRAN($E; i, j$) lies entirely outside $(b_{pq})_{rt}^c$ will have the value 1 in the erosion.

Now, suppose $(e_{pq})_{xy}$ is the *minimal origin-containing bound matrix for E*—that is, $(e_{pq})_{xy}$ is the bound matrix of least dimensions that contains all activated pixels of E and also has the origin within its frame. Suppose also that the dimensions of $(e_{pq})_{xy}$ are u by v. Finally, augment the bound matrix $(b_{pq})_{rt}^c$ by $u - 1$ rows of 1's on top and bottom, and by $v - 1$ columns of 1's on either side. Let this *augmented* bound matrix be denoted by $(b'_{pq})_{r-v+1, t+u-1}^c$. It is clear that if E is translated by any pixel outside the frame of the augmented matrix, then the resulting translate will be a subimage of S^c, and hence all such pixels will be activated in the eroded image. Consequently, we need only check the translations of E by pixels in the augmented minimal complementary bound matrix for S^c. This is a finite quantity that is equal to

$(m + 2u - 2)(n + 2v - 2)$, where the minimal bound matrix of S possesses dimensions m by n.

Example 2.17:

Let S and E be as in Example 2.13. Then S^c is given by

$$S^c = \begin{pmatrix} 1 & * & 1 & * & 1 \\ 1 & * & * & 1 & * \\ ① & * & * & * & 1 \end{pmatrix}^c$$

Since the minimal bound matrix for E is 2 by 2, and since it already contains the origin, the necessary augmented version of S^c is given by

$$S^c = \begin{pmatrix} 1 & 1 & 1 & 1 & 1 & 1 \\ 1 & * & 1 & * & 1 & 1 \\ 1 & * & * & 1 & * & 1 \\ ① & * & * & * & 1 & 1 \\ 1 & 1 & 1 & 1 & 1 & 1 \end{pmatrix}^c$$

Recalling that all pixels external to the frame have gray value 1 (since we are referring to S^c), we successively translate E about the augmented bound matrix and retain the value $S^c(i, j) = 1$ for any (i, j) for which

$$\text{TRAN}(E; i, j) \ll S^c$$

In other words, those pixels (i, j) for which $\text{TRAN}(E; i, j)$ is a subimage of S^c constitute the erosion. For example, pixel $(1, -1)$ is activated in the eroded image since $\text{TRAN}(E; 1, -1)$ is a subimage of S^c. After performing all necessary translations throughout the augmented bound matrix for S^c, we obtain

$$\text{ERODE}(S^c, E) = \begin{pmatrix} 1 & 1 & 1 & 1 & 1 & 1 \\ 1 & * & * & * & * & 1 \\ 1 & * & * & * & * & * \\ ① & * & * & * & * & * \\ 1 & 1 & * & * & * & 1 \end{pmatrix}^c$$

In minimal form containing the origin, we have

$$\text{ERODE}(S^c, E) = \begin{pmatrix} 1 & * & * & * & * & 1 \\ 1 & * & * & * & * & * \\ ① & * & * & * & * & * \\ 1 & 1 & * & * & * & 1 \end{pmatrix}^c$$

Example 2.18:

Using the results of Example 2.17 together with property M-6, we have

$$\text{DILATE}(S, -E) = [\text{ERODE}(S^c, E)]^c$$

$$= \text{COMP}\left[\begin{pmatrix} 1 & * & * & * & * & 1 \\ 1 & * & * & * & * & * \\ ① & * & * & * & * & * \\ 1 & 1 & * & * & * & 1 \end{pmatrix}^c\right]$$

$$= \begin{pmatrix} * & 1 & 1 & 1 & 1 & * \\ * & 1 & 1 & 1 & 1 & 1 \\ ⊛ & 1 & 1 & 1 & 1 & 1 \\ * & * & 1 & 1 & 1 & * \end{pmatrix}$$

Sec. 2.5   Digital Minkowski Algebra

The same result can be obtained by dilating S directly by

$$-E = \text{NINETY}^2(E) = \begin{pmatrix} ① & 1 \\ * & 1 \end{pmatrix}$$

Example 2.19:

Let S once again be the image of Example 2.13; however, let $E_1 = (1 \quad ①)$ and

$$E_2 = \begin{pmatrix} 1 & * \\ 1 & ⊛ \end{pmatrix}$$

Then $E = \text{EXTMAX}(E_1, E_2)$ is the structuring element of Example 2.13. Now,

$$S \boxplus E_1 = \begin{pmatrix} 1 & 1 & 1 & 1 & * \\ 1 & 1 & 1 & 1 & 1 \\ ① & 1 & 1 & 1 & * \end{pmatrix}$$

and

$$S \boxplus E_2 = \begin{pmatrix} 1 & * & 1 & * \\ 1 & 1 & 1 & 1 \\ 1 & 1 & 1 & 1 \\ ① & 1 & 1 & * \end{pmatrix}$$

Thus, operating by EXTMAX yields precisely the result for $S \boxplus E$ that was obtained in Example 2.13. This result is of course guaranteed by property M-15.

Example 2.20:

Let E_1, E_2, and E be as in Example 2.19, and let

$$T = \begin{pmatrix} 1 & 1 & 1 & * \\ 1 & 1 & * & 1 \\ ① & 1 & * & * \end{pmatrix}$$

Then

$$T \boxminus E_1 = \begin{pmatrix} 1 & 1 \\ 1 & * \\ ① & * \end{pmatrix}$$

$$T \boxminus E_2 = \begin{pmatrix} 1 & 1 \\ 1 & 1 \\ * & ⊛ \end{pmatrix}$$

and

$$\text{MIN}(T \boxminus E_1, T \boxminus E_2) = \begin{pmatrix} 1 \\ 1 \\ ⊛ \end{pmatrix} = T \boxminus E$$

This is exactly the result demanded by property M-16.

Example 2.21:

Let E_1, E_2, and T be as in Example 2.20. Then

$$E_1 \boxplus E_2 = \begin{pmatrix} 1 & 1 & * \\ 1 & 1 & ⊛ \end{pmatrix}$$

and

$$(T \boxminus E_1) \boxminus E_2 = \begin{pmatrix} 1 & * \\ 1 & * \\ * & \circledast \end{pmatrix} = T \boxminus (E_1 \boxplus E_2)$$

which illustrates property M-18.

Example 2.22:

In operator terminology, properties M-2, M-4, M-8, M-16, M-17, and M-18 become, respectively,

M-2: DILATE$[A$, DILATE$(B, C)] = $ DILATE$[$DILATE$(A, B), C]$

M-4: DILATE$(A, B) = $ COMP$[$ERODE$($COMP(A), NINETY$^2(B))]$

M-8: ERODE$[A$, NINETY$^2($TRAN$(B; i, j))]$

$\qquad\qquad = $ ERODE$[$TRAN$(A; i, j)$, NINETY$^2(B)]$

$\qquad\qquad = $ TRAN$[$ERODE$(A$, NINETY$^2(B)); i, j]$

M-16: ERODE$[A$, NINETY$^2($EXTMAX$(B, C))]$

$\qquad\qquad = $ MIN$[$ERODE$(A$, NINETY$^2(B))$, ERODE$(A$, NINETY$^2(C))]$

M-17: ERODE$[$MIN(B, C), NINETY$^2(A)]$

$\qquad\qquad = $ MIN$[$ERODE$(B$, NINETY$^2(A))$, ERODE$(C$, NINETY$^2(A))]$

M-18: ERODE$[$ERODE$(A$, NINETY$^2(B))$, NINETY$^2(C)]$

$\qquad\qquad = $ ERODE$[A$, NINETY$^2($DILATE$(B, C))]$

Noting that

$\qquad$ NINETY$^2[$EXTMAX$(B, C)] = $ EXTMAX$[$NINETY$^2(B)$, NINETY$^2(C)]$

and using the fact that NINETY$^2[$NINETY$^2(S)] = S$ (idempotence of NINETY^2), we can rewrite M-16 as

M-16: ERODE$[A$, EXTMAX$(B, C)] = $ MIN$[$ERODE(A, B), ERODE$(A, C)]$

Similarly, the idempotence of NINETY^2 allows us to write M-17 as

M-17: ERODE$[$MIN$(B, C), A] = $ MIN$[$ERODE(B, A), ERODE$(C, A)]$

Other such reductions are possible utilizing the algebraic properties of the morphological primitives in conjunction with the properties of the macro-operators derived from those primitives.

2.6 DIGITAL OPENING AND CLOSING

The digital opening and closing are defined analogously to the definitions of the corresponding Euclidean operators, i.e.,

$$\text{OPEN}(S, E) = [S \boxminus (-E)] \boxplus E = \text{DILATE}[\text{ERODE}(S, E), E]$$

Sec. 2.6 Digital Opening and Closing

and

$$\text{CLOSE}(S, E) = [S \boxplus (-E)] \boxminus E = \text{ERODE}[\text{DILATE}(S, -E), -E]$$

Comments concerning the manner in which the Euclidean opening and closing affect the input image apply without material alteration to the digital versions of the same operators. Figure 2.14 gives black-and-white interpretations of the opening and closing of image S of Figure 2.7 by the structuring element E of that same figure.

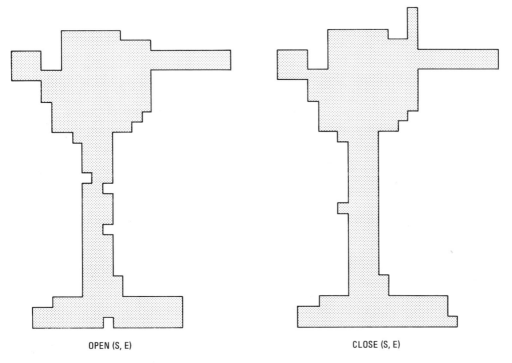

Figure 2.14 Digital opening and closing

The respective block diagrams of OPEN and CLOSE are

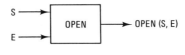

and

The block diagram specifications for OPEN and CLOSE result directly from their respective definitions in terms of DILATE and ERODE. These are given in Figure 2.15.

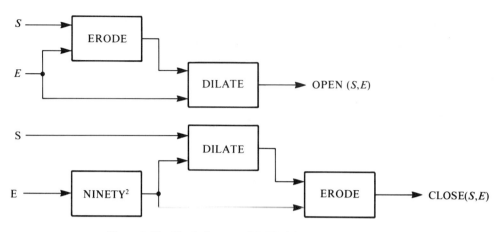

Figure 2.15 Block diagram of OPEN/block diagram of CLOSE

In a form similar to Theorem 1.2, the opening can be represented as an extended maximum (union) of fitted translates of the structuring element:

$$\text{OPEN}(S, E) = \bigvee \{\text{TRAN}(E; i, j): \text{TRAN}(E; i, j) \ll S\}$$

Example 2.23:

Consider the image S and the structuring element E of Example 2.13. There are three translations of E that fit into S: TRAN$(E; 2, 0)$, TRAN$(E; 3, 0)$, and TRAN$(E; 2, 1)$. These are precisely the translations that yielded the erosion given in Example 2.14. The extended maximum of these translations gives the opening:

$$\text{OPEN}(S, E) = \begin{pmatrix} * & 1 & * & * \\ * & 1 & 1 & * \\ \circledast & 1 & 1 & 1 \end{pmatrix}$$

The manner in which the opening reveals textural information is revealed in the next example, where we utilize *digital linear structuring elements*. A digital linear structuring element is called *vertical* if it is a column of 1's with the foot of the column situated at the origin, and *horizontal* if it is a row of 1's with the leftmost entry at the origin. These structuring elements will play a central role in the construction of digital linear granulometries (Section 3.6).

Example 2.24:

Let

$$S = \begin{pmatrix} * & 1 & * & * & 1 & 1 \\ 1 & ① & 1 & * & 1 & 1 \\ 1 & 1 & * & * & 1 & 1 \\ 1 & * & * & * & 1 & 1 \\ 1 & * & * & * & 1 & * \end{pmatrix}$$

$$E = \begin{pmatrix} 1 \\ 1 \\ ① \end{pmatrix} \quad \text{and} \quad E' = (① \quad 1 \quad 1)$$

Sec. 2.6 Digital Opening and Closing

Opening S by E yields

$$\text{OPENS}(S, E) = \begin{pmatrix} * & 1 & * & * & 1 & 1 \\ 1 & ① & * & * & 1 & 1 \\ 1 & 1 & * & * & 1 & 1 \\ 1 & * & * & * & 1 & 1 \\ 1 & * & * & * & 1 & * \end{pmatrix}$$

This follows since OPEN(S, E) is given by the union, or extended maximum, of the following translates of E: TRAN(E; -1, -2), TRAN(E; -1, -3), TRAN(E; 0, -1), TRAN(E; 3, -1), TRAN(E; 3, -2), TRAN(E; 3, -3), TRAN(E; 4, -1), and TRAN (E; 4, -2). On the other hand, opening by E' yields

$$\text{OPEN}(S, E') = (\,1 \quad ① \quad 1 \quad * \quad * \quad *\,)$$

The only translate of E' that fits is TRAN(E'; -1, 0).

Note that the activated pixels of S tend to form vertical strings. Thus, there appears to be vertical elongation. By contrast, there is little or no horizontal elongation. The result is a greatly diminished image when S is opened by a horizontal linear element and little change when it is opened by a vertical element. Obviously, the result depends not only on the shape of the structuring element but also on the size. Had the vertical element been of greater length, it, too, could have resulted in an extensive diminution of the original image. It is precisely this ability of opening to distinguish textural differences that makes it so useful.

Example 2.25:

We compute an opening and a closing directly from the respective definitions in terms of DILATE and ERODE. Let S and E be as in Example 2.13. Then

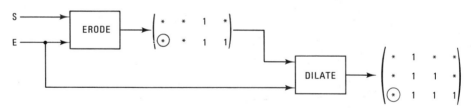

which is the opening of S by E. (See Example 2.23.) As for the closing, Figure 2.16 gives a walk-through.

Theorems 1.3 and 1.4 both carry over to the digital setting. For instance, the three conditions of Theorem 1.3 become

(i) OPEN(S, E) ≪ S
(ii) If S_1 ≪ S_2, then OPEN(S_1, E) ≪ OPEN(S_2, E)
(iii) OPEN[OPEN(S, E), E] = OPEN(S, E)

A similar restatement applies to Theorem 1.4, except in its case, (i) expresses extensivity. Note that the comments regarding the importance of these theorems to the construction of morphological filters apply without alteration.

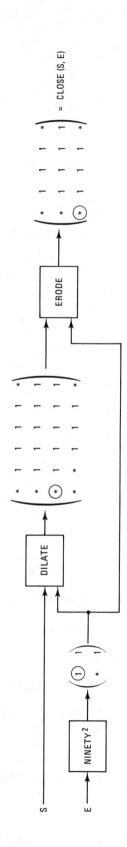

Figure 2.16 Walk-through of closing

Sec. 2.6   Digital Opening and Closing

Just as the properties of Euclidean dilation and erosion had the digital analogues given in Table 2.1, the properties of Euclidean opening and closing have analogues that apply to OPEN and CLOSE. These are summarized in Table 2.2.

TABLE 2.2

M-19: $CLOSE(S, E)^c = OPEN(S^c, E)$

M-20: $[CLOSE(S, E)](p, q) = \begin{cases} 1, & \text{if, for all } (i, j) \text{ such that } [TRAN(E; i, j)](p, q) = 1, \\ & TRAN(E; i, j) \wedge S \neq \emptyset \\ *, & \text{otherwise} \end{cases}$

M-21: $OPEN[TRAN(S; i, j), E] = TRAN[OPEN(S, E); i, j]$

M-22: $CLOSE[TRAN(S; i, j), E] = TRAN[CLOSE(S, E); i, j]$

Property M-20 deserves close attention. According to the set-theoretic interpretation of constant-image operations, pixel (p, q) lies in the closing if and only if every translate of E containing (p, q) intersects the original image S, which is precisely what is stated by the Euclidean version of M-20. To employ M-20 directly upon bound matrices, the structuring element must be translated to all pixels in the grid structure, an impossible task to implement. However, pixels outside the minimal bound matrix frame of S certainly cannot be in the closing. Moreover, since the closing is extensive, we need only check those pixels lying in the bound matrix frame that are deactivated; those that are activated are sure to be in $CLOSE(S, E)$.

Example 2.26:

For the image S and the structuring element E of Example 2.13, $CLOSE(S, E)$ was computed in Figure 2.16. Treating E as a template and successively placing the activated pixels of E over the deactivated pixels in the bound matrix of S, the closing can be obtained by M-20. In finding the closing by this method, each pixel in the bound matrix can occupy only three positions in the template since E possesses only three activated pixels. For example, there are three translates of E that have the value 1 at $(2, 2)$: $TRAN(E; 3, 2)$, $TRAN(E; 2, 2)$, and $TRAN(E; 3, 1)$. Since each of these translates has nonempty intersection with S, the pixel $(2, 2)$ is activated in the closing.

On the other hand, consider the pixel $(4, 2)$. It is not true that each of the three translates of E that has the value 1 at $(4, 2)$ also has nonempty intersection with S. In particular,

$$TRAN(E; 5, 2) \wedge S = \emptyset$$

even though $[TRAN(E; 5, 2)](4, 2) = 1$. Consequently,

$$[CLOSE(S, E)](4, 2) = *$$

Augmentation plays a role in property M-19. To open a complementary bound matrix by fitting structuring element E, we need to augment the matrix by $u - 1$ rows of 1's on top and bottom, and by $v - 1$ rows of 1's on either side, where u and v are the dimensions of the minimal bound matrix representing E. Notice the difference in augmentation in this case as opposed to the situation that occurred with erosion, where the minimal origin-containing bound matrix was used: insofar as the opening is concerned, the position of the origin with respect to the activated pixels of the structuring element is irrelevant. Indeed,

$$\text{OPEN}[S, \text{TRAN}(E; i, j)] = \text{OPEN}(S, E)$$

Example 2.27:

Let S and E be as in Example 2.17. Once again the template E is translated about the augmented minimal bound matrix for S^c. Once this is done, the extended maximum, or union, of all those translates that are subimages of S^c is taken. The result is given by

$$\text{OPENS}(S^c, E) = \begin{pmatrix} 1 & 1 & 1 & 1 & 1 & 1 \\ 1 & * & * & * & 1 & 1 \\ 1 & * & * & * & * & 1 \\ \text{\textcircled{1}} & * & * & * & 1 & 1 \\ 1 & 1 & 1 & 1 & 1 & 1 \end{pmatrix}^c$$

As in the Euclidean case, property M-19 can be written in other forms. However, these are strictly analogous and therefore will not be explicitly stated.

Since the notions of an image being open or closed with respect to another image carry over directly from the Euclidean case, properties M-23, M-24, and M-25 have direct analogues.

2.7 DIGITIZATION

Thus far, we have developed the fundamental morphological operations in two different settings, Euclidean and digital. In order to move from one setting to the other, we must formally characterize the manner in which we will *digitize*, or *sample*, a constant Euclidean image. There are numerous ways of doing so; however, we shall introduce only one way, a way that is in accord with typical grid-type approximations employed in other areas of mathematical analysis.

Consider the square grid of points $(i/2^k, j/2^k)$ in the plane, where k is fixed (thereby determining the *sampling rate*), and where i and j are arbitrary integers. The plane is then the union of the closed squares Q_p centered at p and of edge length 2^{-k}, where p is a grid point. In this setting, given a Euclidean constant image A, we associate with it the Euclidean constant image $\tilde{A}$, where $\tilde{A}$ is the union of all the pixel squares Q_p such that A intersects the interior of Q_p. The intention is to approximate A by $\tilde{A}$. Finally, we define the *digitization* of A to be $\bar{A}$, where $\bar{A}$ is the collection of all grid points p such that $Q_p \subset \tilde{A}$. (See Figure 2.17.) Since $\bar{A}$ is defined on the collection of 2^{-k}-grid points, it can be digitally represented as a subset of the integral lattice $Z \times Z$; in particular, each element of $\bar{A}$ can be represented by a pair (i, j).

Following are some immediate relations among A, $\tilde{A}$, and $\bar{A}$:

(*i*) The interior of A is a subset of $\tilde{A}$.
(*ii*) $m(A) \leq m(\tilde{A})$, where m denotes area (Lebesgue measure).
(*iii*) $m(\tilde{A}) = 2^{-2k} \text{CARD}(\bar{A})$.
(*iv*) If $A_1 \subset A_2$, then $\tilde{A}_1 \subset \tilde{A}_2$ and hence $m(\tilde{A}_1) \leq m(\tilde{A}_2)$.

Sec. 2.7　Digitization

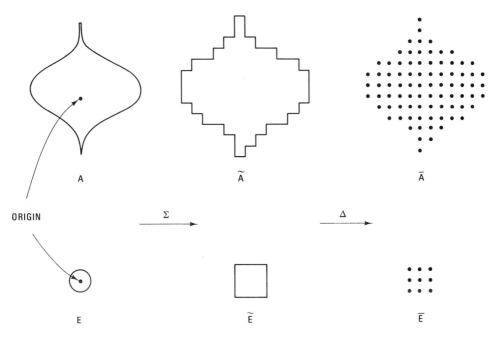

Figure 2.17　Digitization process

Given a Euclidean image A, we denote the mapping $A \to \tilde{A}$ by $\tilde{A} = \Sigma(A)$. Given a Euclidean image $\tilde{A}$, which is a union of pixel squares Q_p, we denote the mapping $\tilde{A} \to \overline{A}$ by Δ. Note that Δ is invertible in that $\Delta^{-1}(\overline{A})$ is well defined.

The *digitization problem (sampling problem)* can be crudely stated as follows: given a Euclidean image mapping Ψ, can one find a digital mapping $\overline{\Psi}$ which in some sense "does essentially the same thing as Ψ"? To solve the problem, we need to specify exactly what the quoted expression means. This is done schematically in Figure 2.18.

In theory, given an image A and a Euclidean image-to-image mapping Ψ, we could simply find $\Psi(A)$. In practice, of course, we wish to accomplish this digitally.

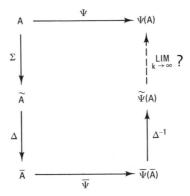

Figure 2.18　Definition of a digitizable operation

We proceed as follows:

1. Digitize A—that is, find $\bar{A}$.
2. Operate on the $Z \times Z$ specification of $\bar{A}$ by some "digital version" of Ψ, called $\overline{\Psi}$.
3. Given the output $\overline{\Psi}(\bar{A})$ of $\overline{\Psi}$, form the Euclidean image $\Delta^{-1}[\overline{\Psi}(\bar{A})]$, which is the union of all squares Q_p such that the center of Q_p lies in the $Z \times Z$ representation $\overline{\Psi}(\bar{A})$.

Letting $\tilde{\Psi}(A) = \Delta^{-1}[\overline{\Psi}(\bar{A})]$, we wish to find out how well $\tilde{\Psi}(A)$ approximates the theoretically desired result $\Psi(A)$.

To be precise, both $\overline{\Psi}$ and $\tilde{\Psi}$ actually depend on k, which is why there is the limit operation in Figure 2.18. Indeed, we say that Ψ is *digitizable* if, for any compact set A,

$$\lim_{k \to \infty} \tilde{\Psi}(A) = \Psi(A).$$

As stated, the limit operation poses serious difficulties: not only must one define in what sense the limit is to be taken, but also, in practice the value of k is fixed by the actual digitizer in use. Insofar as the first problem is concerned, at least three approaches have been taken:

1. Matheron[1] has developed a topology of images in the Euclidean plane and Serra[2] has presented the digitization problem in terms of the Matheron topology.
2. The Hausdorff metric (see Section 3.3) induces a convergence criterion which is a special case of convergence in the Matheron topology. Digitization can be discussed in terms of this metric, thereby avoiding the abstractness of the Matheron topology.[3]
3. Both the Matheron topology and the Hausdorff metric are extremely sensitive to noise. Another approach to the digitization problem is to examine the set-theoretic difference between the outputs $\tilde{\Psi}(A)$ and $\Psi(A)$.[4]

The second difficulty mentioned, that being the fixed size of k, is of a more practical nature, and it is this latter problem on which we wish to focus. To be precise, we desire some estimate of the maximum difference between the outputs of $\tilde{\Psi}$ and Ψ.

Perhaps one of the most important digitizations is that of the opening. Suppose A and E are compact sets and we wish to find $O(A, E)$, but we wish to proceed digitally. We first digitize A and E; that is, we find $\tilde{A} = \Sigma(A)$, $\tilde{E} = \Sigma(E)$, $\bar{A} = \Delta(\tilde{A})$, and $\bar{E} = \Delta(\tilde{E})$. Then, from an inspection of Figure 2.17, we see that we should open $\tilde{A}$ by $\tilde{E}$ by checking the fits of $\tilde{E}$ into $\tilde{A}$. This is accomplished digitally by finding OPEN($\bar{A}$, $\bar{E}$) and then reconstructing the Euclidean image by taking the union of all pixel squares with centers in OPEN($\bar{A}$, $\bar{E}$). Symbolically, the desired approximation is given by $\Delta^{-1}[\text{OPEN}(\bar{A}, \bar{E})]$. The important point is that the computation of OPEN($\bar{A}$, $\bar{E}$) is a purely digital operation on bound matrices. If we fix E, then, using the notation of Figure 2.18, we have $\Psi(A) = O(A, E)$, $\overline{\Psi}(\bar{A}) = \text{OPEN}(\bar{A}, \bar{E})$, and $\tilde{\Psi}(A) = \Delta^{-1}[\text{OPEN}(\bar{A}, \bar{E})]$.

Sec. 2.7 Digitization

Example 2.28:

For A and E as given in Figure 2.17, the bound matrix form of $\bar{A}$ is given in Figure 2.19. Figure 2.20 gives the Euclidean opening $O(A, E)$ together with the approximation that was obtained digitally.

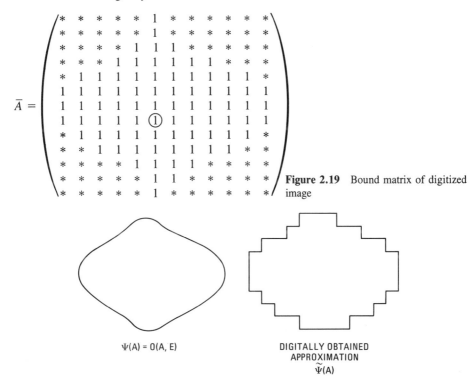

Figure 2.19 Bound matrix of digitized image

Figure 2.20 A digitally obtained approximation

The preceding methodology for digitally generating the opening can also be employed to digitally generate the dilation, erosion, and closing of an image. The mappings $\bar{\Psi}$ are defined by DILATE($\bar{A}$, $\bar{E}$), ERODE($\bar{A}$, $\bar{E}$), and CLOSE($\bar{A}$, $\bar{E}$), respectively.

As we shall see in Sections 3.2 and 3.5, basic morphological feature measurements are obtained by taking the areas of the outputs of operations such as the opening and the erosion. Consequently, an important question arises concerning the extent to which the area of the Euclidean output $\Psi(A)$ compares with the area of the digitally derived Euclidean output $\tilde{\Psi}(A)$. By relation (*iii*) above, for the opening,

$$m[\tilde{\Psi}(A)] = 2^{-2k} \operatorname{CARD}[\operatorname{OPEN}(\bar{A}, \bar{E})]$$

Although we shall not pursue the matter here, for convex A and E, bounds on errors of the form

$$\left| m[\Psi(A)] - m[\tilde{\Psi}(A)] \right|$$

have been found for dilation, erosion, and opening.[5,6]

EXERCISES

2.1. Let
$$f = \begin{pmatrix} 1 & 2 & 3 & -1 \\ 0 & * & 1 & -2 \\ * & 2 & 0 & 5 \end{pmatrix}_{0,2}$$

and

$$g = \begin{pmatrix} 2 & 2 & 0 \\ * & 2 & -2 \\ 4 & 1 & 6 \end{pmatrix}_{1,1}$$

Find:
(a) EXTMAX(f, g)   (e) NINETY^2(g)
(b) MIN(f, g)   (f) DOMAIN(f)
(c) TRAN($f; 2, -4$)   (g) RANGE(f)
(d) NINETY(g)   (h) CREATE([(1, -1), (0, 1), (0, 0)], [2, *, -2])

2.2. Consider the constant images

$$S = \begin{pmatrix} 1 & 1 & 1 & 1 & 1 & * \\ 1 & 1 & 1 & * & * & * \\ 1 & 1 & 1 & 1 & * & * \\ * & * & 1 & 1 & 1 & 1 \\ * & ⊛ & 1 & 1 & 1 & 1 \\ * & * & 1 & 1 & 1 & 1 \end{pmatrix}$$

$$T = \begin{pmatrix} ① & 1 & 1 & 1 & 1 & 1 & 1 \\ 1 & 1 & * & * & 1 & * & 1 \\ * & * & * & * & 1 & 1 & 1 \end{pmatrix}$$

and

$$E = \begin{pmatrix} 1 & 1 \\ ① & 1 \end{pmatrix}$$

Find
(a) S^c (g) ERODE(S, E)
(b) $S \wedge T$ (h) ERODE(T, E)
(c) $S \vee T$ (i) OPEN(S, E)
(d) TRAN($E; 1, 2$)   (j) OPEN(T, E)
(e) DILATE(S, E)   (k) CLOSE(S, E)
(f) DILATE(T, E)   (l) CLOSE(T, E)

2.3. Using S and E of Exercise 2.2, find the following morphological outputs without using any properties mentioned in this chapter:
(a) DILATE(S^c, E)   (c) OPEN(S^c, E)
(b) ERODE(S^c, E)   (d) CLOSE(S^c, E)

2.4. Using results obtained in Exercise 2.2, find the following morphological outputs by using any of the properties mentioned in this chapter.
(a) DILATE(T^c, E)   (c) OPEN(T^c, E)
(b) ERODE(T^c, E)   (d) CLOSE(T^c, E)

Sec. 2.8 Exercises

2.5. Using the images

$$A = \begin{pmatrix} 1 & 1 & 1 \\ 1 & ① & 1 \\ 1 & 1 & 1 \end{pmatrix}$$

$$B = \begin{pmatrix} 1 & * & 1 \\ ① & * & 1 \end{pmatrix}$$

and

$$C = \begin{pmatrix} 1 & 1 \\ ⊛ & * \\ 1 & 1 \end{pmatrix}$$

demonstrate properties M-1 through M-7 of Table 2.1.

2.6. Using the images A and B of Exercises 2.5, and letting $i = 2$ and $j = -3$, demonstrate properties M-8 and M-14 of Table 2.1.

2.7. Let A be the image S of Exercise 2.2, and let

$$B = \begin{pmatrix} 1 & * \\ ⊛ & 1 \end{pmatrix}$$

and

$$C = \begin{pmatrix} 1 & 1 \\ ① & * \end{pmatrix}$$

Using A, B, and C, demonstrate M-15 through M-18 of Table 2.1.

2.8. Let A and E be the images given in Figure 2.21, both appearing on the background grid.
(a) Find $\tilde{A}$, $\bar{A}$, $\tilde{E}$, and $\bar{E}$.
(b) Using the digitization methodology of Section 2.7, find the approximations $\Delta^{-1}[\text{OPEN}(\bar{A}, \bar{E})]$, $\Delta^{-1}[\text{ERODE}(\bar{A}, \bar{E})]$, $\Delta^{-1}[\text{DILATE}(\bar{A}, \bar{B})]$, and $\Delta^{-1}[\text{CLOSE}(\bar{A}, \bar{E})]$ to the respective Euclidean operations $O(A, E)$, $\mathscr{E}(A, E)$, $\mathscr{D}(A, E)$, and $C(A, E)$. In each case find the area of the error.

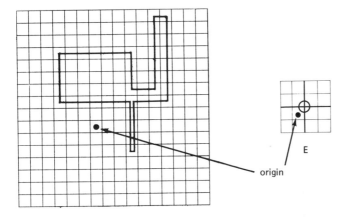

Figure 2.21 Images for Exercise 2.8

2.9. Using A and E of Exercise 2.8, perform the Euclidean operation $\mathcal{D}(\tilde{A}, \tilde{E})$. Note that this Euclidean operation is not identical to $\Delta^{-1}[\text{DILATE}(\overline{A}, \overline{E})]$. This means that digital implementation of dilation through the digitization schema is not equivalent to Euclidean dilation of $\tilde{A}$ by $\tilde{E}$. A similar comment applies to erosion via digitization. On the other hand, note that

$$O(\tilde{A}, \tilde{E}) = \Delta^{-1}[\text{OPEN}(\overline{A}, \overline{E})]$$

In fact, it is easy to show that this equality always holds. More generally, erosion and dilation behave quite differently with respect to digitization than do opening and closing.[7]

FOOTNOTES FOR CHAPTER 2

1. Georges Matheron, *Random Sets and Integral Geometry* (New York: Wiley, 1975), p. 1–16.
2. Jean Serra, *Image Analysis and Mathematical Morphology* (New York: Academic Press, 1983), p. 207–220.
3. E. R. Dougherty and C. R. Giardina, "Sampling Criteria for Euclidean Images," SPSE 39th Annual Conference (1986).
4. E. R. Dougherty and C. R. Giardina, "Error Bounds For Morphologically Derived Feature Measurements," SIAM Journal on Applied Mathematics (in press).
5. Ibid.
6. E. R. Dougherty and C. R. Giardina, *Image Processing: Continuous to Discrete*, Volume I (Englewood Cliffs: Prentice-Hall, 1987).
7. Dougherty and Giardina, "Error Bounds for Morphologically Derived Feature Measurements."

3

Morphological Features

3.1 QUANTITATIVE FEATURE GENERATION

It is common practice in both image and signal processing to operate on an image in such a manner as to produce a new structure to "replace" the original image. This new structure might be the result of a transform technique, or it might simply be a set of *feature* measurements taken on the image. In either case, given an image f, we arrive at a transformed structure $\mathcal{T}(f)$, where the specific transformation employed depends upon the goals we have in mind.

The information revealed by the transformation depends upon the mathematical properties it possesses. So, too, does the amount of information lost in the transformation process—for instance, if no information is lost, then the process is invertible. It might also be that the process preserves certain fundamental mathematical operations, as in the case of linear transformations. Any significance regarding the preservation of information or mathematical structure is relative to the intent of the investigator.

In pattern recognition, the problem is classification. If an operation $\mathcal{T}$ results in a satisfactory recognizer system, then all else is of little interest. For example, if a filter reduces noise to acceptable levels, then whether the filter is linear or not may be of no consequence.

The genesis of the morphological methodology lies in the search for structure within an image. The underlying strategy in the description of structure is to understand the textural or geometric properties of an image by probing the microstructure of the image with various forms. Indeed, our study of the Minkowski algebra has had the purpose of developing a systematic mathematical framework within which the study of such probes can take place. As we have seen in the case of a constant image, the analysis is geometric in character. The intent is to approach image processing from the vantage point of human perception by deriving quantitative measures of natural perceptual categories, thereby exploiting whatever inherent congruences exist between image structure and ordinary human recognition. Necessarily, such an approach must break free of the classical linear-space framework that has so long dominated applied mathematics. Nevertheless, the method is well suited to eventual integration into an artificial intelligence schema: for a computer vision system to yield image-based decisions resembling those that result from direct human understanding, the categories upon which that system operates must correspond well to native human perceptual categories, whether or not the ensuing mathematical apparatus happens to be one that has served well for other classes of problems.

In searching for a given pattern within an image, a person perceives the image through the filter of his or her own motivation. Sensory data are not passively received and acted upon by analytic intelligence; rather, they are organized by the brain into *percepts*, and it is these percepts that are the raw material for analysis. To employ engineering terminology, one might loosely refer to the act of perception (i.e., of rendering data into percepts) as a form of data compression. Such compression involves a choice, prior to sensory reception, as to what manner the compression is to take place in and what end it serves. In addition to this sensory organization, higher level filtering must take place in order to search for desired patterns.

The elaboration of structure, which is, after all, the intent of image processing, involves an analysis of the relationships between the component parts of whatever object is under investigation. While it might be argued that structure inheres within an image, it certainly cannot be maintained that inherent structure is measurable, or even perceivable, while remaining outside the categories of human perception and conception. Consequently, those relationships which comprise (perceived) image structure are imposed upon the image by intelligence and do not exist independently of intelligence. In the words of Matheron,

> In general, the structure of an object is defined as the set of relationships existing between elements or parts of the object. In order to experimentally determine this structure, we must try, one after the other, each of the possible relationships and examine whether or not it is verified. Of course, the image constructed by such a process will depend to the greatest extent on the choice made for the system of relationships considered as possible. Hence, this choice plays a priori a constitutive role (in the Kantian meaning) and determines the relative worth of the concept of structure at which we arrive.[1]

Thus, only those aspects of the image which conform to some predetermined set of relational categories are relevant, and in that sense the image engineer's choice of these categories *constitutes*, or frames, the image. For practical morphological image processing, this means that the type of filtering or probing of an image depends upon

Sec. 3.2 Image Functionals

the particular knowledge desired. Once the image is constituted in terms of the relational base of this desired knowledge, other characteristics of the image are no longer accessible. However, if the base is well chosen relative to one's aims, the other characteristics are irrelevent.

Guided by the foregoing philosophy, morphological feature extraction has tended to center around a particular methodology according to which an image is probed by a family of structuring elements, resulting in a sequence of filtered images. Each of the filtered images is then "measured" in turn, and the sequence of measurements serves as the *signature* of the image. The methodology is summarized in Figure 3.1, where the loop structure is evident. The input to the diagram is an image, and the output is a sequence of parameters.

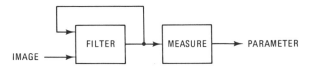

Figure 3.1 Block diagram illustration of parameter extraction

The properties of the foregoing transformation procedure depend not only upon the choice of structuring elements, but also upon the choice of filtering technique. Among such techniques, we shall pay particular attention to the granulometries of Section 3.5. Not only do these have especially pleasing properties, but they also characterize an entire class of one-parameter filters, the so-called *Euclidean granulometries*. Section 5.5 is devoted to this characterization.

3.2 IMAGE FUNCTIONALS

In this section, we shall study some general properties that the numerical parameters associated with images may or may not satisfy. Different classification problems require that one select parameters with different properties. The properties we shall discuss are not restricted in their application to morphology: *any* quantification of structure, whether it results from morphological, statistical, Fourier, or other methods, can be analyzed in terms of the constraints to be considered subsequently.

A numerical function on the set of all Euclidean constant images is called an *image functional* or *feature parameter*. A general image functional is a mapping

$$q: 2^{R \times R} \longrightarrow R$$

In practice, this definition is often too general, so we usually restrict our attention to some subclass, say X, of the class of all images and speak of an image functional $q: X \to R$. For instance, when we are concerned with topological properties of constant Euclidean images, we customarily restrict our attention to the class $\mathcal{H}$ of compact images. When employing concepts from integral geometry, we often make the further restriction that the images come from the class $\mathcal{H}_c$ of convex compact images. The important point to remember is that when we speak of some image-functional property, we often speak of the satisfaction of that property relative to some subclass of images, which could be the entire collection of constant Euclidean images.

Not all image functionals are suitable for morphological analysis. For specific situations, appropriate requirements must be imposed. Also as mentioned earlier, a given functional might behave differently on some classes of images than it does on others. It is the characterization of image-specific functional properties that will concern us in this section.

The most obvious requirement that might be imposed on an image functional q is invariance under translation. Indeed, suppose a recognition algorithm is operating on some input image in order to detect some object, such as a bridge, within the image. Whatever feature parameters might be involved in the recognition procedure, it is very likely that we do not wish the position of the bridge and its background within the actual image to be of consequence. In a similar vein, if the image is to be used to characterize regions of uniform textural variation, the positions of the regions within the overall grid structure should not influence textural parameters. Accordingly, we have the following property:

Property Q1: An image functional $q: X \to R$ is said to be *translationally invariant* on the class of images X if, for any $A \in X$ and for any $x \in R^2$,

$$q(A + x) = q(A)$$

Some examples of translationally invariant functionals are the area (Lebesque measure) $m(A)$, the perimeter $\Lambda(A)$, and the length of any projection of A onto a line. The last functional, which has proven useful in morphological analysis, is defined as follows: if **u** is a unit vector, then Proj(A, **u**) denotes the length of the projection of A onto any line perpendicular to **u**. (See Figure 3.2.) Since each unit vector **u** is determined by an angle θ with the x-axis, where $0 \leq \theta < 2\pi$, we sometimes write Proj(A, θ).

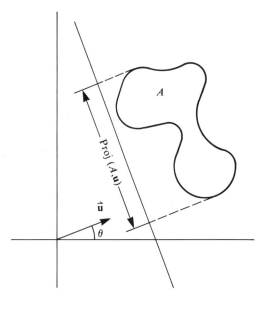

Figure 3.2 Length of projection

Sec. 3.2 Image Functionals

In addition to translational invariance, rotational invariance is another requirement often imposed on image functionals:

Property Q2: $q: X \to R$ is *rotationally invariant* if

$$q[\Omega(A)] = q(A)$$

where $\Omega(A)$ is any rotation of A.

The desirability of rotational invariance is not as clear cut as that of translational invariance. For instance, suppose that recognition depends upon the vertical and horizontal relation between component parts of an image. Then rotation of the image would change these relationships. On the other hand, if a missile is using image information to home in on a target, the polar orientation of the object will sometimes be of no consequence. It should be obvious that m and Λ are rotationally invariant but that Proj is not. However, if the projections are averaged over all unit vectors, then the resulting average projection, given by

$$\frac{1}{2\pi} \int_0^{2\pi} \text{Proj}(A, \theta) \, d\theta = \frac{1}{\pi} \int_0^{\pi} \text{Proj}(A, \theta) \, d\theta$$

is rotationally invariant.

The third property we wish to consider is *homogeneity*. Suppose that a toxicological slide is viewed under different magnifications. Certainly we do not wish intelligence-based decisions regarding the microscopic structure of the substance on the slide to be dependent upon the degree of magnification. Consequently, consider a scalar multiple of the set A, say tA, for $t > 0$. It is well known that area satisfies the relationship $m(tA) = t^2 m(A)$. This equation can be used to relate the operation of the functional under different magnifications. It is described by saying that area is homogeneous of degree 2. More generally, we have property Q3:

Property Q3:   The image functional q is said to be *homogeneous of degree k* if

$$q(tA) = t^k q(A)$$

Now consider the perimeter Λ. To avoid undue analytical difficulties, suppose that the boundary of the set A is piecewise smooth and is given by the parametric representation $x = x(v)$, $y = y(v)$, for $v_1 \leq v \leq v_2$. Then the boundary consists of a finite number of arcs on which both x and y have continuous derivatives. Moreover, tA has boundary given by $tx(v)$ and $ty(v)$. Hence,

$$\Lambda(tA) = \int_{v_1}^{v_2} [(tx'(v))^2 + (ty'(v))^2]^{1/2} \, dv = t\Lambda(A)$$

so that Λ is homogeneous of degree 1.

In order to examine the homogeneity of the projection onto a line, consider the projection of A vertically onto the x-axis. Assume that A consists of a single connected component and is compact. Then there exist points $z_1 = (x_1, y_1)$ and $z_2 = (x_2, y_2)$ in A such that x_1 is the minimum x-coordinate for all z in A and x_2 is the maximum x-coordinate for all z in A, and it follows that

$$\text{Proj}(A, \mathbf{u}) = x_2 - x_1$$

and

$$\text{Proj}(tA, \mathbf{u}) = tx_2 - tx_1 = t\,\text{Proj}(A, \mathbf{u})$$

Since the length of the projection in any direction can be found by rotating and then projecting vertically, we conclude that Proj is homogeneous of degree 1 whenever it is applied to compact sets consisting of a single component. A similar argument applies to compact sets consisting of a finite number of components.

To this point, we have not emphasized the fact that the validity of any image-functional property depends upon the class over which it is to be applied. The matter is of some concern, however, in regard to the remaining properties to be discussed. The first of these is additivity. (See Figure 3.3).

Property Q4: q is said to be *additive* if

$$q(A \cup B) = q(A) + q(B) - q(A \cap B)$$

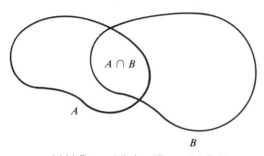

$m(A \cup B) = m(A) + m(B) - m(A \cap B)$
(Area$(A \cup B)$ = Area(A) + Area(B) - Area$(A \cap B)$)   **Figure 3.3** Additivity of area

Those familiar with measure theory are aware that m satisfies the additivity property. Implicit, however, in this statement is the measure-theoretic proposition that if A and B are measurable, then so are $A \cup B$ and $A \cap B$. (If A and B have areas, then so do $A \cup B$ and $A \cap B$.) In other words, the area m satisfies property Q4 provided that the sets under consideration are measurable. (A similar remark, unmentioned earlier, in fact applies to Q2: if A is measurable, then tA is measurable.)

The difficulty inherent in the underlying collection of sets (images) comes to the fore in the morphological study of convex sets: simply because A and B are convex does not guarantee that $A \cup B$ is convex. Thus, to apply Q4 to the collection of convex compact sets, we must state additivity under the assumption that $A \cup B$ is convex. The result is the property of *C-additivity*, which states that q satisfies Q4 as long as A, B, and $A \cup B$ are convex. (If A and B are convex, then $A \cap B$ is automatically convex.) In effect, C-additivity is a weaker condition than measure additivity since a functional may be C-additive but not additive over arbitrary unions of measurable sets. It is of consequence that Λ is C-additive. The notion of the underlying collection of images upon which an image functional may be considered is not purely academic; in fact, it determines the domain over which certain artificial intelligence decision techniques may be applied.

The next property of functionals to be considered is monotonicity.

Sec. 3.2 Image Functionals

Property Q5: q is (monotonically) *increasing* if, whenever $A \subset B$, then $q(A) \leq q(B)$.

Because of their favorable behavior, functionals that are both increasing and translationally invariant are employed extensively in morphological analysis. Indeed, we have already seen that, as image-to-image mappings, for a fixed structuring element E, $\mathcal{D}(\cdot, E)$, $\mathcal{E}(\cdot, E)$, $O(\cdot, E)$, and $C(\cdot, E)$ are increasing. Also, insofar as image functionals are concerned, m and Proj are increasing. And, though Λ is not increasing in general, it is increasing on the class $\mathcal{H}_c$ of convex compact sets.

The last two image-functional properties to be presented are topological in that they involve limits. A notion of "distance" between two nonempty compact sets shall be introduced. For any nonempty compact set $A \in \mathcal{H}$ and for any point $x \in R^2$, we define the *distance* from x to A as

$$d(x, A) = \min\{|x - a|: a \in A\}$$

Using d, we define the *Hausdorff metric* on the class $\mathcal{H}$ of nonempty compact sets: for any two sets A and B in $\mathcal{H}$,

$$h(A, B) = \max(\max_{b \in B} d(b, A), \max_{a \in A} d(a, B))$$

Intuitively, the Hausdorff metric measures the greatest distance by which A differs from B. (See Figure 3.4.)

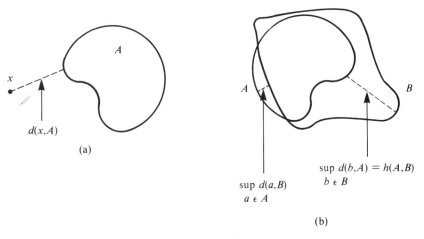

Figure 3.4 Hausdorff metric

The Hausdorff metric plays a fundamental role as regards the behavior of image functionals. Intuitively, to say that $\lim_{n \to \infty} h(A_n, A) = 0$, where $A_n, A \in \mathcal{H}$, is to say that for large n, A_n is nearly identical to A. Hence, as images, A_n and A appear very much alike. This is why the Hausdorff metric is often employed to define image convergence for compact sets. That is, we define $\lim_{n \to \infty} A_n = A$ if $\lim_{n \to \infty} h(A_n, A) = 0$. (See Figure 3.5.)

The concept of continuity in calculus says that a real-valued function f is continuous at a point a if $\lim_{x \to a} f(x) = f(a)$. This notion of a limit is generalized

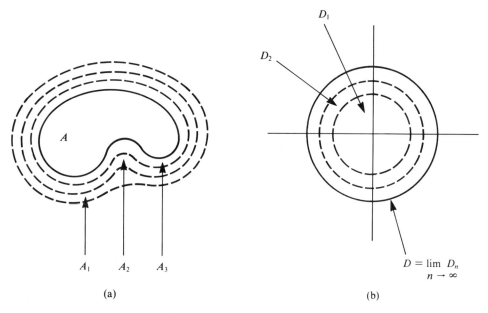

Figure 3.5 Convergence of sets

topologically into many areas of mathematics. Insofar as image functionals are concerned, it enters into continuity as specified by property Q6:

Property Q6: q is *continuous* if $\lim_{n\to\infty} q(A_n) = q(A)$ whenever $\{A_n\}$ is a sequence of compact sets such that $\lim_{n\to\infty} A_n = A$ in the Hausdorff metric.

Intuitively, the image functional q is continuous if, whenever the compact images A_n approach the compact image A, the functional values of A_n approach the functional value of A. But care must be taken: once again, the class of images under consideration is vitally important. For instance, if we restrict our attention to convex compact sets, then Λ is continuous: that is, $\lim_{n\to\infty} A_n = A$ implies that $\lim_{n\to\infty} \Lambda(A_n) = \Lambda(A)$. However, this is not true for compact sets in general.

Example 3.1:

Consider the sequence of unit disks of radius $1 - 1/2n$ and centered at the origin. (See Figure 3.5(b).) Call these disks D_n, and let D denote the unit disk centered at the origin. Then $D_n \to D$, and also,

$$\Lambda(D_n) = 2\pi(1 - 1/2n) \longrightarrow 2\pi = \Lambda(D)$$

as it must since Λ is continuous on the class $\mathcal{H}_c$ of nonempty convex compact sets.

Example 3.2:

Consider again the closed unit disk D. Suppose a grid with mesh $1/2^k$ is placed over D in such a way that the origin of the grid is situated at the center of D. Let A_k be the union of all grid squares whose interiors intersect D (See Figure 3.6.) Using the notation of Section 2.7, we have $A_k = \bar{D}$ for the $(1/2^k)$-grid. As grids of finer and finer mesh size

Sec. 3.2 Image Functionals

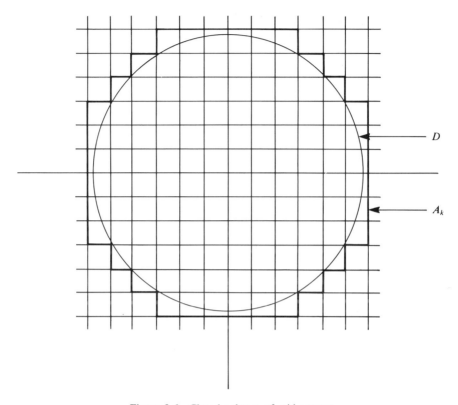

Figure 3.6 Closed polygon of grid squares

are taken (i.e., as $k \to \infty$), $\lim_{k \to \infty} A_k = D$, using the Hausdorff metric. But $\Lambda(A_k) \geq 8$ for all k, while $\Lambda(D) = 2\pi$. Hence, Λ is not continuous over the collection $\mathcal{H}$ of nonempty compact sets.

It turns out that a weaker condition than image-functional continuity is suitable for morphological analysis. Consider a collection of nonempty compact sets A_n that are *nested*—i.e., for which $A_1 \supset A_2 \supset A_3 \supset \ldots$. For such a collection, the Cantor Intersection Theorem, a well-known theorem of advanced calculus, states that there is a nonempty intersection A that is also compact. An example of a sequence of this sort is given in Figure 3.5(a). Another well-known theorem, this time from measure theory, states that for such a nested collection, $\lim_{n \to \infty} m(A_n) = m(A)$. The next image-functional property is formulated in accordance with these theorems.

Property Q7: If q is increasing and, for any nested sequence of compact nonempty sets,

$$\lim_{n \to \infty} q(A_n) = q\left[\bigcap_{n=1}^{\infty} A_n\right]$$

then q is said to be *continuous from above*.

As with the other image-functional properties presented, one might choose to reduce the class of sets over which Q7 is to be employed. For instance, one might consider an image functional's being continuous from above only over convex compact sets. Note that if q is increasing and continuous, then it must be continuous from above.

As noted in Section 3.1, morphological feature generation often involves the application of some image functional to the output of some family of image-to-image transformations. These transformations normally consist of dilations, erosions, openings, or closings. For instance, the granulometries of Section 3.5 are, in accordance with Figure 3.1, generated by successive openings of an image by ever larger structuring elements. Area measurements are then taken on the outputs. Consequently, we are utilizing functionals of the form

$$A \longrightarrow O(A, E) \longrightarrow m[O(A, E)]$$

An important question is how the properties Q1 through Q7 apply to such composition-type image functionals. For example, according to Theorem 1.3(ii), $O(\cdot, E)$ is increasing in the first variable. Given that m is increasing, can we then conclude that the composition $m[O(\cdot, E)]$ is also increasing? Theorem 3.1 answers this question, as well as some related ones, in the affirmative.

Theorem 3.1.[2] Let $\Psi(A)$ denote either dilation, erosion, opening, or closing of a compact image A by a fixed compact structuring element, and let the image functional q be defined on the class $\mathcal{H}$. Then if q is increasing, translationally invariant, or continuous from above, the composition functional $q \circ \Psi$ is respectively increasing, translationally invariant, or continuous from above.

Note that Theorem 3.1 is stated in the context of compact sets only. Although it could be generalized to other settings, the compact set model suffices for the purposes of both Euclidean modeling and digital implementation. The salient point in the stated circumstances is that if A and B are compact, then so, too, are $\mathcal{D}(A, B)$, $\mathcal{E}(A, B)$, $O(A, B)$, and $C(A, B)$.

Example 3.3:

Since the length of a projection in a given direction, Proj(A, **u**), is translationally invariant, then so is an opening followed by a projection; indeed, according to Theorem 3.1,

$$\text{Proj}[O(A + x, B), \mathbf{u}] = \text{Proj}[O(A, B), \mathbf{u}]$$

3.3 EUCLIDEAN IMAGE MODELING

In his seminal work on mathematical morphology, Matheron focuses attention on closed subsets of the Euclidean plane. In other words, he restricts his purview to those constant Euclidean images which contain their topological boundaries. His view is that in the Euclidean model the boundary of a set lacks empirical meaning. In his own words,

Does a point of the boundary of A belong to A or not? From an experimental point of view, this question is absolutely senseless, because no physical reality corresponds perfectly to the notion of a point belonging to the boundary of the grains [image]. Indeed, for an experimentalist, there exist no true points but rather spots, with small but nonzero dimensions and rather poorly defined boundaries.[3]

Thus, given that the boundary of a Euclidean set is unimportant for experimental reasons, Matheron chooses, for mathematical reasons, to assume that all images contain their boundaries, i.e., that all images are closed.

From the perspective of image processing, Matheron's model presents no difficulty. Since we prefer to consider a Euclidean image as any subset of R^2, we need only be careful to specify the restriction to closed sets whenever it applies, which is only when Euclidean topological considerations are of interest. When we are not concerned with topological questions, we prefer to view an image simply as a subset of the plane.

Yet certain important modeling questions remain to be addressed. For example, since our purpose is electronic digital image processing, one might question the relevance of topological questions that involve limiting notions. Such a questioning is not frivolous; indeed, it forces us to come to grips with crucial modeling decisions—for instance, Should we concern ourselves with the Euclidean model when, in fact, we never intend to implement it? We believe the answer is clear cut: as long as our geometric intuition resides in the Euclidean world, then so must our underlying modeling. As noted in Section 3.1, it is our own categories which lead to the manner in which the relevant nature of the image is constituted. Thus, intelligence must be brought to bear on image features that conform in a reasonable manner to our own perceptual categories.

A good example is the entire notion of image-functional continuity. It is obvious from the definition that the Hausdorff metric is far too sensitive to noise to be of direct practical value. Nevertheless, it certainly conforms to our geometric intuition insofar as that intuition relates to image convergence. Hence, it is natural to focus attention on image functionals that are continuous relative to the Hausdorff metric. Such functionals behave well from the point of view of continuous change. Hopefully, so, too, will their digital counterparts. It is of little concern that noise will almost assuredly prevent us from ever utilizing the notion of continuity for practical implementation. That is not the point; rather, it is that our knowledge concerning ideal behavior will allow us to choose for use those digital functionals that derive from well-behaved Euclidean functionals.

3.4 INTEGRAL GEOMETRY AND IMAGE FUNCTIONALS

Image functionals that have been adopted from the area of integral geometry have proven to be instrumental in the morphological analysis of images. Five salient reasons for the success of these functionals are the following:

1. They represent quantifications of geometrically intuitive, and hence perceptually relevant, image characteristics.

2. They are closely related to each other, and their properties have been extensively investigated, especially by the German mathematician H. Hadwiger.
3. Fundamental theorems of integral geometry relate parameters as they apply to different-dimensional spaces. Consequently, integral geometric image functionals are *stereological* in that parameters computed for two-dimensional images taken by sectioning three-dimensional bodies may be utilized to estimate related three-dimensional parameters of the original body, a problem central to biology and the material sciences. Since the exposition at hand is concerned solely with image processing, stereological considerations will not be specifically pursued.
4. Image functionals from integral geometry often have representations in terms of the morphological operations of Minkowski algebra.
5. Upon digitization of an image, the digital versions of the Minkowski algebra operations can be implemented in a highly parallel fashion, and hence digital approximations of many integral geometric functionals are obtainable under acceptable time contraints.

In this section, we shall continue discussing the image functionals m, Λ, and Proj, which are parameters central to integral geometry. We shall also present some basic results concerning integral geometry and morphological analysis.

In the preceding section, the projection length $\text{Proj}(A, \theta)$ of A onto a line perpendicular to the direction θ was mentioned frequently. We also mentioned the rotational invariance of the average of these projections. In fact, for a convex compact set $A \in \mathcal{K}_c$,

$$\Lambda(A) = \pi \frac{1}{2\pi} \int_0^{2\pi} \text{Proj}(A, \theta) \, d\theta$$

In words, the length of the boundary equals π times the average length of the projections. This is the well-known Cauchy projection theorem for convex compact sets in R^2. Note that because of symmetry, the average need be taken only over 0 to π. In such a case, the denominator of the coefficient of the integral is π.

Example 3.3:

Let A be a square of edge length 1. Then $\Lambda(A) = 4$. Since the orientation of A is inconsequential when averaging the projections, suppose it is situated with one side horizontal as in Fig 3.7. Then, from the figure, it can be seen that whatever the value of θ between 0 and π, the length of the projection is given by $|\cos \theta| + |\sin \theta|$. Therefore, the average projection length is given by

$$\frac{1}{\pi} \int_0^{\pi} [|\cos \theta| + |\sin \theta|] \, d\theta = \frac{4}{\pi} = \frac{1}{\pi} \Lambda(A)$$

as the Cauchy theorem states it must.

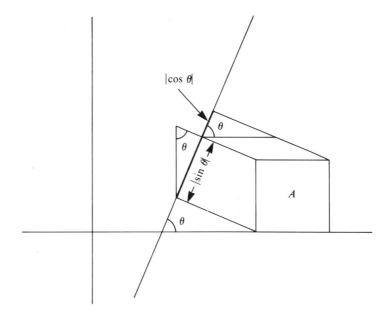

Figure 3.7 Square with edge 1

Numerous image-functional properties relating to m and Λ have been mentioned thus far. If the underlying collection of sets is $\mathcal{K}_c$, so that m and Λ are restricted to convex compact sets, then both image functionals satisfy Q1 through Q6, where homogeneity for m is of degree 2 and for Λ is of degree 1. (Note that Q7 is also satisfied since the stronger continuity condition Q6 is satisfied.) A fundamental characterization theorem has been given by Hadwiger which in essence states a converse in terms of Q1, Q2, Q4, and Q6.

Theorem 3.2[4] (Hadwiger). Suppose q is an image functional on $\mathcal{K}_c$ that is invariant with respect to translation and rotation, is C-additive, and is continuous. Then q can be represented as a linear combination of m, Λ, and 1; that is, there exist constants a, b, and c such that for any convex compact set A,

$$q(A) = am(A) + b\Lambda(A) + c$$

Moreover, if the continuity assumption is replaced by an assumption that q is increasing, then not only is q also continuous on $\mathcal{K}_c$, but the constants a, b, and c are nonnegative.

Several comments concerning Hadwiger's theorem are in order. We shall concentrate on the form where q is assumed to be increasing, since this a more natural requirement than is continuity.

The Hadwiger Theorem limits the form of those image functionals for which we require monotonicity (that the functional be increasing), additivity, and translational and rotational invariance. Given any image functional q on some class of compact sets, say $\mathcal{H}$, which contains $\mathcal{H}_c$, the restriction of q to $\mathcal{H}_c$ (not q itself!) must be of the form given in the theorem. While this is quite a strong conclusion, one must be careful not to overplay the consequences of the Hadwiger Theorem. First, note that it requires rotational invariance, which, as has been previously mentioned, is certainly not desirable in all circumstances. Moreover, the assumption of rotational invariance cannot be dropped from the hypothesis. Second, the characterization of an image functional on $\mathcal{H}_c$ does not uniquely determine it on some larger domain such as $\mathcal{H}$ itself. For consider the class $\mathcal{R}_c$ consisting of finite unions of convex compact sets. This class is known as the *convex ring* since if A, $B \in \mathcal{R}_c$ then also $A \cap B$, $A \cup B \in \mathcal{R}_c$ (as long as we allow the null set to be an element of $\mathcal{R}_c$). For any functional of the form $am + b\Lambda + c$ on $\mathcal{H}_c$, there exist numerous extensions to $\mathcal{R}_c$[5]. Finally, the theorem demands the additivity requirement. Nonadditive functionals are not restricted by Hadwiger's theorem. Indeed, continuity from above, such as that mentioned in Theorem 3.1 for compositions including erosion, opening, and closing, appears to be a much more natural requirement than additivity. This coincides with a current trend in AI where the notion of a "fuzzy measure" is nothing more than a generalization of continuity from above.

Before we leave Hadwiger's theorem, its relation to the Cauchy theorem should be noted. The two functionals in the Hadwiger characterization are related in that $\Lambda(A)$ can be obtained by averaging lengths of projections of A, each of which is actually a one-dimensional measure parameter, whereas $\Lambda(A)$ is a two-dimensional parameter. This is one of the stereological relations mentioned earlier. If we were to consider three-dimensional parameters, Hadwiger's theorem would generalize and each of the relevant parameters would be an average of two-dimensional parameters, in one case of areas and in another of perimeters. In fact, the theorem is generalizable to n-dimensional Euclidean space, and the parameters involved are known as the *Minkowski functionals*.[6]

It was previously mentioned that the Minkowski algebra operators can be employed to find geometric parameters. As an illustration, let us suppose that A consists of a single simply connected component. (That is, A is connected and has no holes, but is not necessarily convex.) Let B be the unit disk centered at the origin. Then $A \oplus tB$ equals the set of all points within t of the set A; graphically, it is A together with a band C of width t about it. Letting m denote the area (measure), we have

$$m(A \oplus tB) = m(A) + m(C)$$

By means of the unit normal $\mathbf{N}$, the unit tangent $\mathbf{T}$, and the curvature κ, we see (Figure 3.8) that

$$m(C) = \int_0^{\Lambda(A)} \left[t\, ds + \frac{1}{2} \operatorname{sign}(\kappa)\, t^2 \sin \theta \right]$$

But

$$\sin \theta = |\mathbf{T}(s) \times \mathbf{T}(s + ds)| = |\mathbf{T}(s) \times [\mathbf{T}(s) + \kappa\, ds\, \mathbf{N}(s)]|$$
$$= |\kappa\, ds|$$

Sec. 3.4 Integral Geometry and Image Functionals

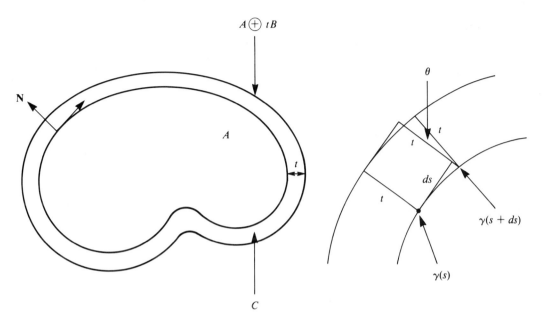

Figure 3.8 Parameters in calculation of area of dilation

since **T** is perpendicular to **N**. Consequently,

$$m(C) = \int_0^{\Lambda(A)} \left[t\, ds + \frac{1}{2} t^2\, \kappa(s)\, ds \right]$$
$$= t \int_0^{\Lambda(A)} ds + \frac{1}{2} t^2 \int_0^{\Lambda(A)} \kappa(s)\, ds$$
$$= t\Lambda(A) + \pi t^2$$

(Note that the above demonstration holds with only slight modification as long as there exists a tangent at all but a finite number of points of the boundary of A.) As a result,

$$\lim_{t \to 0} \frac{m(A \oplus tB) - m(A)}{t} = \lim_{t \to 0} \frac{t\Lambda(A) + \pi t^2}{t} = \Lambda(A)$$

In other words, using the notion of a one-sided derivative, we have

$$\Lambda(A) = \frac{d}{dt}[m(A \oplus tB)]\Big|_{t=0}$$

Though the derivation has been carried out for a set having a single simply connected component, it can be carried out for reasonably well-behaved sets that consist of more than one component or are mutliply connected (have holes). In sum, the total perimeter of a set can be found by using dilation. In practice, this methodology must be accomplished digitally.

Example 3.4:

Let A be a 2 by 3 rectangle. (See Figure 3.9.) Then $m(A \oplus tB) = 6 + 10t + \pi t^2$. Therefore,

$$\lim_{t \to 0} \frac{1}{t}[m(A \oplus tB) - m(A)] = \lim_{t \to 0} [10 + \pi t] = 10$$

which is the perimeter of the rectangle.

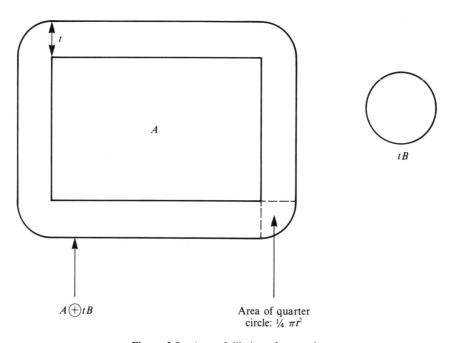

Figure 3.9 Area of dilation of rectangle

It is also possible to find a projection length of a simply connected component by use of the dilation.[7] Consider the Minkowski sum $A \oplus tE$, where E is a unit interval starting at the origin and in the same direction as a unit vector **u**. Although we shall not go through the theoretical details, as long as A is sufficiently regular, the length Proj$(A, \mathbf{u})$ of the projection can be obtained as a one-sided derivative of $m(A \oplus tE)$, viz.,

$$\text{Proj}(A, \mathbf{u}) = \frac{d}{dt}[m(A \oplus tE)]|_{t=0}$$

$$= \lim_{t \to 0} \frac{m(A \oplus tE) - m(A)}{t}$$

Example 3.5:

Let A be a square of edge length 1 sitting with its edges parallel to the coordinate axes.

Sec. 3.4   Integral Geometry and Image Functionals   95

Let **u** be the unit vector pointing in the 45° direction. (See Figure 3.10.) Then

$$m(A \oplus tE) = 1 + t\sqrt{2}$$

Hence,

$$\lim_{t \to 0} \frac{m(A \oplus tE) - m(A)}{t} = \sqrt{2}$$

which is exactly equal to the length of the diagonal, which is in turn equal to Proj(A, **u**).

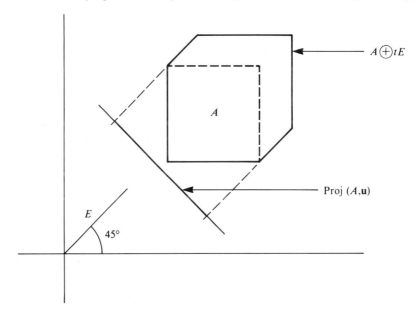

Figure 3.10 Projection of square

Many other morphological formulations of geometric parameters are possible. Whereas the preceding methodologies have involved dilation by a parameterized family of balls and a dilation by a parameterized family of line segments, the next one involves erosion by a family of two point structuring elements generated by some set $\{0, w\}$. Indeed, we consider the eroded sets $A \ominus (-tE) = A \cap (A - tw)$, where $E = \{0, w\}$ and $|w| = 1$. We define the *covariance* of A to be

$$[\text{Cov}(A, w)](t) = m[A \ominus (-tE)]$$

When no confusion can arise, we write $C(t)$ to denote the covariance.

Though no proof will be given, it is known that for sufficiently regular compact sets, the perimeter can be found by using averages of the derivative of the covariance at 0.[8] Indeed,

$$\frac{1}{2\pi} \int_0^{2\pi} C'(0) \, d\theta = \frac{-\Lambda(A)}{\pi}$$

where the average is taken over all Cov (A, w) such that $|w| = 1$.

Example 3.6:

Consider a unit square with sides parallel to the coordinate axes. For a given angle θ and sufficiently small t (see Figure 3.11),

$$[\text{Cov}(A, \theta)](t) = 1 - [t|\sin\theta| + t|\cos\theta| - t^2|\cos\theta\sin\theta|]$$

Hence,

$$[\text{Cov}(A, \theta)]'(0) = -[|\sin\theta| + |\cos\theta|]$$

Averaging over all θ between 0 and 2π yields

$$\frac{1}{2\pi}\int_0^{2\pi} C'(0)\,d\theta = \frac{-1}{2\pi}\int_0^{2\pi}[|\sin\theta| + |\cos\theta|]\,d\theta$$

$$= \frac{-4}{\pi} = \frac{-\Lambda(A)}{\pi}$$

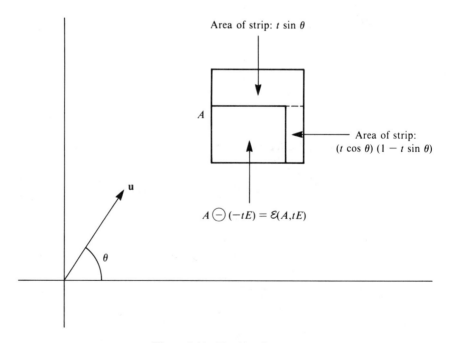

Figure 3.11 Erosion of square

3.5 GRANULOMETRIES

Having investigated some of the ways in which geometric feature parameters are generated morphologically, we return to the general feature methodology outlined in Figure 3.1. We discuss the implementation of that methodology through the use of families of openings to accomplish the filtering denoted in the middle block of the figure.

Given a fixed convex compact set E, we consider the functional composition sequence

$$A \longrightarrow O(A, tE) \longrightarrow m[O(A, tE)]$$

where A is a compact set, $t > 0$, and m denotes area (Lebesque measure). According to Theorem 1.5, if $t \geq t'$, then $O(A, tE) \subset O(A, t'E)$. Consequently, the real-valued function

$$t \longrightarrow u_A(t) = m[O(A, tE)]$$

is decreasing. Moreover, as long as E consists of more than a single point, the opening of A by tE will be empty for sufficiently large t. Hence, $u_A(t) = 0$ for sufficiently large t. Under these conditions, the mapping $t \to O(A, tE)$ is known as a *granulometry* and the function $u_A(t)$ is known as the *size distribution* generated by the granulometry. As regards the latter, when no possible confusion can arise, we simply write $u(t)$.

Example 3.7:

Consider the image A which is a rectangle of length 2 and width 3. Suppose E is a square of edge length 1. Then

$$O(A, tE) = \begin{cases} A & \text{if } t \leq 2 \\ \emptyset & \text{if } t > 2 \end{cases}$$

Hence,

$$u(t) = \begin{cases} 6 & \text{if } t \leq 2 \\ 0 & \text{if } t > 0 \end{cases}$$

The graph of $u(t)$ is given in Figure 3.12(b).

Example 3.8:

Consider the same image as in the previous example, only this time let E be the closed unit disk. A typical opening of A by E is given in Figure 3.12(a). In general,

$$u(t) = \begin{cases} \pi t^2 + 2(3 - 2t) + 2t(2 - 2t) & \text{if } t \leq 1 \\ 0 & \text{if } t > 1 \end{cases}$$

The graph of $u(t)$ for this example is given in Figure 3.12(c).

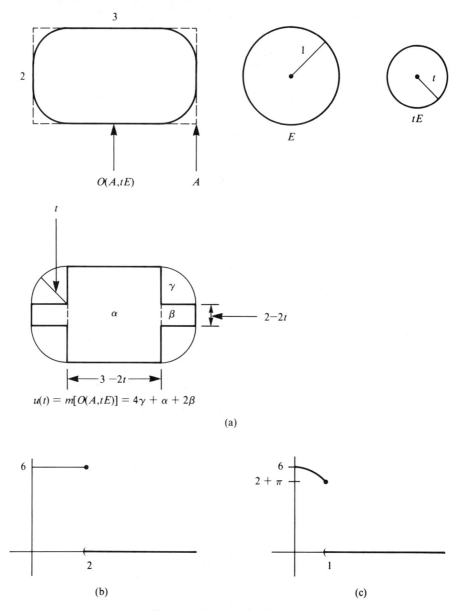

Figure 3.12 Size distributions

The preceding examples demonstrate the manner in which the shape of an object can lead to different size distributions depending upon the structuring element employed. What is of interest is that different shapes lead to different size distributions when opened by the same one-parameter family of structuring elements. Should an image contain intricate textural properties, these can be revealed by granulometric analysis. Because of the complexity of the calculations involved, digital morphological processing needs to be utilized. This will be discussed in Section 3.6.

Sec. 3.5   Granulometries

From an intuitive standpoint, the granulometric process represents a type of filtering in which those sections of the image which are not sufficiently large to hold the structuring element tE are removed from the image. Indeed, Figure 3.13 gives an illustration of just such a filtering when the structuring elements E, $2E$, and $3E$ are applied to the image in the figure. Put in classical engineering terminology, a granulometry is a one-parameter family of low-pass filters. The filters are low pass because it is high-frequency fluctuation between a set and its complement which is attenuated in the output images when the structuring element is sufficiently smooth.

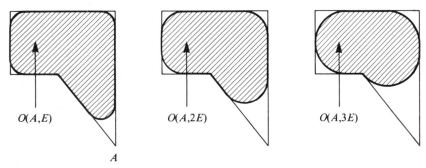

Figure 3.13 Granulometry as a form of filtering

Returning to the granulometries themselves, let $\Psi_t(A)$ denote the opening $O(A, tE)$. Then the family $\{\Psi_t\}$ has numerous algebraic properties. These will be discussed in detail from an axiomatic point of view in Section 5.5; however, the following will be listed here without proof for those who are not concerned with the theoretical underpinnings:

(i) $r \geq s$ implies that $\Psi_r(A) \subset \Psi_s(A)$.
(ii) $A \subset B$ implies that $\Psi_t(A) \subset \Psi_t(B)$.
(iii) For any $r, s \geq 0$, $\Psi_r[\Psi_s(A)] = \Psi_s[\Psi_r(A)] = \Psi_{\max(r,s)}(A)$.
(iv) $\Psi_t(A + x) = \Psi_t(A) + x$.

In brief, (i) has already been mentioned, (ii) states the fact that opening by tE is an increasing transformation, and (iii) states that opening by an iteration of rE and sE is equivalent to opening solely by the scalar mutliple of the maximum of r and s times E. Property (iv) says that opening a translate is the same as translating the corresponding opening.

Now, for the moment, let us fix t and consider $A \rightarrow u_A(t)$ as an image functional. According to Theorem 3.1, since m is increasing, translationally invariant, and continuous from above, so is the image functional $u_A(t) = m[O(A, tE)]$. That is, mathematically, if $A \subset B$, then $u_A(t) \leq u_B(t)$; if $x \in R^2$, then $u_{A+x}(t) = u_A(t)$, and if $\{A_n\}$ is a nested sequence, then $\lim_{n \to \infty} u_{A_n}(t) = u_A(t)$, where $A = \bigcap_{n=1}^{\infty} A_n$. Notice that $u_A(t)$ is not rotationally invariant unless the structuring element is the closed unit disk. This lack of rotational invariance means that granulometric size distributions are sensitive to orientation; hence, an image must usually be rotationally normalized prior to feature classification based on granulometric analysis.

Before illustrating how one might employ a size distribution for image recognition, let us return momentarily to Examples 3.7 and 3.8. In both, the graph of $u(t)$ is continuous from the left. Although we shall not prove the statement, as long as A is compact and the structuring element is compact and convex, as is the case in these examples, the resulting size distribution will be continuous from the left.

Now instead of working directly with a size distribution $u(t)$, it is convenient to normalize it by introducing the function

$$F(t) = 1 - \frac{u(t)}{u(0)} = 1 - \frac{u(t)}{m(A)}$$

$F(t)$ is a true distribution in the probabilistic sense: it is increasing from 0 to 1 and continuous from the left. Since $F(t)$ is a probability distribution function, its derivative $F'(t)$ is a probability density. (We assume that $F'(t)$ might well include some delta functions in its makeup). Using $F'(t)$, we can compute the moments

$$m^k = \int_0^T t^k F'(t)\, dt$$

where we assume that $t = T$ is the point where $F'(t)$ vanishes. The moments m^k can be employed as feature parameters.

Example 3.9:

Let A be a square of edge length 1, and let E be the closed unit disk. Then the orientation of A is irrelevant to the size distribution $u(t)$; i.e., for fixed t, $u(t)$ is rotationally invariant. Moreover,

$$u(t) = \begin{cases} 1 - (4 - \pi)t^2, & \text{if } t \le \tfrac{1}{2} \\ 0, & \text{if } t > \tfrac{1}{2} \end{cases}$$

and $m(A) = 1$. Hence,

$$F(t) = \begin{cases} 0, & \text{if } t \le 0 \\ (4 - \pi)t^2, & \text{if } 0 < t \le \tfrac{1}{2} \\ 1, & \text{if } t > \tfrac{1}{2} \end{cases}$$

and $F'(t) = g(t) + \frac{\pi}{4}\delta(t - \tfrac{1}{2})$, where

$$g(t) = \begin{cases} 2(4 - \pi)t, & \text{if } 0 \le t < \tfrac{1}{2} \\ 0, & \text{otherwise} \end{cases}$$

and δ denotes the delta function. Therefore, for $k = 1, 2, \ldots$, the kth moment m^k is given by

$$m^k = \int_0^\infty t^k \left[g(t) + \frac{\pi}{4} \delta(t - \tfrac{1}{2}) \right] dt$$

$$= \int_0^{1/2} 2(4 - \pi)t^{k+1}\, dt + \frac{\pi}{2^{k+2}}$$

$$= \left[\frac{4 - \pi}{k + 2} + \frac{\pi}{2} \right] 2^{-k-1}$$

In the computation of feature parameters, one must usually employ some normalization procedure to correct size (i.e., magnification) differences. Due to properties M-12 and M-13, the granulometric size distributions behave very well with respect to magnification (scalar multiplication) of the image A. As a result, there is a close relationship between the moments of A and those of a scalar multiple tA of A.

To see this relationship, employing M-12 and M-13, we have

$$O(rA, tE) = [rA \ominus (-tE)] \oplus tE$$

$$= \left[r\left(A \ominus \frac{-t}{r}E\right)\right] \oplus tE$$

$$= r\left[\left(A \ominus \frac{-t}{r}E\right) \oplus \frac{t}{r}E\right]$$

$$= rO[A, (t/r)E]$$

Consequently, by the 2-homogeneity of m,

$$u_{rA}(t) = m[O(rA, tE)] = m[rO(A, (t/r)E)] = r^2 u_A(t/r)$$

Thus,

$$F_{rA}(t) = 1 - \frac{u_{rA}(t)}{m(rA)} = 1 - \frac{r^2 u_A(t/r)}{r^2 m(A)} = 1 - \frac{u_A(t/r)}{m(A)}$$

and

$$F'_{rA}(t) = -\frac{u'_A(t/r)}{rm(A)} = r^{-1} F'_A(t/r)$$

Now suppose $F'_A(t)$ vanishes at $t = T$. Then $F'_A(t/r)$ vanishes at $t = rT$. Hence, using the substitution $s = t/r$, we obtain

$$m^k_{rA} = \int_0^{rT} t^k F'_{rA}(t)\, dt = r^{-1} \int_0^{rT} t^k F'_A(t/r)\, dt$$

$$= r^{-1} \int_0^T (rs)^k F'_A(s)\, r\, ds = r^k m^k_A$$

In sum, the kth moment generated by the granulometry $\{O(rA, tE)\}$ equals r^k times the kth moment generated by the granulometry $\{O(A, tE)\}$.

Example 3.10:

In Example 3.9, we employed the square of edge length 1 and found the resulting moments. If we now employ the square of edge length 2, the new moments should be 2^k times the original moments. A direct calculation will verify this.

3.6 DIGITAL SIZE DISTRIBUTIONS

The granulometric size distributions provide textural information by measuring the manner in which opening by structuring elements of ever-increasing size shrinks the activated portion of the image. In this section we shall examine how the entire procedure can be implemented in a digital setting, as well as look at some digital size distributions that are generated by erosion instead of opening.

A difficulty immediately arises in moving the granulometric methodology from the Euclidean to the digital setting: the granulometry $t \to O(A, tE)$ involves a scalar multiplication, and, as noted in Section 2.5, we do not employ a digital analogue to that operation.

One way to proceed is to use the digitization metholodogy of Section 2.7: given a compact set A, and a convex compact set E, approximate

$$u_A(t) = m[\Psi_t(A)] = m[O(A, tE)]$$

by

$$\tilde{u}_A(t) = m[\tilde{\Psi}_t(A)] = 2^{-2k} \text{CARD}[\text{OPEN}(\overline{A}, \overline{tE})]$$

It can be shown that for convex images, the approximation of $u_A(t)$ by $\tilde{u}_A(t)$ is good in the L_1 sense.[9]

Now suppose A and E are the Euclidean images given in Figure 3.14. Note that tE is simply a stretched version of E; for any $t > 0$, $\widetilde{tE}$ is a vertical string of pixel squares, and $\overline{tE}$, in bound matrix form, is simply a column of 1's that grows in length as t grows. Accordingly, the preceding digitization methodology results in the digital

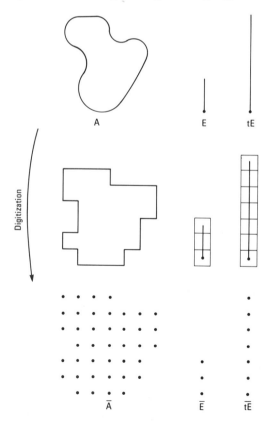

NOTE: $\overline{tE}$ CONSISTS OF k ACTIVATED PIXELS

Figure 3.14 Digitization process

Sec. 3.6   Digital Size Distributions   103

opening of $\bar{A}$ by successive structuring elements E_k, for $k = 1, 2, \ldots$, where E_k is a vertical string of 1-valued pixels starting at the origin pixel $(0, 0)$ and ending at (and including) the pixel $(0, k - 1)$. According to the methodology of Figure 2.18, instead of employing the Euclidean granulometry $t \to O(A, tE)$, we would in fact be using a digital granulometry of the form $k \to \text{OPEN}(\bar{A}, E_k)$. Of course, the size distribution $\bar{u}_A(t)$ would still be a function of t. Nevertheless, the operative relation would be $k \to \text{OPEN}(\bar{A}, E_k)$.

Using the preceding discussion as our motivation, we define the *linear digital granulometries*

$$k \longrightarrow \text{OPEN}(S, V(k))$$

and

$$k \longrightarrow \text{OPEN}(S, H(k))$$

where $k = 1, 2, 3, \ldots$,

$$V(k) = \begin{pmatrix} 1 \\ \vdots \\ 1 \\ \textcircled{1} \end{pmatrix}$$

and

$$H(k) = (\textcircled{1} \; 1 \; \ldots \; 1)$$

each with k activated pixels. In the Euclidean case, the size distributions were created by finding the areas that resulted from the granulometries. If we were to employ the digitization methodology, we would simply count, at each stage, the number of activated pixels and then find the approximate area by multiplying by 2^{-2k}. Thus, rather than reconvert to the Euclidean scheme, we can simply stay digital. Doing so, we generate the *digital linear granulometric size distributions*

$$\omega_1(k) = \text{CARD}[\text{OPEN}(S, V(k))]$$

and

$$\omega_2(k) = \text{CARD}[\text{OPEN}(S, H(k))]$$

For the sake of notational clarity, we have omitted mention of the original set in the size distribution notation. This should present no great difficulty since the set S under consideration should be clearly known from the context of the discussion. Recall that CARD outputs the cardinality of the domain of a digital image.

Example 3.11:

Let S be the image given in Figure 3.15. Then the size distributions ω_1 and ω_2 are given by

$k =$	1	2	3	4	5	6	7	8	9	10	11	12	13
$\omega_1(k) =$	80	76	70	70	66	66	54	54	54	45	35	24	0
$\omega_2(k) =$	80	66	52	22	10	0	0	0	0	0	0	0	0

For $k > 13$, $\omega_1(k) = \omega_2(k) = 0$.

$$S = \begin{pmatrix} * & * & * & 1 & 1 & 1 & * & * & 1 & 1 & 1 & * \\ 1 & 1 & * & 1 & 1 & 1 & 1 & * & 1 & 1 & 1 & 1 \\ 1 & * & 1 & 1 & 1 & * & * & * & 1 & 1 & 1 & * \\ 1 & * & 1 & 1 & 1 & * & * & * & 1 & 1 & 1 & * \\ 1 & * & 1 & 1 & 1 & * & * & * & 1 & 1 & 1 & * \\ 1 & * & 1 & 1 & 1 & * & * & * & 1 & 1 & 1 & * \\ 1 & * & 1 & 1 & * & 1 & 1 & 1 & 1 & 1 & * & * \\ 1 & 1 & 1 & 1 & * & 1 & 1 & 1 & 1 & 1 & * & * \\ 1 & * & 1 & 1 & * & * & 1 & * & 1 & 1 & * & * \\ 1 & * & 1 & 1 & * & * & 1 & * & 1 & 1 & * & * \\ 1 & * & 1 & 1 & * & * & * & * & 1 & * & * & * \\ 1 & * & * & 1 & * & * & * & * & 1 & * & * & * \end{pmatrix}$$

Figure 3.15 Image S

Note in the preceding example that the vertical size distribution $\omega_1(k)$ is less affected by small k than is the horizontal size distribution $\omega_2(k)$. This means that $H(k)$ has a greater filtering effect than $V(k)$. The difference in the behavior of the size distributions results from the fact that the image contains a greater distribution of linear size in the vertical direction than it does in the horizontal direction. The size distributions provide some sort of quantification of textural information. In general, the measure of a granulometry yields a quantification of the manner in which activated pixels are clustered relative to the shape of the structuring element.

Just as we normalized the Euclidean granulometric size distributions to create true probability distributions, so too can we normalize the digital granulometries. In the latter case, each normalized distribution is a step-type probability distribution function, and hence its derivative consists of a sum of delta functions. Nevertheless, the feature analysis methodology using the method of moments can be employed once again. Because of the similarity of the techniques, we shall forego the details.

Whereas the granulometric size distributions result from the dimunition of an image due to successive openings by structuring elements of increasing size, other size distributions result from a corresponding dimunition due to erosion by increasingly large structuring elements. Instead of finding $m[O(A, tE)]$, we can compute $m[\mathcal{E}(A, tE)]$. We have paid particular attention to the granulometric procedure because of the excellent mathematical properties possessed by the granulometries; however, one might, for heuristic reasons, choose to employ size distributions resulting from erosion. Rather than proceed through a full explication of the Euclidean erosion methodology, we shall content ourselves with a brief discussion of the digital case, as well as the digital version of the morphological covariance. (See Section 3.3.)

The digital size distributions generated by linear erosion result from the mappings

$$k \longrightarrow \text{ERODE}(S, V(k))$$

and

$$k \longrightarrow \text{ERODE}(S, H(k))$$

The corresponding size distributions are defined by

$$\mu_1(k) = \text{CARD}[\text{ERODE}(S, V(k))]$$

and
$$\mu_2(k) = \text{CARD}[\text{ERODE}(S, H(k))]$$

Example 3.12:

Let S be the image given in Figure 3.15. Then the size distributions μ_1 and μ_2 are given by

$k =$	1	2	3	4	5	6	7	8	9	10	11	12	13
$\mu_1(k) =$	80	65	54	46	38	31	24	19	14	9	5	2	0
$\mu_2(k) =$	80	44	22	7	2	0	0	0	0	0	0	0	0

For $k > 13$, $\mu_1(k) = \mu_2(k) = 0$.

One can also utilize a digital version of the morphological covariance to create size distributions. One essential difference here is that the resulting distributions are not decreasing functions of k. To proceed, define the digital structuring elements

$$v(k) = \begin{pmatrix} 1 \\ * \\ \vdots \\ * \\ \textcircled{1} \end{pmatrix} \quad \text{and} \quad h(k) = (\textcircled{1} * * \ldots * 1)$$

each of which, except for $k = 1$, consists of two activated pixels at either end of a string of k pixels. For the case $k = 1$, $v(k) = h(k) = (\textcircled{1})$. The *vertical digital covariance function* and the *horizontal digital covariance function* are then respectively defined by

$$\nu_1(k) = \text{CARD}[\text{ERODE}(S, v(k))]$$

and

$$\nu_2(k) = \text{CARD}[\text{ERODE}(S, h(k))]$$

Some of the important properties of the digital covariance, which of course are analogues of the corresponding Euclidean properties, are:

(i) $\nu_1(1) = \text{CARD}(S)$
(ii) $\nu_1(k) = \text{CARD}[\text{ERODE}(S, -v(k))]$
(iii) $\nu_1(k) = \text{CARD}[S \wedge \text{TRAN}(S; 0, -k + 1)]$

The last equality follows directly from the fact that

$$S \wedge \text{TRAN}(S; 0, -k + 1) = \text{ERODE}(S, v(k))$$

Analogous properties hold for $\nu_2(k)$.

Example 3.13:

Let S be the image given in Figure 3.16. Then the covariance functions $\nu_1(k)$ and $\nu_2(k)$

$$S = \begin{pmatrix} 1 & 1 & 1 & * & * & * & 1 & 1 \\ 1 & 1 & 1 & * & * & * & 1 & 1 \\ 1 & 1 & 1 & * & * & * & 1 & 1 \\ 1 & 1 & 1 & 1 & 1 & 1 & 1 & 1 \\ 1 & 1 & 1 & 1 & 1 & 1 & 1 & 1 \\ 1 & 1 & * & * & * & 1 & 1 & * \\ 1 & 1 & * & * & * & 1 & 1 & * \\ 1 & 1 & * & * & * & 1 & 1 & * \end{pmatrix}$$

Figure 3.16 Image S for which covariance function is to be found

are given by

$k =$	1	2	3	4	5	6	7	8	9
$\nu_1(k) =$	43	35	27	21	15	9	6	3	0
$\nu_2(k) =$	43	29	15	8	14	18	13	5	0

For $k > 9$, $\nu_1(k) = \nu_2(k) = 0$. Notice how there is a fairly steady decline in the vertical covariance. This reflects the lack of deactivated pixels separating strings of activated pixels in the vertical direction. On the other hand, there are six different strings of deactivated pixels separating strings of activated pixels in the horizontal direction. This gives rise to two local maxima for the function $\nu_2(k)$, one at $k = 1$ and another at $k = 6$.

Insofar as interpretation of any of the aforementioned size distributions is concerned, one should always be aware that such interpretation depends heavily on the experience of the investigator relative to the class of images under discussion. While it has been used to identify different images on the basis of texture, the behavior of the size distributions is not describable in elementary geometric terms. It is true that size distributions provide quantitative feature information, but it is also true that their utility depends upon a deep understanding of the scientific structures to which they are to be applied.

3.7 A STOCHASTIC APPROACH

The genesis of the morphological methodology lies in the search for structure within an image and the concomitant quantification of that structure. In his early work, Matheron saw that quantification as essentially statistical in nature. In this section, we hope to give some insight into the stochastic model and to illustrate how the probabilistic outlook leads to certain structural parameters associated with an image.

Given a subset A of the Euclidean plane, one can imagine throwing a small image E containing the origin onto the plane in such a manner that the origin ends up in A. One can then raise the question as to what the probability is that all of E lies within A. Referring to the discussion in Section 3.1, E can be viewed as a random probe, and the probability of whether or not E falls entirely within A depends upon the shape of the probe and the shape (or structure) of A. In other words, the probability serves as a statistical descriptor, and, since it is dependent upon the geometries of A and E, it is called a *geometric probability*.

Sec. 3.7 A Stochastic Approach

For those used to deterministic features, the preceding analysis might appear somewhat strange. One has the inclination to ask what the "meaning" is of the resulting probability, which we shall henceforth denote by $P(E \subset A)$. Unfortunately, such a query can only be answered by saying that the meaning of $P(E \subset A)$ lies within the methodology for finding $P(E \subset A)$. Indeed, if one attempts to dig too deeply into the meaning of any statistical descriptor, one finds oneself in precisely the same quandary. This is because statistical meaning is by nature methodological: while it is true that the scientist might choose for heuristic reasons to employ one probe over another, the choice reflects his or her own motivation; it does not impart deterministic meaning to the outcome, or possible outcomes, of the experiment. Thus, using the terminology of Section 3.1, in choosing the method and the probe the image engineer *constitutes* the meaning of structure upon which intelligence must focus.

Now suppose we restrict our attention to a finite image S in the grid and a discrete structuring element E containing the origin $(0, 0)$. If we assume, as we did above, that the tossing of E always results in the origin's lying in S, and if we assume that the origin can fall with equal likelihood on any pixel in S, then the probability $P(E \subset S)$ is given by the number of pixels (i, j) in S for which $\text{TRAN}(E; i, j) \leqslant S$ divided by the total number of pixels in S. Consequently,

$$P(E \subset S) = \frac{\text{CARD}[\text{ERODE}(S, E)]}{\text{CARD}(S)}$$

In other words, the geometric probability $P(E \subset S)$ depends upon the erosion.

Let us now proceed to investigate how we are naturally led to a feature-parameter methodology based on the preceding stochastic approach. In order to search for structure within an image S, we could calculate a vector of probabilities of the form $P(E^k \subset S)$, where $E^1, E^2, \ldots, E^m$ are a predetermined set of structuring elements, the collection of which, in Matheron's words, plays an "a priori constitutive role" in the definition of structure. Once this choice of structuring elements has been made, S can be eroded by the various elements, each erosion resulting in a geometric probability

$$P_k = \text{Prob}\,[\text{TRAN}\,(E^k; i,j) \leqslant S \mid (i, j) \in S]$$
$$= \frac{\text{CARD}[\text{ERODE}(S, E^k)]}{\text{CARD}(S)}$$

The vector

$$\mathbf{P} = \begin{pmatrix} P_1 \\ P_2 \\ \vdots \\ P_m \end{pmatrix}$$

serves as a measure of the manner in which the archetypal geometric forms $E^1, E^2, \ldots, E^m$ fit into the image, and the distribution of the P_k can serve as a signature of the image.

Example 3.14:

Consider the archetypal collection $\{E^1, E^2, E^3\}$, where

$$E^1 = \begin{pmatrix} 1 & 1 & 1 \end{pmatrix}_{0,0}$$

$$E^2 = \begin{pmatrix} 1 & 1 & 1 \\ * & 1 & * \\ 1 & * & * \end{pmatrix}_{0,2}$$

$$E^3 = \begin{pmatrix} 1 & 1 & 1 \\ * & * & 1 \\ * & * & 1 \end{pmatrix}_{0,2}$$

Intuitively, in a digital sense E^1 represents an infinitesimal straight line, E^2 an infinitesimal 45° angle, and E^3 an infinitesimal 90° angle. Keep in mind that we are looking for infinitesimal structure; if we were looking for macrostructures, we would employ much larger structuring elements, recognizing of course that this would *eo ipso* result in lower probabilities P_k.

Now suppose

$$S = \begin{pmatrix} 1 & 1 & * & 1 & * & * & 1 \\ * & 1 & * & 1 & 1 & * & 1 \\ * & 1 & * & * & 1 & * & * \\ * & 1 & * & 1 & * & 1 & 1 \\ * & * & * & 1 & 1 & * & * \\ 1 & 1 & * & * & * & 1 & 1 \end{pmatrix}_{0,5}$$

Then $\text{ERODE}(S, E^1) = \text{ERODE}(S, E^2) = \text{ERODE}(S, E^3) = \emptyset$, and hence $P_1 = P_2 = P_3 = 0$.

Also, suppose

$$T = \begin{pmatrix} 1 & 1 & 1 & 1 & 1 & 1 & 1 \\ 1 & 1 & 1 & 1 & 1 & 1 & 1 \\ * & * & * & * & * & 1 & 1 \\ * & * & * & * & * & 1 & 1 \\ * & * & * & * & * & 1 & 1 \\ * & * & * & * & * & 1 & 1 \end{pmatrix}_{0,5}$$

Then

$$\text{ERODE}(T, E^1) = \begin{pmatrix} 1 & 1 & 1 & 1 & 1 \\ 1 & 1 & 1 & 1 & 1 \end{pmatrix}_{0,5}$$

$$\text{ERODE}(T, E^2) = \emptyset$$

and

$$\text{ERODE}(T, E^3) = \begin{pmatrix} 1 & 1 \\ 1 & 1 \end{pmatrix}_{3,5}$$

Hence,

$$\mathbf{P} = \begin{pmatrix} \frac{5}{11} \\ 0 \\ \frac{2}{11} \end{pmatrix}$$

Clearly, P_1 has picked up the high degree of linear structure, while P_3 has detected a trace of infinitesimal 90° structure.

Numerous variations and extensions of the stochastic methodology can be employed. Our intent, however, has been to introduce the probabilistic model while at the same time illustrating further the manner in which structuring elements can be employed to yield textural information. Indeed, one could utilize the methodology of this section without ever being aware that the parameters were derived on the basis of statistical considerations.

EXERCISES

3.1. Let A and B be the sets given in Exericse 1.2. Find and graph the size distribution generated by the granulometry $t \to O(A, tB)$.

3.2. Let A and C be the sets given in Exercise 1.2. Find and graph the size distribution generated by the granulometry $t \to O(A, tC)$.

3.3. Using the integral formula that gives the length of a piecewise smooth boundary, show that the boundary length Λ is translationally and rotationally invariant.

3.4. Let
$$E = \{(x, y): 0 \le y \le 2x \text{ and } 0 \le x \le 3\}$$
The perimeter of E is $9 + \sqrt{45}$. For all θ, find $\text{Proj}(E, \theta)$ and show that the Cauchy projection theorem holds, i.e., show that the average over $0 \le \theta \le 2\pi$ of the projections is equal to $\Lambda(E)/\pi$.

3.5. Give an example to show that Λ is not, in general, increasing (does not satisfy Q5), even though it is increasing on convex compact sets. Use the Cauchy projection theorem to show that Λ is increasing on compact convex sets.

3.6. For $n = 1, 2, 3, \ldots$, let
$$A_n = \left\{(x, y): 0 \le y \le \frac{1}{n}x \text{ and } 0 \le x \le 1\right\}$$
and
$$A = \{(x, y): y = 0 \text{ and } 0 \le x \le 1\}$$
Show that $\lim_{n \to \infty} A_n = A$ in the Hausdorff metric by actually computing $h(A_n, A)$. Note the extreme sensitivity of the Hausdorff metric to noise by considering $h(B_n, A)$, where $B_n = A_n \cup \{(0, 2)\}$.

3.7. Using the sets B_n from Exercise 3.6, show that $\lim_{n \to \infty} m(B_n) = 0$, where m denotes the Lebesgue measure (area). This result should be expected since m is continuous from above and, using set A of Exercise 3.6, we have
$$\bigcap_{n=1}^{\infty} B_n = A \cup \{(0, 2)\}$$
But $m[A \cup \{(0, 2)\}] = 0$.

3.8. Let E be the triangle given in Exercise 3.4. In Section 3.4 it was shown that the measure of E can be found morphologically by using a one-sided derivative. Using that methodology, find the area of E.

3.9. Using the triangle E of Exercise 3.4 and the dilation technique of Section 3.4, find Proj$(E, \mathbf{u})$, where $\mathbf{u}$ is the unit vector pointing in the 0° direction. Repeat for the 90° direction.

3.10. Let

$$R = \begin{pmatrix} 1 & 1 & 1 & * & 1 & 1 & 1 & 1 & 1 \\ 1 & * & * & 1 & * & * & * & * & 1 \\ 1 & 1 & 1 & 1 & 1 & 1 & 1 & * & * & * \\ * & * & * & 1 & 1 & 1 & 1 & 1 & 1 \\ 1 & 1 & 1 & * & 1 & * & 1 & 1 & * & * \\ * & * & * & * & * & 1 & * & 1 & 1 & 1 \\ \textcircled{1} & 1 & 1 & 1 & 1 & 1 & 1 & 1 & 1 & 1 \end{pmatrix}$$

For the image R, find:

(a) $\omega_1(k)$ (d) $\mu_2(k)$
(b) $\omega_2(k)$ (e) $v_1(k)$
(c) $\mu_1(k)$ (f) $v_2(k)$

FOOTNOTES FOR CHAPTER 3

1. Georges Matheron, *Random Sets and Integral Geometry* (New York: Wiley, 1975) p. ix.
2. E. R. Dougherty and C. R. Giardina, *Image Processing: Continuous to Discrete*, Volume I (Englewood Cliffs: Prentice-Hall, 1987), p. 130–131.
3. Matheron, *Random Sets and Integral Geometry,* ix–x.
4. H. Hadwiger, *Vorslesungen Uber Inhalt, Oberflache Und Isoperimetric* (Berlin: Springer, 1957), p. 221.
5. Matheron, *Random Sets and Integral Geometry,* p. 119.
6. Hadwiger, *Vorslesungen Uber Inhalt, Oberflache Und Isoperimetric,* p. 221.
7. G. Watson, "Mathematical Morphology," Tech. Report No. 21, Dept. of Stat., 1973: 11.
8. Ibid., p. 13.
9. E. R. Dougherty and C. R. Giardina, "Error Bounds for Morphologically Derived Feature Measurements," SIAM Journal on Applied Mathematics (in press).

4

Topological Processing

4.1 TOPOLOGICAL PRELIMINARIES

Numerous classes of topological procedures exist in image processing that are concerned with the underlying region upon which an image is defined. Some of these generic methodologies are region growing, curve filling, tracking, thinning, and, in some cases, edge detection. In this chapter we shall investigate some morphological techniques that can be employed to accomplish topological processing on constant (1–*) images, especially digital images. Because of the identification of constant images with subsets of $Z \times Z$ or $R \times R$, topological processing of such images is most natural.

A digital morphological transformation is a transformation that employs only operators formed from either the augmented morphological basis or the structural basis. Although we shall concentrate on algorithms that are solely morphological, in practice one makes no such restriction. The procedures to be outlined can be used in conjunction with procedures from the general image algebra to produce suitable algorithms, or they may be altered through the introduction of nonmorphological operations.

We begin by defining certain concepts from digital topology that correspond to concepts from Euclidean topology. Included will be digital variants of connectivity, the closure, the boundary, and convexity.

Two pixels are called *direct* (*strong*) *neighbors* if they share a common side. The direct neighbors of pixel (i, j) are pixels $(i, j - 1)$, $(i, j + 1)$, $(i + 1, j)$, and $(i - 1, j)$. Two pixels are called *indirect* (*weak*) *neighbors* if they share exactly one common corner. The indirect neighbors of pixel (i, j) are pixels $(i - 1, j - 1)$, $(i - 1, j + 1)$, $(i + 1, j - 1)$, and $(i + 1, j + 1)$. (See Figure 4.1.) Two pixels are simply called *neighbors* if they are either direct or indirect neighbors.

O	X	O
X	(i, j)	X
O	X	O

X DIRECT NEIGHBOR
O INDIRECT NEIGHBOR   **Figure 4.1**   Neighbors of a pixel

A *path* is a sequence of pixels $p_1, p_2, \ldots, p_n$ such that p_k is a neighbor of p_{k+1} for $k = 1, 2, \ldots, n - 1$. Pixel p_1 is called the *initial* pixel and p_n is the *terminal* pixel. A path is called *simple* if no pixel is repeated, except perhaps the first and last, and if no pixel has more than two direct neighbors. It is called *closed* if the first and last pixels coincide.

Example 4.1

The labeled pixels in the following grid can be sequenced to form several paths.

```
- - b - - -
- a - c e -
- h - f d -
- - g - - -
```

Some examples are

P_1: a, b, c, d, e, c, f, g

P_2: a, b, c, d, e, f, g

P_3: $a, b, c, d, e, c, f, g, h, a$

P_4: $a, b, c, d, e, f, g, h, a$

P_2 is simple, P_3 is closed, and P_4 is simple and closed. For those familiar with the definition of a simple Euclidean path, notice that, if the lattice points of P_2 were joined in order by straight lines, the resulting path would not be simple in the Euclidean sense. Once again, digitization has created an apparent anomaly.

A collection of pixels is said to be *connected* if for any two pixels in the collection there exists a path of pixels in the collection such that one is the initial pixel and the other is the terminal pixel of the path. A connected set of pixels is usually called a *region*. A constant (1–*) image is called *connected* if its activated pixels form a connected collection.

Example 4.2

Let

$$S = \begin{pmatrix} * & * & 1 & * & * & * & 1 \\ * & * & 1 & * & 1 & 1 & * \\ * & * & 1 & * & 1 & 1 & 1 \\ * & * & * & 1 & 1 & * & * \end{pmatrix}_{3,-1}$$

and

$$T = \begin{pmatrix} * & * & 1 & * & * & * & * \\ 1 & 1 & 1 & * & * & * & * \\ 1 & 1 & * & * & 1 & * & * \\ * & * & * & 1 & 1 & 1 & * \\ * & * & * & * & 1 & 1 & 1 \end{pmatrix}_{-3,2}$$

Then S is connected and T is disconnected.

Given a digital constant image S, the *boundary* of S, BOUND(S), is the constant image whose domain is such that each pixel within it has a neighbor in the domain of S but is not in the domain of S itself. Two points should be noted. First, for those who are familiar with the boundary of a Euclidean set, BOUND is a digital version of that notion; however, it is only one of several possible digital alternatives, each suffering from its own digitization shortcomings. Second, the concept of boundary, as defined herein, is often called the *external* boundary, since BOUND(S) has a domain of pixels that are external to the domain of S.

Example 4.3

Let

$$S = \begin{pmatrix} * & * & * & * & * & * & * & * & * \\ * & * & * & 1 & * & 1 & * & * & * \\ * & * & * & 1 & * & 1 & * & * & * \\ * & * & 1 & 1 & 1 & 1 & 1 & * & * \\ * & 1 & 1 & 1 & 1 & 1 & 1 & * \\ * & * & * & * & * & 1 & 1 & * & * \\ * & 1 & 1 & * & * & * & 1 & * & * \\ * & 1 & 1 & 1 & * & * & * & * & * \\ * & * & * & * & * & * & * & * & * \end{pmatrix}_{0,8}$$

Then

$$\text{BOUND}(S) = \begin{pmatrix} * & * & 1 & 1 & 1 & 1 & 1 & * & * \\ * & * & 1 & * & 1 & * & 1 & * & * \\ * & 1 & 1 & * & 1 & * & 1 & 1 & * \\ 1 & 1 & * & * & * & * & * & 1 & 1 \\ 1 & * & * & * & * & * & * & * & 1 \\ 1 & 1 & 1 & 1 & 1 & * & * & 1 & 1 \\ 1 & * & * & 1 & 1 & 1 & 1 & * & 1 \\ 1 & * & * & * & 1 & * & 1 & 1 & 1 \\ 1 & 1 & 1 & 1 & 1 & * & * & * & * \end{pmatrix}_{0,8}$$

The boundary (external boundary) of a constant image *encloses* that image. Due to the discreteness of the grid, this enclosing at times appears paradoxical. For instance, in Example 4.3 the image S can intuitively be viewed as the union of two disjoint connected subimages S_1 and S_2, where S_1 is the image consisting of five black pixels in the lower left corner of the bound matrix for S, and S_2 consists of the remaining activated pixels of S. Rigorously,

$$S = S_1 \vee S_2 = \text{EXTMAX}(S_1, S_2)$$

In accordance with Euclidean intuition, the boundaries of S_1 and S_2 should be disjoint. This is not the case. As is often the problem, digitization yields thick sets where one would desire a thin set. More precisely, Euclidean points have no diameter, whereas pixels are squares with edges having a fixed positive length. Redefining the boundary in some other manner might solve this particular anomaly; however, the new definition would have its own problems.

BOUND is specified morphologically by the following block diagram:

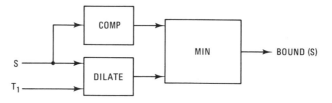

where

$$T_1 = \begin{pmatrix} 1 & 1 & 1 \\ 1 & \textcircled{1} & 1 \\ 1 & 1 & 1 \end{pmatrix}$$

Given this definition of the boundary, we define the digital closure by

$$\text{CLOSURE}(S) = \text{EXTMAX}[S, \text{BOUND}(S)]$$

An issue often addressed in image processing is the *filling* of a region determined by a given simple closed curve. The problem, in its most general form, is extremely delicate due to the intricacies of digitization. Consider the image S in Figure 4.2, which has a domain that is a simple closed curve; the symbol # is temporarily used to denote the value at a pixel where the gray value is undefined. It appears that the #-value pixels are *interior* to the curve S. Yet if I denotes the constant image with those pixels activated, it is certainly not true that $S = \text{BOUND}(I)$. Suppose we were to argue that the notion of *internal* boundary should be applied, where the internal

$$S = \begin{pmatrix} 1 & * & * & * & 1 & * & * & * \\ 1 & 1 & * & 1 & 1 & * & * & * \\ 1 & \# & 1 & \# & \# & 1 & 1 & * \\ 1 & \# & \# & \# & 1 & 1 & \# & 1 \\ * & 1 & 1 & 1 & 1 & * & 1 & * \end{pmatrix}_{0,4}$$

Figure 4.2 Simple closed curve-type image

boundary of $S \vee I$ is the collection of pixels in the domain of $S \vee I$ that have neighbors outside the domain of $S \vee I$. Would S then be the boundary of $S \vee I$? Certainly not! Pixel (1, 1) is in the domain of $S \vee I$ and it has a neighbor outside that domain. Yet it is not in S. Another argument might be to use the internal boundary notion but include a pixel if and only if it has a direct neighbor outside the domain. However, this would also fail since pixel (4, 1) is in the domain of $S \vee I$ and does not have a direct neighbor outside that domain, but is nevertheless in the domain of S! It is possible to point out many such seeming pathologies. Yet the problem is inherent in the digitization scheme. To obtain rigorous results, precise definitions must be generated, and theorems must be carefully checked against those definitions. There is certainly no definitive, universally accepted approach to the matter. In this text, we shall demonstrate one algorithm for filling a particular type of curve and shall leave the multitude of other approaches to the literature.

An image is said to be *directly connected* if, for any two pixels in its domain, there is a path $p_1, p_2, \ldots, p_n$ between the two given pixels such that p_k is a direct neighbor of p_{k+1} for $k = 1, 2, \ldots, n - 1$. Intuitively, for a directly connected image, it is possible to go from any one pixel in the domain to any other by a sequence of horizontal and vertical steps.

A constant image C is said to be *strongly grid convex* if three conditions are satisfied:

1. C is directly connected.
2. If two pixels in the same column of the minimal bound matrix representing C are activated, then so are all pixels in the column that lie between them (vertical convexity).
3. If two pixels in the same row of the minimal bound matrix representing C are activated, then so are all pixels in the row that lie between them (horizontal convexity).

Figure 4.3 gives a strongly grid convex image. From this example, it should be clear that strong grid convexity and Euclidean convexity are quite different notions. Yet there is one sense in which they are the same. If P and Q are any two activated pixels of a grid convex image that lie in the same column or the same row, then all pixels between them are activated. Since there are only two principal directions on the grid, vertical and horizontal, the definition of grid convexity remains essentially true to the underlying notion of convexity. Figure 4.4 presents an image that fails to be

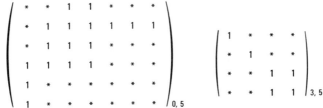

Figure 4.3 Strongly convex image

Figure 4.4 Weakly convex image

strongly grid convex because it violates condition 1 of the definition, even though it is connected. Such an image is said to be *weakly grid convex*. If only conditions 2 and 3 are satisfied, then the image is called *pseudoconvex*.

4.2 REGION PROCESSING

Rather than find the boundary proper of an image, we might simply be interested in an edge that approximates the boundary. In image processing, the term *edge* is not so well defined as the term *boundary*. It has the intuitive meaning of boundary, but an edge need not be rigorously specified. Indeed, an image can have numerous "edges," each one being defined by the methodology by which it has been derived. The intent of an edge-detection methodology is to compress the image, not to output a specific mathematical entity.

Figure 4.5(a) depicts the image A, its dilation by a small disk B, and the set-theoretic subtraction of A from $\mathcal{D}(A, B)$. It is clear that the subtraction

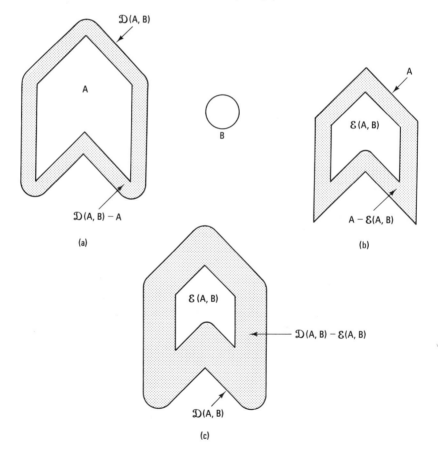

Figure 4.5 Euclidean edge images

Sec. 4.2 Region Processing

$\mathcal{D}(A, B) - A$ is an approximation of the true Euclidean boundary of A. The thickness of that approximation results from the thickness of B. If the diameter of B is small, we might certainly wish to call $\mathcal{D}(A, B) - A$ an *edge* of A. In fact, we would probably call $\mathcal{D}(A, B) - A$ an *external edge*.

Figure 4.5(b) depicts $\mathcal{E}(A, B)$ and $A - \mathcal{E}(A, B)$ for the images A and B of part (a) of the figure. This difference, too, gives an approximation of the boundary, only this time the edge is *internal*. If, on the other hand, one desired an edge that "straddled" the true boundary, he or she might consider the difference $\mathcal{D}(A, B) - \mathcal{E}(A, B)$. This image is depicted in Figure 4.5(c).

The specification of BOUND presented earlier is an exact digital analogue of the edge detection methodology resulting from the subtraction of the input image from a dilate of that image, where the *square neighbor mask*

$$T_1 = \begin{pmatrix} 1 & 1 & 1 \\ 1 & \textcircled{1} & 1 \\ 1 & 1 & 1 \end{pmatrix}$$

has been employed. If we consider the same specification but allow the second image T to be a variable input, then we arrive at a binary morphological edge operation which we shall call MOREDGE(S, T).

Example 4.4:

Let S be the image given in Example 4.3, and let T_2 be the *strong neighbor mask*

$$T_2 = \begin{pmatrix} * & 1 & * \\ 1 & \textcircled{1} & 1 \\ * & 1 & * \end{pmatrix}$$

Then

$$\text{DILATE}(S, T_2) = \begin{pmatrix} * & * & * & 1 & * & 1 & * & * & * \\ * & * & 1 & 1 & 1 & 1 & 1 & * & * \\ * & * & 1 & 1 & 1 & 1 & 1 & * & * \\ * & 1 & 1 & 1 & 1 & 1 & 1 & 1 & * \\ 1 & 1 & 1 & 1 & 1 & 1 & 1 & 1 & 1 \\ * & 1 & 1 & 1 & 1 & 1 & 1 & 1 & * \\ 1 & 1 & 1 & 1 & * & 1 & 1 & 1 & 1 \\ 1 & 1 & 1 & 1 & * & * & 1 & * \\ * & 1 & 1 & 1 & * & * & * & * & * \end{pmatrix}_{0,8}$$

and

$$\text{MOREDGE}(S, T_2) = \begin{pmatrix} * & * & * & 1 & * & 1 & * & * & * \\ * & * & 1 & * & 1 & * & 1 & * & * \\ * & * & 1 & * & 1 & * & 1 & * & * \\ * & 1 & * & * & * & * & * & 1 & * \\ 1 & * & * & * & * & * & * & * & 1 \\ * & 1 & 1 & 1 & 1 & * & * & 1 & * \\ 1 & * & * & 1 & * & 1 & 1 & * & 1 \\ 1 & * & * & * & 1 & * & * & 1 & * \\ * & 1 & 1 & 1 & * & * & * & * & * \end{pmatrix}_{0,8}$$

Note that MOREDGE(S, T_2) is sparser than BOUND(S). Indeed, if MOREDGE(S, T_2) provides sufficient information regarding S, then it is preferable to BOUND(S) since it requires less computation to obtain and less storage to save. Put succinctly, MOREDGE($\cdot$, T_2) provides more compression than does BOUND.

Other structuring elements besides T_1 and T_2 can be employed in MOREDGE. A larger structuring element will produce a thicker edge. In essence, MOREDGE is a *thinning* algorithm in that it produces a thin replica of the input image. Since it removes the original image from a dilated version of the image, it yields an external edge. If we define a dual edge operator by subtracting ERODE(S, T) from S, then an internal edge is produced. There are many variations on the theme.

We now consider the notion of a *digital convex hull*. In a manner analogous to the Euclidean definition, for any constant digital image S, we define HULL(S) to be the *smallest* pseudoconvex image such that $S \ll \text{HULL}(S)$. Whereas MOREDGE and its variants are thinning algorithms, HULL is a thickening algorithm. The next example shows why we do not define the convex hull in terms of strong or weak convexity: there can be more than one "smallest" strongly convex or weakly convex image containing a given image.

Example 4.5:

Let

$$S = \begin{pmatrix} 1 & * & * & * & * & * \\ * & 1 & * & 1 & * & * \\ * & * & 1 & 1 & * & 1 \\ * & * & * & * & * & * \\ * & * & 1 & * & * & 1 \end{pmatrix}_{0,4}$$

Then

$$\text{HULL}(S) = \begin{pmatrix} 1 & * & * & * & * & * \\ * & 1 & 1 & 1 & * & * \\ * & * & 1 & 1 & 1 & 1 \\ * & * & 1 & 1 & 1 & 1 \\ * & * & 1 & 1 & 1 & 1 \end{pmatrix}_{0,4}$$

and HULL(S) is not strongly convex. Note that the images

$$S' = \begin{pmatrix} 1 & 1 & * & * & * & * \\ * & 1 & 1 & 1 & * & * \\ * & * & 1 & 1 & 1 & 1 \\ * & * & 1 & 1 & 1 & 1 \\ * & * & 1 & 1 & 1 & 1 \end{pmatrix}_{0,4}$$

and

$$S'' = \begin{pmatrix} 1 & * & * & * & * & * \\ 1 & 1 & 1 & 1 & * & * \\ * & * & 1 & 1 & 1 & 1 \\ * & * & 1 & 1 & 1 & 1 \\ * & * & 1 & 1 & 1 & 1 \end{pmatrix}_{0,4}$$

Sec. 4.2 Region Processing

are minimal strongly convex images containing S: if any pixel in either of them is deactivated, the resulting image is either not strongly convex or does not contain S.

The connectivity problem cannot be solved by defining the hull to be the smallest weakly convex image containing S. For consider

$$T = \begin{pmatrix} 1 & * & * & * \\ * & * & * & 1 \end{pmatrix}_{0,0}$$

Both

$$T' = \begin{pmatrix} 1 & 1 & * & * \\ * & * & 1 & 1 \end{pmatrix}_{0,0}$$

and

$$T'' = \begin{pmatrix} 1 & * & * & * \\ * & 1 & 1 & 1 \end{pmatrix}_{0,0}$$

are minimal weakly convex images containing T.

We shall not give a morphological specification of HULL at this time. HULL is an example of a 0-bounded increasing τ-mapping. We shall see in Section 5.3 that such mappings can be readily computed employing EXTMAX and ERODE.

The hull methodology can be useful for filling a region enclosed by a path. If the path should happen to be the boundary of a weakly convex region, then applying HULL to the image whose activated pixels form the path will produce the desired result. Extensions of the method are possible if we rigorously define the interior of a simple closed path.

In the next example, we consider a closed path with a point lying within the region enclosed by the path. We shall apply a dilation methodology to fill the enclosed region.

Example 4.6:

Let

$$S = \begin{pmatrix} 1 & 1 & 1 & * & * & * & * & * \\ 1 & * & * & 1 & 1 & * & * & * \\ 1 & * & * & * & * & 1 & 1 & * \\ 1 & * & * & 1 & * & * & * & 1 \\ * & 1 & * & * & * & * & * & 1 \\ * & 1 & * & * & * & * & 1 & * \\ * & * & 1 & 1 & 1 & 1 & 1 & * \end{pmatrix}_{2,5}$$

$$T_2 = \begin{pmatrix} * & 1 & * \\ 1 & \textcircled{1} & 1 \\ * & 1 & * \end{pmatrix}$$

$S^{(1)} = (1)_{5,2}$, and form

$$S^{(2)} = \text{MIN}[\text{DILATE}(S^{(1)}, T_2), S^c],$$

The net effect is to dilate $S^{(1)}$ by T_2 and then set-theoretically subtract from the dilation any pixel that also happens to be in S. Proceeding inductively, form

$$S^{(3)} = \text{MIN}[\text{DILATE}(S^{(2)}, T_2), S^c]$$

which is the set-theoretic subtraction of S from the dilation of $S^{(2)}$ by T_2. Continue the procedure until $S^{(k+1)} = S^{(k)}$. Then $S^{(k)}$ is the output of the algorithm. We obtain, in order,

$$S^{(2)} = \begin{pmatrix} * & * & * & * & * & * & * & * \\ * & * & * & * & * & * & * & * \\ * & * & * & 1 & * & * & * & * \\ * & * & 1 & * & 1 & * & * & * \\ * & * & * & 1 & * & * & * & * \\ * & * & * & * & * & * & * & * \\ * & * & * & * & * & * & * & * \end{pmatrix}_{2,5}$$

$$S^{(3)} = \begin{pmatrix} * & * & * & * & * & * & * & * \\ * & * & * & * & * & * & * & * \\ * & * & 1 & 1 & 1 & * & * & * \\ * & 1 & 1 & * & 1 & 1 & * & * \\ * & * & 1 & 1 & 1 & * & * & * \\ * & * & * & 1 & * & * & * & * \\ * & * & * & * & * & * & * & * \end{pmatrix}_{2,5}$$

$$S^{(4)} = \begin{pmatrix} * & * & * & * & * & * & * & * \\ * & * & 1 & * & * & * & * & * \\ * & 1 & 1 & 1 & 1 & * & * & * \\ * & 1 & 1 & * & 1 & 1 & 1 & * \\ * & * & 1 & 1 & 1 & 1 & * & * \\ * & * & 1 & 1 & 1 & * & * & * \\ * & * & * & * & * & * & * & * \end{pmatrix}_{2,5}$$

$$S^{(5)} = \begin{pmatrix} * & * & * & * & * & * & * & * \\ * & 1 & 1 & * & * & * & * & * \\ * & 1 & 1 & 1 & 1 & * & * & * \\ * & 1 & 1 & * & 1 & 1 & 1 & * \\ * & * & 1 & 1 & 1 & 1 & 1 & * \\ * & * & 1 & 1 & 1 & 1 & * & * \\ * & * & * & * & * & * & * & * \end{pmatrix}_{2,5}$$

with $S^{(5)}$ being the output image. If one wishes, the starting pixel, (5, 2), can be activated at the completion of the recursion.

Note that had we employed

$$T_1 = \begin{pmatrix} 1 & 1 & 1 \\ 1 & \textcircled{1} & 1 \\ 1 & 1 & 1 \end{pmatrix}$$

instead of T_2, pixel (7, 4), which lies outside the desired region, would have been activated in $S^{(4)}$.

4.3 SKELETON

Whereas edge detection methods thin a figure by producing a facsimile of the boundary, the *skeleton* or *medial axis* algorithm thins by creating an archetypal stick figure that internally locates the central axis of the figure. To begin with, we shall consider

Sec. 4.3   Skeleton

the skeleton of a set in the Euclidean plane. The Euclidean skeleton of a set S is defined in the following manner. For each x in S, let $D(x)$ denote the largest disk centered at x such that $D(x)$ is a subset of S. Then x is in the skeleton of S if there does not exist a disk D_1, not necessarily centered at x, such that D_1 properly contains $D(x)$ and such that D_1 is contained in S. For example, consider the isosceles triangle in Figure 4.6(a). The skeleton is drawn in dark lines. Note that, whereas the point x lies in the skeleton, since $D(x)$ cannot be included in a larger disk still within the triangle [Figure 4.6(b)], the point w does not lie in the skeleton, since $D(w)$ is a subset of the disk D', which is itself a subset of the triangle [Figure 4.6(c)].

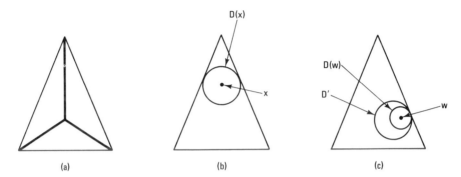

Figure 4.6 Determining skeleton for isosceles triangle

Figure 4.7 gives a good indication of some of the intuitive notions concerning the Euclidean skeleton. While the skeleton gives a decent replication of the shape of a figure that is already somewhat thin [Figure 4.7(a)], it is far less appropriate when applied to a *thick* figure [Figure 4.7(b)]. Moreover, different geometric figures may possess the same skeleton [Figures 4.7(c) and (d)]. Perhaps most importantly, the skeleton is extremely sensitive to noise. An infinitesimal distortion of the original shape can result in a vastly altered skeleton. For instance, in Figures 4.7(e) and (f), notice how the removal of a tiny section of the figure results in a drastically changed skeleton.

In proceeding to a digital definition of the skeleton, we are immediately confronted with the impossibility of finding an exact analogue to a Euclidean disk. While there are several ways to give a digital version of Euclidean disks, we shall content ourselves with the collection of "square disks" given in Figure 4.8. Each of these is a constant image, or template, in which the origin is near the center of the domain. One might legitimately argue that a proper extension of the disk notion would require that we omit the even-numbered digital disks D_2, D_4, D_6, ..., since for these the center of the template is not a true center in the sense of symmetry. If we were to do this, however, the resulting thinning procedure would often result in skeletons that were not sufficiently thin. In any event, note that for the even-numbered disks the center has been defined in a consistent fashion.

The definition of the digital skeleton can now be stated in a manner analogous to the corresponding Euclidean definition. Let T be a constant image (pixel values 1 or $*$). For any pixel (i, j) in the domain of T, the maximal disk for (i, j), MAXDISK(i, j), is the highest-numbered disk D_k, translated so that its new center is

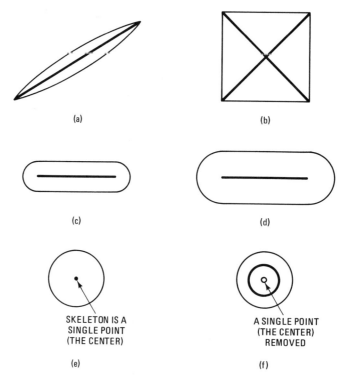

Figure 4.7 Skeleton for various pictures

$$D_1 = \begin{pmatrix} \textcircled{1} \end{pmatrix}$$

$$D_2 = \begin{pmatrix} 1 & 1 \\ \textcircled{1} & 1 \end{pmatrix}$$

$$D_3 = \begin{pmatrix} 1 & 1 & 1 \\ 1 & \textcircled{1} & 1 \\ 1 & 1 & 1 \end{pmatrix}$$

$$D_4 = \begin{pmatrix} 1 & 1 & 1 & 1 \\ 1 & 1 & 1 & 1 \\ 1 & \textcircled{1} & 1 & 1 \\ 1 & 1 & 1 & 1 \end{pmatrix}$$

$$D_5 = \begin{pmatrix} 1 & 1 & 1 & 1 & 1 \\ 1 & 1 & 1 & 1 & 1 \\ 1 & 1 & \textcircled{1} & 1 & 1 \\ 1 & 1 & 1 & 1 & 1 \\ 1 & 1 & 1 & 1 & 1 \end{pmatrix}$$

etc.

Figure 4.8 Square disks of increasing size

Sec. 4.3 Skeleton

at (i, j), such that $\text{TRAN}(D_k; i, j)$ is a subimage of T. The skeleton of T, $\text{SKEL}(T)$, is a constant image (1's and *'s) such that a pixel lies within the domain of $\text{SKEL}(T)$ if and only if its maximal disk is not a proper subimage of any other translated disk that is itself a subimage of T. Intuitively, (i, j) is in the digital skeleton if and only if its maximal disk is not a proper subset of some other disk lying within T.

Example 4.7:

For the image T of Figure 4.9(a), the maximal disk for the pixel (1, 8) is given by

$$\text{MAXDISK}(1, 8) = \begin{pmatrix} 1 & 1 \\ 1 & 1 \end{pmatrix}_{1, 9}$$

This is schematically indicated in Figure 4.9(b). Also shown are the illustrations of the maximal disks

$$\text{MAXDISK}(5, 1) = \begin{pmatrix} 1 & 1 & 1 & 1 \\ 1 & 1 & 1 & 1 \\ 1 & 1 & 1 & 1 \\ 1 & 1 & 1 & 1 \end{pmatrix}_{4, 3}$$

and

$$\text{MAXDISK}(6, 5) = \begin{pmatrix} 1 & 1 & 1 \\ 1 & 1 & 1 \\ 1 & 1 & 1 \end{pmatrix}_{5, 6}$$

Pixels (1, 8) and (5, 1) lie in the domain of the skeleton of T, but pixel (6, 5) does not. This latter claim follows from the fact that the domain of $\text{MAXDISK}(6, 5)$ lies properly within the domain of $\text{TRAN}(D_4; 5, 4)$, which is itself a subimage of T.

Finally, Figure 4.9(c) gives the skeleton $\text{SKEL}(T)$. Assuming the underlying image to be the letter Y, the skeleton gives a fairly good replication. Unfortunately, the pixel (8, 3), which appears to be affected by noise, results in an extraneous activated pixel in the skeleton. Moreover, the skeleton is not connected: the fifth row of the bound matrix contains no activated pixels. As a result, the top section of the letter Y is separated from the bottom section. Such connectivity problems are common for the skeleton. They make its utilization highly problematic.

The implementation of the operator SKEL is given in Figure 4.10. In that figure, the number m is given by the minimum between the number of rows and columns of the minimal bound matrix for T. In essence, the block diagram of Figure 4.10 works by finding the skeleton pixels that have maximal disk of edge length 1, then those with maximal disk of edge length 2, and so on. It then takes the set-theoretic union of those pixel classes.

Figure 4.11 provides a walk-through of the first three branches of Figure 4.10 with input image R. For the remaining branches, $k = 4, 5$, and 6, the first erosion yields the empty image. This will usually be the case for values of k near to the value m unless the minimal bound matrix consists of mostly activated pixels, with all deactivated pixels near the outer edge of the matrix. Notice also that the image $S \wedge T^c$ is the image whose activated pixels are those contained in the domain of S minus (set-theoretic difference) the domain of T.

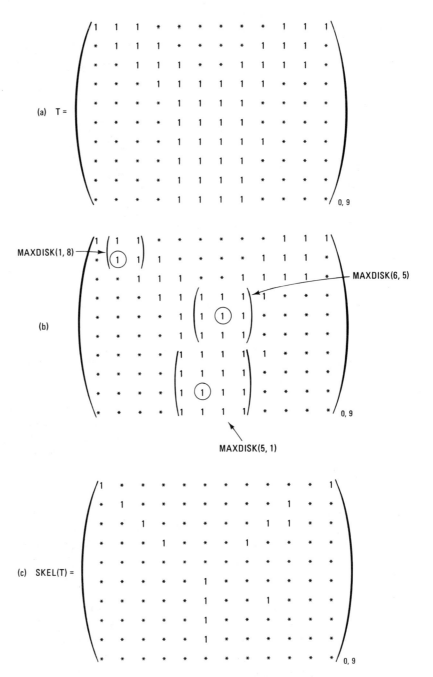

Figure 4.9 Image and associated skeleton image

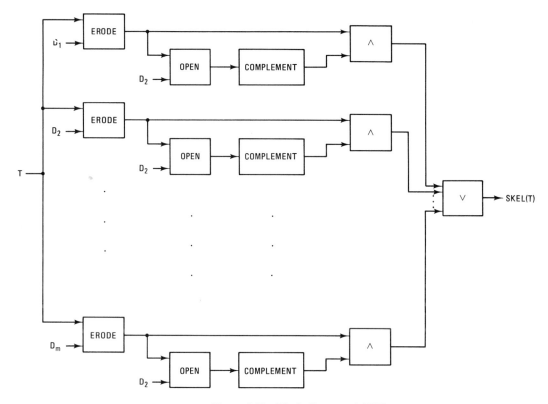

Figure 4.10 Block diagram of SKEL

4.4 HIT-AND-MISS OPERATOR

In this section we discuss the thinning and thickening procedures that result from the use of the *hit-and-miss* operator of Serra.[1] For an input image S, the hit-and-miss operator is defined by

$$\text{HIT}(S; T_1, T_2) = \text{MIN}[\text{ERODE}(S, T_1), \text{ERODE}(S^c, T_2)]$$

where T_1 and T_2 are bounded structuring elements that are disjoint. By the definition of erosion, $\text{HIT}(S; T_1, T_2)$ consists of all pixels (i, j) such that $\text{TRAN}(T_1; i, j) \leqslant S$ and $\text{TRAN}(T_2; i, j) \leqslant S^c$. For convenience of notation, we let T denote the pair (T_1, T_2) and write $\text{HIT}(S; T)$. Figuratively speaking, $\text{TRAN}(T_1; i, j)$ "hits" S and $\text{TRAN}(T_2; i, j)$ "misses" S.

Because it operates by fitting structuring elements into both the image and the complement of the image, the hit-and-miss operator probes the relation between the image and its complement relative to the structuring pair.

One of the applications of HIT involves a type of thinning. Given a constant image S and a pair $T = (T_1, T_2)$ of disjoint structuring elements, we define

$$\text{THIN}(S; T) = \text{MIN}[S, \text{HIT}(S; T)^c]$$

In other words, $\text{THIN}(S; T)$ is formed by taking the set-theoretic difference

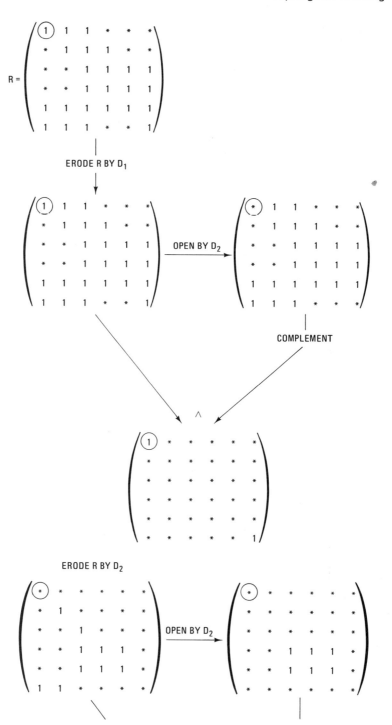

Figure 4.11 Walk-through for SKEL operation

Sec. 4.4 Hit-and-Miss Operator

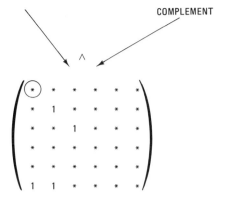

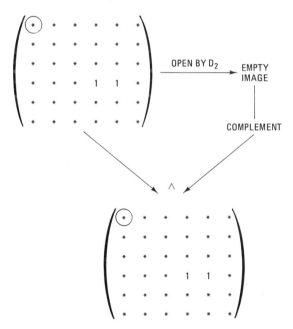

Figure 4.11 (cont.)

$S - \text{HIT}(S; T)$. In practice, a sequence of structuring element pairs $T^{(1)}, T^{(2)}, \ldots$ is employed and the following sequence of outputs results:

$$S^{(1)} = \text{THIN}(S; T^{(1)})$$
$$S^{(2)} = \text{THIN}(S^{(1)}; T^{(2)})$$
$$S^{(3)} = \text{THIN}(S^{(2)}; T^{(3)})$$
$$\vdots$$

Thus, we proceed to inductively generate a sequence of ever-thinner sets. If the original input is finite, the procedure eventually halts and a final thin, or skeletal-type, image results. Note that in the sequential thinning process one might simply employ the same structuring pair repeatedly.

Example 4.8:

Let

$$T_1 = \begin{pmatrix} * & 1 & * \\ * & \textcircled{1} & 1 \\ * & * & * \end{pmatrix}$$

$$T_2 = \begin{pmatrix} * & * & * \\ * & \textcircled{*} & * \\ 1 & * & * \end{pmatrix}$$

and

$$S = \begin{pmatrix} 1 & 1 & 1 & 1 & * & * & * & * \\ 1 & 1 & 1 & 1 & * & * & * & * \\ * & 1 & 1 & 1 & 1 & * & * & * \\ * & * & * & * & 1 & 1 & 1 & 1 \\ * & * & * & * & 1 & 1 & 1 & 1 \\ * & * & * & * & 1 & 1 & 1 & 1 \end{pmatrix}_{0,5}$$

We apply sequential thinning to S with repeated use of the structuring pair $T = (T_1, T_2)$:

$$\text{HIT}(S; T) = \begin{pmatrix} * & * & * & * & * & * & * & * \\ 1 & 1 & * & * & * & * & * & * \\ * & 1 & 1 & 1 & * & * & * & * \\ * & * & * & * & 1 & * & * & * \\ * & * & * & * & 1 & * & * & * \\ * & * & * & * & 1 & 1 & 1 & * \end{pmatrix}_{0,5}$$

$$S^{(1)} = \begin{pmatrix} 1 & 1 & 1 & 1 & * & * & * & * \\ * & * & 1 & 1 & * & * & * & * \\ * & * & * & * & 1 & * & * & * \\ * & * & * & * & * & 1 & 1 & 1 \\ * & * & * & * & * & 1 & 1 & 1 \\ * & * & * & * & * & * & * & 1 \end{pmatrix}_{0,5}$$

Sec. 4.4 Hit-and-Miss Operator

$$\text{HIT}(S^{(1)}; T) = \begin{pmatrix} * & * & * & * & * & * & * & * \\ * & * & 1 & * & * & * & * & * \\ * & * & * & * & * & * & * & * \\ * & * & * & * & * & * & * & * \\ * & * & * & * & * & 1 & 1 & * \\ * & * & * & * & * & * & * & * \end{pmatrix}_{0,5}$$

$$S^{(2)} = \begin{pmatrix} 1 & 1 & 1 & 1 & * & * & * & * \\ * & * & * & 1 & * & * & * & * \\ * & * & * & * & 1 & * & * & * \\ * & * & * & * & * & 1 & 1 & 1 \\ * & * & * & * & * & * & * & 1 \\ * & * & * & * & * & * & * & 1 \end{pmatrix}_{0,5}$$

Since $\text{HIT}(S^{(2)}; T)$ is the null image, the sequential thinning algorithm halts at $S^{(2)}$, which is the desired thin figure. Note that $S^{(2)}$ is a path that does tend to track the body of S. In that sense it is skeletal in nature.

The output of the sequential thinning methodology is highly dependent upon the choice of the structuring pair. Whereas the pair in Example 4.8 revealed a skeletal figure, a different pair might reveal an entirely different thin figure. In essence, the method remains highly heuristic as opposed to morphological procedures in general, which tend to be well grounded in a structured theory.

The preceding thinning methodology resulted from taking the set-theoretic difference of the output of HIT from the input. A corresponding thickening methodology results from forming the union (EXTMAX) of the output of HIT with the input, thus:

$$\text{THICK}(S; T) = \text{EXTMAX}[S, \text{HIT}(S; T)]$$

Once again we assume that T_1 and T_2 are disjoint; however, in the thickening methodology we also assume that the origin is not activated in T_1, or else we would have $\text{THICK}(S; T) = S$. We employ the thickening sequence

$$S_{(1)} = \text{THICK}(S; T^{(1)})$$
$$S_{(2)} = \text{THICK}(S_{(1)}; T^{(2)})$$
$$S_{(3)} = \text{THICK}(S_{(2)}; T^{(3)})$$
$$\vdots$$

The following derivation shows that there is a duality between the operators THICK and THIN:

$$\text{THICK}(S; T_1, T_2)^c = [S \vee \text{HIT}(S; T_1, T_2)]^c$$
$$= S^c \wedge \text{HIT}(S; T_1, T_2)^c$$
$$= S^c \wedge [(S \boxminus (-T_1)) \wedge (S^c \boxminus (-T_2))]^c$$
$$= S^c \wedge [(S^c \boxminus (-T_2)) \wedge (S \boxminus (-T_1))]^c$$
$$= S^c \wedge \text{HIT}(S^c; T_2, T_1)^c$$
$$= \text{THIN}(S^c; T_2, T_1)$$

Example 4.9:

Let

$$T_1 = \begin{pmatrix} 1 & 1 \\ \circledast & 1 \end{pmatrix}$$

$$T_2 = \begin{pmatrix} * & * \\ \textcircled{1} & * \end{pmatrix}$$

and

$$S = \begin{pmatrix} 1 & 1 & 1 & 1 & 1 \\ * & * & * & * & 1 \\ * & * & * & * & 1 \\ * & 1 & * & * & 1 \end{pmatrix}_{0,3}$$

Two sequential thickenings with the pair (T_1, T_2) gives

$$S_{(1)} = \begin{pmatrix} 1 & 1 & 1 & 1 & 1 \\ * & * & * & 1 & 1 \\ * & * & * & * & 1 \\ * & 1 & * & * & 1 \end{pmatrix}_{0,3}$$

and

$$S_{(2)} = \begin{pmatrix} 1 & 1 & 1 & 1 & 1 \\ * & * & 1 & 1 & 1 \\ * & * & * & 1 & 1 \\ * & 1 & * & * & 1 \end{pmatrix}_{0,3}$$

After six thickenings, we obtain

$$S_{(6)} = \begin{pmatrix} 1 & 1 & 1 & 1 & 1 \\ 1 & 1 & 1 & 1 & 1 \\ 1 & 1 & 1 & 1 & 1 \\ 1 & 1 & 1 & 1 & 1 \end{pmatrix}_{0,3}$$

which is the minimal fully activated bound matrix containing S. Not too much should be made of this result, however. Indeed, if we let S' be the same as S except that we deactivate pixel $(4, 3)$, then $\text{THICK}(S'; T) = S'$. In this case, thickening with T produces no new activated pixels.

Not only is the preceding sequential thickening procedure highly heuristic, but it need not terminate after a finite number of steps. In general, then, it appears that sequential hit-and-miss thinning is superior to sequential hit-and-miss thickening, but both methodologies should be used with care. Each takes a great deal of practice and intuition upon the part of the user. Serra provides a guide to the use of structuring pairs; however, a hexagonal grid is employed, which makes the presentation difficult to interpret.[2]

EXERCISES

4.1. Let

$$S = \begin{pmatrix} * & * & * & 1 & * & * & 1 & * \\ * & * & * & 1 & * & * & 1 & * \\ 1 & 1 & 1 & 1 & 1 & 1 & 1 & 1 \\ * & * & * & 1 & 1 & 1 & 1 & * \\ * & * & * & 1 & 1 & 1 & 1 & * \\ \circledast & * & * & * & * & 1 & 1 & 1 \end{pmatrix}$$

Find:

(a) BOUND(S)
(b) CLOSURE(S)
(c) MOREDGE(S, T_2), where T_2 is the mask given in Example 4.4
(d) MOREDGE(S, T_3), where

$$T_3 = \begin{pmatrix} * & * & 1 \\ * & ① & * \\ 1 & * & * \end{pmatrix}$$

(e) HULL(S)

4.2. Define the edge operator

$$\text{INEDGE}(S, T)) = \text{MIN}[S, \text{ERODE}(S, T)^c]$$

which is simply the set-theoretic difference between S and the erosion of S by T. Find INEDGE(S, T), where

$$S = \begin{pmatrix} * & 1 & 1 & 1 & 1 & 1 & * \\ * & 1 & 1 & 1 & 1 & * & * \\ * & 1 & 1 & 1 & * & * & * \\ * & 1 & 1 & 1 & * & * & * \\ 1 & 1 & 1 & 1 & * & 1 & 1 & 1 \\ 1 & 1 & 1 & * & * & 1 & 1 & 1 \\ ① & 1 & 1 & * & * & 1 & 1 & * \end{pmatrix}$$

and T is

(a) the square neighbor mask
(b) the strong neighbor mask
(c) the weak neighbor mask
(d) T_3 of Exercise 4.1

4.3. Find two minimal strongly convex images that contain the image

$$S = \begin{pmatrix} 1 & * \\ * & ① \end{pmatrix}$$

4.4. Employ the dilation methodology of Example 4.6 to fill the path defined by the image

$$S = \begin{pmatrix} 1 & 1 & 1 & * & * & * & * & * \\ 1 & * & * & 1 & 1 & * & * & * \\ 1 & * & * & * & * & 1 & * & * \\ 1 & * & * & * & * & 1 & * & * \\ 1 & * & * & * & * & * & 1 & 1 \\ * & 1 & 1 & * & * & * & * & 1 \\ * & * & * & 1 & 1 & 1 & * & 1 \\ * & * & * & * & * & * & 1 & * \end{pmatrix}_{0,0}$$

4.5. Employ the dilation methodology of Example 4.6 on the image S of Exercise 4.4, except employ the square neighbor mask instead.

4.6. Apply SKEL to the image

$$U = \begin{pmatrix} 1 & 1 & 1 & 1 & * & * & * & * \\ 1 & 1 & 1 & * & * & * & * & * \\ 1 & 1 & 1 & * & * & * & * & * \\ 1 & 1 & 1 & * & * & * & * & * \\ 1 & 1 & 1 & * & * & * & * & * \\ 1 & 1 & 1 & * & * & * & * & * \\ 1 & 1 & 1 & 1 & 1 & 1 & * & * \\ 1 & 1 & 1 & 1 & 1 & 1 & 1 & 1 \\ 1 & 1 & 1 & 1 & 1 & 1 & 1 & 1 \end{pmatrix}_{-7,8}$$

4.7. Apply the thinning methodology of Example 4.8 to the image U of Exercise 4.6.

4.8. Apply the thinning methodology of Example 4.8 to the image S of that example; however, employ the mask pair

$$T_1 = \begin{pmatrix} * & 1 & * \\ 1 & \textcircled{1} & * \\ * & * & * \end{pmatrix}$$

$$T_2 = \begin{pmatrix} * & * & * \\ * & \circledast & * \\ * & * & 1 \end{pmatrix}$$

4.9. Find THICK(S; T_1, T_2) for the image S of Exercise 4.4 and the masks T_1 and T_2 of Example 4.9.

FOOTNOTES FOR CHAPTER 4

1. Jean Serra, *Image Analysis and Mathematical Morphology* (New York: Academic Press, 1983), p. 390.
2. Ibid., p. 392.

5
Morphological Filters for Two-valued Images

5.1 INCREASING τ-MAPPINGS

Perhaps the most promising area of morphological image processing is that dealing with morphological filters. In this chapter we consider the constant-image case, and in a later chapter we generalize to gray-scale images.

We have seen that erosion and opening have a filtering effect from within a constant image, with opening giving a much finer result than erosion. By duality, dilation and closing filter externally by internally filtering the complement. Moreover, all four of the mappings are translation invariant and increasing. It is upon these latter characteristics that we shall focus here.

The presentation will not be theoretical. Rather, we shall give a brief explication of the fundamental Euclidean theory, citing proofs in the appendix, and then proceed to the digital setting, which is of more practical concern. Examples will be presented through the use of bound matrices.

A mapping $\Psi: 2^{R \times R} \to 2^{R \times R}$, where $2^{R \times R}$ denotes the set of all subsets of R^2, is said to be *increasing* if, whenever $A \subset B$, then $\Psi(A) \subset \Psi(B)$. In other words, Ψ preserves the subset relation. Properties M-9 and M-10 respectively state that, for a fixed structuring element, dilation and erosion are increasing mappings. Theorems 1.3 (*ii*) and 1.4 (*ii*) state respectively that, for a fixed structuring element, the opening and the closing are increasing. Figure 5.1 illustrates the concept of an increasing mapping.

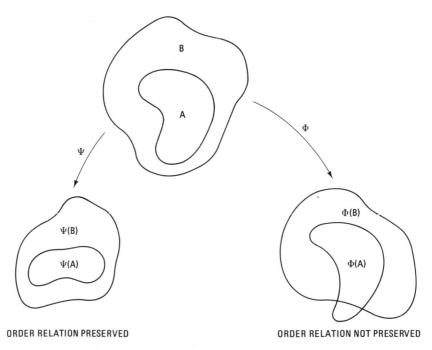

Figure 5.1 Increasing and nonincreasing mappings

A mapping $\Psi: 2^{R \times R} \to 2^{R \times R}$ is called a τ-*mapping* if it is *translation invariant*—i.e., if

$$\Psi(A + x) = \Psi(A) + x$$

for any subset A of R^2 and for any point $x \in R^2$. Properties M-3 and M-8 respectively state that, for a fixed structuring element, dilation and erosion are translation invariant. So are the opening and the closing.

Mappings that are both increasing and translation invariant are called *morphological filters*. Or, put another way, the most general morphological filter is an increasing τ-mapping. Morphological filters preserve the natural set-theoretic ordering and are space invariant. Examples of such filters are the convex hull operator, the topological closure, the umbra transform (see Section 6.2), and various other topological algorithms.

While the notion of an increasing τ-mapping might seem rather general, Matheron has shown that all such mappings can be characterized by erosions and dilations. The representations he has developed (in the Euclidean setting) prove to be pragmatic when adapted to digital images.

An important property of morphological filters is that if Φ and Ψ are increasing τ-mappings, then the composition $\Phi \circ \Psi$ is also an increasing τ-mapping. Indeed, if $A \subset B$, then $\Psi(A) \subset \Psi(B)$, and hence

$$[\Phi \circ \Psi](A) = \Phi[\Psi(A)] \subset \Phi[\Psi(B)] = [\Phi \circ \Psi](B)$$

Sec. 5.1 Increasing τ-Mappings

Also

$$[\Phi \circ \Psi](A + x) = \Phi[\Psi(A + x)] = \Phi[\Psi(A) + x]$$
$$= \Phi[\Psi(A)] + x = [\Phi \circ \Psi](A) + x$$

The key to the theory of morphological filters is the kernel of an increasing τ-mapping. The *kernel* of such a mapping Ψ, Ker $[\Psi]$, is defined by

$$\text{Ker } [\Psi] = \{A: \overline{0} \in \Psi(A)\}$$

where $\overline{0}$ denotes the origin $(0, 0)$.

Example 5.1:

Define the mapping Γ as follows: the point $(x, y) \in \Gamma(A)$ if and only if there exist points (x', y) and (x, y') in A such that $x' \leq x$ and $y' \leq y$. (See Figure 5.2.) Graphically, $\Gamma(A)$ is constructed by taking the union of all horizontal rays in the positive direction emanating from points in A, taking the union of all vertical rays in the positive direction emanating from points of A, and then taking the intersection of the two unions. A little examination shows that Γ is an increasing τ-mapping. Moreover, $A \in \text{Ker } [\Gamma]$ if and only if either A contains the origin or A contains points on each of the negative coordinate axes.

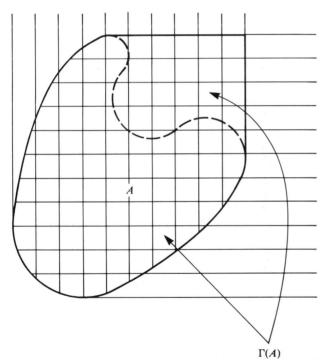

Figure 5.2 Graphical construction of $\Gamma(A)$

Knowledge of its kernel gives one full knowledge of an increasing τ-mapping; indeed, we have the following theorem, which states that two such mappings that have the same kernel must be identical.

Theorem 5.1[1]**.** Suppose Ψ_1 and Ψ_2 are increasing τ-mappings. Then Ker $[\Psi_1] \subset$ Ker $[\Psi_2]$ if and only if $\Psi_1(A) \subset \Psi_2(A)$ for any image $A \subset R^2$. In particular, if Ker $[\Psi_1]$ = Ker $[\Psi_2]$, then $\Psi_1 = \Psi_2$.

The next theorem, which is perhaps the most important one concerning morphological filters, states that every increasing τ-mapping can be represented as a union of erosions. In other words, in theory, all morphological filters are generated by erosions.

Theorem 5.2 (Matheron Representation Theorem)[2]**.** If Ψ is an increasing τ-mapping, then for any image $A \subset R^2$,

$$\Psi(A) = \bigcup_{B \in \text{Ker}[\Psi]} \mathscr{E}(A, B)$$

Theorem 5.2 is highly theoretical. Although it gives a characterization of an increasing τ-mapping in terms of erosions, the union operation is of arbitrary index size. In fact, while the kernel is in general uncountably infinite, there is great redundancy in the expansion as given by Matheron. In Section 5.3 we shall turn more pragmatic and eliminate that redundancy.

Given an increasing τ-mapping Ψ, we define its dual by

$$\Psi^*(A) = [\Psi(A^c)]^c$$

The notion of a dual is a formalization of the duality relations satisfied, for example, by the dilation and erosion (M-4 and M-5), and the opening and closing (M-19). The point is that, given an increasing τ-mapping Ψ, *ipso facto* there exists a related mapping Ψ^* which, as can be shown[3], must also be an increasing τ-mapping. Moreover, the dual of the dual is the original mapping, i.e., $\Psi^{**} = \Psi$.

The next theorem states that once the kernel of Ψ is known, so is the kernel of its dual.

Theorem 5.3[4]**.** If Ψ is an increasing τ-mapping, then

$$\text{Ker}\,[\Psi^*] = \{A: A^c \text{ is not an element of Ker } [\Psi]\}$$

Given the close relationship between a mapping Ψ and its dual Ψ^*, and given the fact that the dual of the erosion is a dilation, it is not surprising that the Matheron Representation Theorem takes another form, one where the expansion is in terms of dilations. This dual form utilizes the kernel of Ψ^*.

Theorem 5.4 (Matheron)[5]**.** If Ψ is an increasing τ-mapping, then for any image $A \subset R^2$,

$$\Psi(A) = \bigcap_{B \in \text{Ker}[\Psi^*]} \mathscr{D}(A, -B)$$

5.2 DIGITAL INCREASING τ-MAPPINGS

In this section we bring the theory of the previous section down to practical, digitally implementable terms. We consider mappings of the form $\Psi: 2^{Z \times Z} \to 2^{Z \times Z}$, the so-called constant-image-to-constant-image mappings. For theoretical reasons, we shall not restrict ourselves to images with finite domains; however, in practice, it is just those images in which we are interested.

The definitions of increasing monotonicity, translation invariance, the kernel, and dual mapping go over to the digital domain without change, as do Theorems 5.1, 5.2, 5.3, and 5.4, except that 5.2 and 5.4 naturally take the forms

$$\Psi(S) = \bigvee_{T \in \text{Ker}[\Psi]} \text{ERODE}(S, T)$$

and

$$\Psi(S) = \bigwedge_{T \in \text{Ker}[\Psi^*]} \text{DILATE}(S, -T)$$

respectively, where, operationally, $-T = \text{NINETY}^2(T)$. No doubt these somewhat abstruse notions will be illuminated by considering some concrete situations.

For a fixed structuring element E, let

$$\Psi(S) = \text{DILATE}(S, E)$$

$\Psi(S)$ contains the origin $(0, 0)$ if and only if there exists some pixel (i, j) activated in S such that

$$[\text{TRAN}(E; i, j)](0, 0) = 1$$

In other words, the domain of $\text{TRAN}(E; i, j)$ contains the origin. Of course, we already know, by M-7, that

$$\Psi^*(S) = \text{DILATE}(S^c, E)^c = \text{ERODE}(S, -E)$$

Example 5.2:

Let

$$E = \begin{pmatrix} 1 & 1 \\ \textcircled{1} & * \end{pmatrix}$$

and $\Psi(S) = \text{DILATE}(S, E)$. For S to be in Ker $[\Psi]$, there must exist a translation of E by some element in S such that the translate contains the origin in its domain. Certainly, any image containing the origin is in the kernel. Some other images in the kernel as well are

$$S_1 = \begin{pmatrix} \textcircled{*} & * & 1 \\ 1 & 1 & * \end{pmatrix}$$

and

$$S_2 = \begin{pmatrix} * & \textcircled{*} & 1 \\ 1 & * & 1 \end{pmatrix}$$

Not in the kernel is the image

$$S_3 = \begin{pmatrix} 1 & 1 \\ \circledast & * \\ * & * \\ 1 & 1 \end{pmatrix}$$

Indeed, an image S is in Ker $[\Psi]$ if and only if one of the three pixels $(0, 0)$, $(-1, -1)$, or $(0, -1)$ is activated in S. But those three pixels constitute the image

$$\begin{pmatrix} * & ① \\ 1 & 1 \end{pmatrix} = -E$$

In other words, S is in the kernel of Ψ if and only if

$$S \wedge (-E) \neq \varnothing$$

Indeed, in general,[6]

$$\text{Ker } [\text{DILATE}(\cdot, E)] = \{S: S \wedge (-E) \neq \varnothing\}$$

Now, for a fixed structuring element E, let

$$\Psi(S) = \text{ERODE}(S, E)$$

$\Psi(S)$ contains the origin if and only if E is a subimage of S. Hence,

$$\text{Ker } [\text{ERODE}(\cdot, E)] = \{S: E \ll S\}$$

Furthermore, by M-6,

$$\Psi^*(S) = \text{DILATE}(S, -E)$$

For fixed structuring element E, consider

$$\Psi(S) = \text{OPEN}(S, E)$$

Then, according to M-19,

$$\Psi^*(S) = \text{CLOSE}(S, E)$$

Moreover, $(0, 0) \in \Psi(S)$ if and only if there exists some (i, j) such that $(0, 0)$ is activated in TRAN$(E; i, j)$ and TRAN$(E; i, j)$ is a subimage of S.

Example 5.3:

Let Ψ be defined as the opening by the structuring element E of Example 5.2. Then the images

$$S_1 = \begin{pmatrix} 1 & 1 & 1 & ① \\ 1 & * & 1 & 1 \end{pmatrix}$$

and

$$S_2 = \begin{pmatrix} 1 & 1 & * \\ ① & 1 & 1 \\ 1 & * & * \end{pmatrix}$$

Sec. 5.2 Digital Increasing τ-Mappings

are in Ker $[\Psi]$. On the other hand, the image

$$S_3 = \begin{pmatrix} 1 & 1 & \textcircled{1} \\ 1 & * & 1 \\ 1 & 1 & 1 \end{pmatrix}$$

is not in Ker $[\Psi]$. It can be seen that

$$\text{TRAN}(E; -1, -1) \ll S_1$$

and

$$\text{TRAN}(E; 0, 0) \ll S_2$$

with the origin pixel $(0, 0)$ being activated in each translate. In each case there exists some pixel (i, j) in the domain of E such that

$$\text{TRAN}(E; -i, -j) \ll S_r$$

$r = 1$ or 2. On the other hand, there does not exist a pixel in the domain of E for which the preceding subimage relation is satisfied for S_3. Indeed, none of the following three images is a subimage of S_3: $\text{TRAN}(E; 0, 0)$, $\text{TRAN}(E; 0, -1)$, $\text{TRAN}(E; -1, -1)$. In general, it can be shown that[7]

$$\text{Ker}\,[\text{OPEN}(\cdot, E)] = \bigcup_{(i,j)\in\text{DOMAIN}(E)} \{S: \text{TRAN}(E; -i, -j) \ll S\}$$

In other words, $S \in \text{Ker}\,[\text{OPEN}(\cdot, E)]$ if and only if there exists a pixel (i, j) activated in E such that the translation of E by $-i$ and $-j$ is a subimage of S.

Since $\text{CLOSE}(S, E) = \Psi^*(S)$, according to Theorem 5.3 the kernel of the closing is equal to the set of all images S for which S^c is not in the kernel of the opening—i.e., for which there does not exist any pixel (i, j) in the domain of E with $\text{TRAN}(E; -i, -j)$ being a subimage of S^c.

In Section 4.2 we discussed the pseudoconvex hull operator HULL, which is an increasing τ-mapping. To describe the kernel of HULL, we utilize Figure 5.3. In that figure there are four activated grid points, P_1, P_2, P_3, and P_4, each in a different quadrant. We need to list the conditions under which, for a given input image S, $(0, 0)$ lies in the domain of $\text{HULL}(S)$. Certainly this is true in the trivial case where S is activated at $(0, 0)$. More importantly, $S \in \text{Ker}\,[\text{HULL}]$ if (1) S has an activated pixel

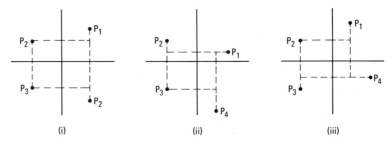

Figure 5.3 Kernel of HULL

in each quadrant (refer to Figure 5.3), (2) at least two of the pixels P_1, P_2, P_3, and P_4 lie in a vertical or horizontal line, and (3) one of the following conditions holds (where for sake of discussion we assume that P_2 and P_3 lie on a vertical line):

(i) P_1 and P_4 lie on a vertical horizontal line.

(ii) P_1 lies (inclusively) between the horizontal lines running through P_2 and P_3, and P_4 lies to the left of the vertical line through P_1 or above the horizontal line through P_3.

(iii) P_4 lies (inclusively) between the horizontal lines running through P_2 and P_3, and P_1 lies to the left of the vertical line through P_4 or below the horizontal line through P_2.

More generally, S is in the kernel of HULL if, upon activating all pixels lying between any two pixels of S that lie on a horizontal or vertical line, the resulting image satisfies (1) through (3). (Note the degenerate cases where there are pixels either above and below, or to the left and right of the origin.)

The next two examples consider in detail two mappings that are closely related to the pseudoconvex hull.

Example 5.4:

Define $J(S)$ to be the *horizontal hull* of S, that is, $[J(S)](i, j) = 1$ if and only if $(i, j) \in \text{DOMAIN}(S)$ or there exist activated pixels to the right and left of (i, j). The latter condition is rigorously specified by saying that there exist (i_1, j) and (i_2, j) in the domain of S such that $i_1 < i$ and $i_2 > i$. For instance, if

$$S = \begin{pmatrix} * & * & * & 1 & * \\ * & 1 & * & * & 1 \\ 1 & * & 1 & * & * \\ 1 & * & * & * & 1 \end{pmatrix}_{0,3}$$

then

$$J(S) = \begin{pmatrix} * & * & * & 1 & * \\ * & 1 & 1 & 1 & 1 \\ 1 & 1 & 1 & * & * \\ 1 & 1 & 1 & 1 & 1 \end{pmatrix}_{0,3}$$

Ker $[J]$ consists of all images that contain the origin or that possess at least one activated pixel to the left of the origin and at least one to the right. For instance,

$$S_1 = \begin{pmatrix} 1 & 1 & * & 1 & 1 \\ 1 & * & * & 1 & 1 \\ 1 & * & \circledast & 1 & 1 \end{pmatrix}$$

is in the kernel, whereas

$$S_2 = \begin{pmatrix} 1 & 1 & 1 \\ 1 & 1 & \circledast \\ 1 & 1 & 1 \end{pmatrix}$$

is not.

Again let S be as above. According to Theorem 5.2, $J(S)$ is equal to the union (extended maximum) of all erosions of S by elements in the kernel of J. But Ker $[J]$ is

Sec. 5.2 Digital Increasing τ-Mappings

infinite, and hence the theorem appears to be of little practical use. However, we note that only seven elements of the kernel are actually required for computation; that is,

$$J(S) = \bigvee_{k=1}^{7} \text{ERODE}(S, T_k)$$

where

$$T_1 = (\textcircled{1})$$
$$T_2 = (1 \; \circledast \; 1)$$
$$T_3 = (1 \; \circledast \; * \; 1)$$
$$T_4 = (1 \; \circledast \; * \; * \; 1)$$
$$T_5 = (1 \; * \; \circledast \; 1)$$
$$T_6 = (1 \; * \; * \; \circledast \; 1)$$
$$T_7 = (1 \; * \; \circledast \; * \; 1)$$

This reduction in required size is no accident. We shall return to the problem of reducing the size of the indexing set in the Matheron theorem in the next section.

Example 5.5:

In this example we consider a digital version of the mapping Γ discussed in Example 5.1. Given a constant image S, (i, j) is activated in $\Gamma(S)$ if and only if (i, j) is activated in S or there exists an activated pixel of S below (i, j) and an activated pixel of S to the left of (i, j). For instance, if

$$T = \begin{pmatrix} * & 1 & 1 & * & * \\ * & * & 1 & 1 & 1 \\ * & 1 & * & \textcircled{1} & * \\ * & 1 & 1 & 1 & * \\ * & * & * & * & * \end{pmatrix}$$

then

$$\Gamma(T) = \begin{pmatrix} * & 1 & 1 & 1 & 1 \\ * & * & 1 & 1 & 1 \\ * & 1 & 1 & \textcircled{1} & * \\ * & 1 & 1 & 1 & * \\ * & * & * & * & * \end{pmatrix}$$

The kernel of Γ is the set of all constant images for which the origin is activated or for which there are activated pixels to the left and beneath the origin. Kernel elements are of the form

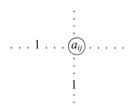

According to Theorem 5.2, Γ can be expressed as a union of erosions by kernel elements. However, as in the previous example, only a small number are actually required to express $\Gamma(S)$ for a given S. For instance, for the preceding image T,

$$\Gamma(T) = \text{ERODE}(T, B_{00}) \vee \text{ERODE}(T, B_{11}) \vee \text{ERODE}(T, B_{12})$$

where the kernel elements B_{00}, B_{11}, and B_{12} are given by

$$B_{00} = (\textcircled{1})$$

$$B_{11} = \begin{pmatrix} 1 & \circledast \\ * & 1 \end{pmatrix}$$

and

$$B_{12} = \begin{pmatrix} 1 & * & \circledast \\ * & * & 1 \end{pmatrix}$$

and where the reason for the notation B_{rs} will become clear in the next section. Note that whereas Theorem 5.2 involves an infinite union, only three kernel elements are required for the expansion of $\Gamma(T)$.

5.3 BASIS FOR THE KERNEL

According to the digital version of Theorem 5.2, an increasing τ-mapping $\Psi: 2^{Z \times Z} \to 2^{Z \times Z}$ possesses the representation

$$\Psi(S) = \bigvee_{T \in \text{Ker}[\Psi]} \text{ERODE}(S, T)$$

Now suppose T_1 and T_2 are kernel elements with $T_1 \ll T_2$. Then, using M-11, we obtain

$$\text{ERODE}(S, T_2) \ll \text{ERODE}(S, T_1)$$

so that $\text{ERODE}(S, T_2)$ is redundant in the EXTMAX expansion of $\Psi(S)$. Hence, it can be eliminated from the expansion. It is precisely this form of elimination that motivates the following definition of the basis for the kernel.

A collection $\mathcal{M}$ of images in the kernel of an increasing τ-mapping Ψ is called a *basis for the kernel* if

1. No element of $\mathcal{M}$ is a proper subimage of any other element of $\mathcal{M}$.
2. For any element $T \in \text{Ker}[\Psi]$, there exists $T' \in \mathcal{M}$ such that $T' \ll T$.

It can be shown that if there exists a basis for the kernel, then the basis is unique.[8] Therefore, we can employ the notation $\text{Bas}[\Psi]$ for the basis without ambiguity. Using the reasoning of the preceding paragraph, we have the following theorem:

Theorem 5.5: If Ψ is an increasing τ-mapping that possesses a basis for its kernel, then, for any image S,

$$\Psi(S) = \bigvee_{T \in \text{Bas}[\Psi]} \text{ERODE}(S, T)$$

Example 5.6:

Consider the mapping Γ of Example 5.5. Γ has basis elements of the following form:

$$B_{00} = (\textcircled{1})$$

$$B_{11} = \begin{pmatrix} 1 & \circledast \\ * & 1 \end{pmatrix}$$

$$B_{12} = \begin{pmatrix} 1 & * & \circledast \\ * & * & 1 \end{pmatrix}$$

$$B_{21} = \begin{pmatrix} 1 & \circledast \\ * & * \\ * & 1 \end{pmatrix}$$

$$B_{22} = \begin{pmatrix} 1 & * & \circledast \\ * & * & * \\ * & * & 1 \end{pmatrix}$$

$$\vdots$$

Thus, a general element in Bas $[\Gamma]$ has the form

$$B_{rs} = \begin{pmatrix} 1 & * & * & \cdots & \circledast \\ * & * & * & \cdots & * \\ * & * & * & \cdots & * \\ \vdots & \vdots & \vdots & & \vdots \\ * & * & * & \cdots & 1 \end{pmatrix}$$

where the bound matrix B_{rs} has $r + 1$ rows and $s + 1$ columns. Theorem 5.5 guarantees the representation of Γ in terms of erosions by the images B_{rs}. Inspection of the general form of the kernel elements that was shown in Example 5.5 reveals that the collection $\{B_{rs}\}$ constitutes a basis.

Several points should be observed in the preceding example. First, the elements of the kernel were not necessarily finite images, whereas all elements B_{rs} are finite. Second, while the size of the expansion has been cut considerably with the elimination of redundancy, the general form of Theorem 5.5 still involves an infinite operation, which, from a practical standpoint, is unacceptable. Third, for the image T of Example 5.5, only three terms of the basis are actually required in the EXTMAX expansion.

Before proceeding to the manner in which the Matheron expansion can be reduced to a finite expansion, we give two sets of conditions under which the existence of a basis for the kernel is guaranteed. Whereas the first is theoretical, the second is of a practical nature and applies to mappings such as those given in Examples 5.4 and 5.5.

Theorem 5.6[9]. Suppose Ψ is an increasing τ-mapping that satisfies the condition of *continuity from above*; i.e., if

$$T_1 \gg T_2 \gg T_3 \gg \cdots$$

then

$$\Psi\left[\bigwedge_{k=1}^{\infty} T_k\right] = \bigwedge_{k=1}^{\infty} \Psi(T_k)$$

Then there exists a basis for the kernel.

Theorem 5.7[10]. Suppose Ψ is an increasing τ-mapping, and suppose every image in the kernel possesses a finite subimage in the kernel. Then there exists a basis for the kernel, and all elements of the basis are finite.

Example 5.6 illustrated the consequences of Theorem 5.7. Not only is the theorem theoretical in that it posits the existence of a basis, but it is also pragmatic in that it guarantees that all the basis elements are bound matrices. Unfortunately, it does nothing to reduce, in theory, the size of the expansion. Indeed, as the following example illustrates, unless some restriction is placed on the mapping Ψ, the situation can still be quite pathological.

Example 5.7:

Consider the mapping Φ defined by $\Phi(S) = Z \times Z$ for any nonempty $S \subset Z \times Z$. Φ has a basis consisting of all singleton images $\{(i, j)\}$, where (i, j) is any pixel in $Z \times Z$. While this basis is arguably superior to the kernel, which is $2^{Z \times Z}$ itself, it nevertheless results in an infinite number of erosions in the basis expansion of any digital image. Although this example is rather absurd from a practical standpoint, it certainly points up the need for some further honing of the Matheron theorem.

In order to get a constraint on the number of terms in the expansion in Theorem 5.5, we place a limit on the degree to which an increasing τ-mapping can expand an input image. If S is a finite constant image, let $\rho(S)$ denote the maximum dimension of the minimal bound matrix representing S, and let $\kappa(S)$ denote the maximum dimension of the minimal origin-containing bound matrix representing S. Moreover, let S_0 denote the constant image which has gray value 1 at each pixel in the frame of the minimal bound matrix and which is $*$ elsewhere. Finally, for $\lambda \geq 0$, let S_λ be the constant image whose minimal bound matrix is obtained from S_0 by extending S_0 λ pixels up, down, to the left, and to the right, and such that S_λ is identically 1 on its minimal bound matrix. Example 5.8 shows these elements for a particular image S.

Example 5.8:

Let

$$S = \begin{pmatrix} * & * & 1 & * & * \\ 1 & * & * & * & * \\ * & 1 & 1 & * & * \\ * & * & * & * & \circledast \end{pmatrix}$$

Then $\rho(S) = 3$, $\kappa(S) = 5$,

$$S_0 = \begin{pmatrix} 1 & 1 & 1 \\ 1 & 1 & 1 \\ 1 & 1 & 1 \end{pmatrix}_{-4, 3}$$

Sec. 5.3   Basis for the Kernel

and

$$S_1 = \begin{pmatrix} 1 & 1 & 1 & 1 & 1 \\ 1 & 1 & 1 & 1 & 1 \\ 1 & 1 & 1 & 1 & 1 \\ 1 & 1 & 1 & 1 & 1 \\ 1 & 1 & 1 & 1 & 1 \end{pmatrix}_{-5,4}$$

In general, if S has a minimal bound matrix of dimensions m by n, then S_λ has dimensions $m + 2\lambda$ by $n + 2\lambda$.

An increasing τ-mapping Ψ is said to be λ-*bounded* if λ is a nonnegative integer such that $\Psi(S) \ll S_\lambda$ for any finite digital image S. For dilation, erosion, opening, and closing relative to some fixed structuring element E, DILATE is $[\kappa(E) - 1]$-bounded, ERODE is $[\kappa(E) - \rho(E)]$-bounded, OPEN is 0-bounded, and CLOSE is 0-bounded. Also, HULL is 0-bounded, the topological closure is 1-bounded, the mapping J of Example 5.4 is 0-bounded, and the mapping Γ of Example 5.5 is 0-bounded. The reason λ-bounded mappings are of interest is that they constitute a class of morphological filters for which the sizes of the output images are bounded in terms of the sizes of the input images. The next theorem gives the expansion in Theorem 5.5 in such a manner that the number of terms in the expansion is bounded by a quantity that is a function of λ and the size, $\rho(S)$, of the input minimal bound matrix. The theorem thus places implicit bounds on the computational complexity of the Matheron representation.

Theorem 5.8[11]**.**   Suppose Ψ is a λ-bounded increasing τ-mapping. Then

$$\Psi(S) = \bigvee_{\substack{T \in \mathrm{Bas}[\Psi] \\ \kappa(T) \leq \rho(S) + \lambda}} \mathrm{ERODE}(S, T)$$

Example 5.9:

Let Γ be the mapping of Example 5.5, and let S be the image of Example 5.8. Then, using the information of Examples 5.6 and 5.8,

$$\Gamma(S) = \bigvee_{\substack{T \in \mathrm{Bas}[\Gamma] \\ \kappa(T) \leq 3 + 0}} \mathrm{ERODE}(S, T)$$

$$= \mathrm{ERODE}(S, B_{00}) \vee \mathrm{ERODE}(S, B_{11}) \vee \mathrm{ERODE}(S, B_{12})$$

$$\vee \mathrm{ERODE}(S, B_{21}) \vee \mathrm{ERODE}(S, B_{22})$$

Thus, the representation of $\Gamma(S)$, for this particular S, requires at most five terms. In fact, only three terms, B_{00}, B_{11}, and B_{12}, are required. Nevertheless, computing even five erosions is certainly far superior to computing an infinite number of erosions.

Example 5.10:

Consider the horizontal hull mapping J of Example 5.4 and the image S of the same example. For that image, $\rho(S) = 5$ and, according to Theorem 5.8,

$$J(S) = \bigvee_{\substack{T \in \mathrm{Bas}[J] \\ \kappa(T) \leq 5 + 0}} \mathrm{ERODE}(S, T)$$

But the collection of images in Bas[J] with $\kappa(T) \leq 5$ consists of precisely the images $T_1, T_2, \ldots, T_7$ that were required for the representation of $J(S)$. In other words, in this instance the bound $\lambda + \rho(S)$ cannot be improved. This shows that the bound in Theorem 5.8 is tight.

In Example 5.2 it was noted that the kernel for the dilation by the structuring element E consists of the set of all images S such that $S \wedge (-E) \neq \emptyset$. This means that every element of the kernel possesses at least one activated pixel of $-E$. As a result, the basis for the kernel of DILATE($\cdot, E$) consists of all singleton images $\{(i, j)\}$ such that $(i, j) \in -E$, or equivalently, all singleton images $\{(-i, -j)\}$ such that $(i, j) \in E$. If E is finite, then the basis itself must be finite, with cardinality the cardinality of E. In any case, Theorem 5.5 applies; that is,

$$\text{DILATE}(S, E) = \bigvee_{T \in \text{Bas}[\text{DILATE}(\cdot, E)]} \text{ERODE}(S, T)$$

$$\bigvee_{(i, j) \in E} \text{ERODE}(S, \{(-i, -j)\})$$

$$\bigvee_{(i, j) \in E} \text{TRAN}(S; i, j)$$

which is precisely the definition of the dilation!

5.4 ALGEBRAIC OPENINGS AND CLOSINGS OF CONSTANT IMAGES

In Section 1.4 we discussed the filtering effects of the opening and the closing, the essential properties of which were given in Theorems 1.3 and 1.4, respectively. For a fixed structuring element E, $O(\cdot, E)$ is antiextensive, increasing, and idempotent, while $C(\cdot, E)$ is extensive, increasing, and idempotent. Moreover, both operations are translation invariant, and hence are increasing τ-mappings. Using Matheron's terminology, they are idempotent morphological filters, one antiextensive and the other extensive. The antiextensive–extensive relationship results from the duality between the filters.

Before formalizing these remarks into a general algebraic theory of filters, it is important to recognize the geometric significance of the properties with which we are concerned. As was demonstrated in the examples of Section 1.4, filters that are antiextensive (or extensive), increasing, and idempotent preserve fundamental perceptual characteristics in such a manner that the input image is distorted in a well-defined manner by the filter and, in addition, the distortion is not compounded by repeated filtering. In the case of gray-scale filters, this perceptual invariance channels the distortion due to filtering.

In general, a mapping $\Psi: 2^{R \times R} \to 2^{R \times R}$ is called an *algebraic opening* if it is antiextensive, increasing, and idempotent; and an *algeraic closing* if it is extensive, increasing, and idempotent. It is called a *τ-opening* (*τ-closing*) if it is a translation-invariant algebraic opening (closing). For a given structuring element E, $O(\cdot, E)$ is

Sec. 5.4   Algebraic Openings and Closings of Constant Images   **147**

a τ-opening and $C(\cdot, E)$ is a τ-closing. This duality between $O(\cdot, E)$ and $C(\cdot, E)$ is not accidental. It is a specific instance of the next theorem.

Theorem 5.9[12]. Ψ^*, the dual of Ψ, is an algebraic closing if and only if Ψ is an algebraic opening, and vice versa.

From a practical point of view, Matheron's theorem to be presented shortly (Theorem 5.10) characterizes all τ-openings and τ-closings in terms of elementary openings and closings, respectively. Thus, the theorem is definitive with regard to the construction of filters that are antiextensive (extensive), increasing, idempotent, and translation invariant.

Just as a certain class of images called the kernel played a central role in the study of increasing τ-mappings, a specific class of sets will be important in the study of algebraic openings and closings. Given an algebraic opening or closing Ψ, the *invariant class*, denoted Inv $[\Psi]$, is the set of all images A such that $\Psi(A) = A$. For the opening $O(\cdot, E)$, the invariant images are those which are E-open; for the closing $C(\cdot, E)$, the invariant images are those which are E-closed.

Now suppose that Ψ is an algebraic opening. Then a subcollection $\mathcal{B}$ of images in Inv $[\Psi]$ is called a *base* for Inv $[\Psi]$ if every image in Inv $[\Psi]$ can be represented as a union of translations of elements in $\mathcal{B}$. In mathematical terminology, a base for Inv $[\Psi]$ *generates* Inv $[\Psi]$. Note that the concept of a base is *not* the same as that of a basis for the kernel. The terms *base* and *basis* are overworked in mathematics and are often employed interchangeably. We do not follow this practice herein.

We now state the Matheron Representation Theorem for τ-openings. It asserts that whereas increasing τ-mappings are characterized by erosions and dilations, τ-openings and τ-closings are characterized by openings and closings, respectively.

Theorem 5.10 (Matheron). A constant-image-to-constant-image mapping Ψ is a τ-opening if and only if there exists a class of sets $\mathcal{B}$ such that[13]

$$\Psi(A) = \bigcup_{B \in \mathcal{B}} O(A, B)$$

Moreover, $\mathcal{B}$ is a base for Inv $[\Psi]$. Furthermore, the dual of Ψ has the representation[14]

$$\Psi^*(A) = \bigcap_{B \in \mathcal{B}} C(A, B)$$

In practice, Theorem 5.10 gives a prescription for the construction and implementation of certain classes of morphological filters, the τ-openings and τ-closings. For instance, to construct a τ-opening, we must find a base of structuring elements that yield openings $O(\cdot, B)$ which, when employed as inputs to a union operation, produce the desired filtering effect. Conversely, should one have a filter that is a τ-opening, it can be implemented in terms of elementary openings.

Rather than give an abstruse example from the Euclidean setting, we shall go straight to the digital case, where all of the preceding definitions hold, except that

$$\Psi: 2^{Z \times Z} \longrightarrow 2^{Z \times Z}$$

For the digital case, Theorem 5.9 holds without change and Theorem 5.10 holds with $\vee$ and $\wedge$ in place of $\cup$ and $\cap$, respectively, and OPEN and CLOSE in place of O and C, respectively.

Example 5.11:

Suppose we wish to filter from an image (deactivate) any pixel that does not have two adjacent activated neighbors. One might wish to engage in such a process under the assumption that the activation of any such pixel is likely due to noise. Let Φ be the mapping that produces the desired output. For instance, if

$$S = \begin{pmatrix} * & 1 & * & 1 & 1 & 1 \\ * & 1 & 1 & 1 & * & 1 \\ \circledast & 1 & * & 1 & * & 1 \\ 1 & * & 1 & * & * & * \end{pmatrix}$$

then

$$\Phi(S) = \begin{pmatrix} * & 1 & * & 1 & 1 & 1 \\ * & 1 & 1 & 1 & * & 1 \\ \circledast & 1 & * & 1 & * & * \end{pmatrix}$$

Since Φ is a τ-opening, the discrete form of Theorem 5.10 applies.

To implement the theorem, we need a description of Inv $[\Phi]$. It should be clear that a digital image (finite or infinite) is invariant under Φ if and only if it does not possess any activated pixels not having at least two adjacent activated neighbors. What is required is a base for Inv $[\Phi]$ that consists of a small number of elements, since this will make the representation expansion small. For the mapping Φ, the following four images constitute a base $\mathcal{B}$:

$$B_1 = \begin{pmatrix} 1 & 1 \\ \circledast & 1 \end{pmatrix} \quad B_2 = \begin{pmatrix} 1 & 1 \\ \circledcirc & * \end{pmatrix} \quad B_3 = \begin{pmatrix} 1 & * \\ \circledcirc & 1 \end{pmatrix} \quad B_4 = \begin{pmatrix} * & 1 \\ \circledcirc & 1 \end{pmatrix}$$

To show that $\mathcal{B}$ is a base for Inv $[\Phi]$, we need to show that given an image $T \in $ Inv $[\Phi]$, T can be represented as a union of translations of elements in $\mathcal{B}$. The problem is essentially one of exhausting the possibilities. If (i, j) is activated in T and (i, j) has at least two adjacent activated neighbors, then at least one translate of one of the B_k's will fit into the image in such a manner as to contain (i, j). For instance, if $(i - 1, j + 1)$ and $(i, j + 1)$ are activated, then

$$\text{TRAN}(B_1; i - 1, j) \ll T$$

If (i, j) does not contain two adjacent activated neighbors, then no B_k will fit. Consequently, T is invariant under Φ if and only if it can be represented as an extended maximum of translates of elements from $\mathcal{B}$.

According to Theorem 5.10,

$$\Phi(S) = \bigvee_{k=1}^{4} \text{OPEN}(S, B_k)$$

which for the image S above becomes

Sec. 5.4 Algebraic Openings and Closings of Constant Images

$$\begin{pmatrix} * & * & * & * & 1 & 1 \\ * & * & 1 & 1 & * & 1 \\ \circledast & * & * & 1 & * & * \end{pmatrix} \vee \begin{pmatrix} * & * & * & 1 & 1 & * \\ * & * & 1 & 1 & 1 & * & * \\ \circledast & * & 1 & * & * & * \end{pmatrix}$$

$$\vee \begin{pmatrix} * & 1 & * & * & * & * \\ * & 1 & 1 & * & * & * \\ \circledast & * & * & * & * & * \end{pmatrix} \vee \begin{pmatrix} * & * & * & 1 & * & * \\ * & * & 1 & 1 & * & * \\ \circledast & * & * & * & * & * \end{pmatrix}$$

which agrees with the output shown previously.

In Example 5.11 it was assumed that we knew the invariant class and wished to find a base that would yield it. Once the base is found, the filter is expressed as an extended maximum of openings over it. From a practical standpoint, the determination of the desired invariant class together with the requirements of antiextensivity, increasing monotonicity, idempotence, and translation invariance determines a unique filter, which is constructible by using EXTMAX and OPEN.

Because Theorem 5.10 is an equivalence, we can also interpret it from the converse perspective. Suppose, for heuristic reasons, that we wish to link a number of elementary openings OPEN($\cdot$, B_1), OPEN($\cdot$, B_2), . . . , OPEN($\cdot$, B_m) to form a filter that filters only those pixels which are removed by all of the individual openings. Then the desired filter takes the form

$$\Psi(S) = \bigvee_{k=1}^{m} \text{OPEN}(S, B_k)$$

According to Theorem 5.10, Ψ is a τ-opening with Inv [Ψ] being made up of unions of translates of the B_k. In other words, $\mathcal{B} = \{B_k\}$ forms a base for Inv [Ψ].

Example 5.12:

Suppose

$$B_1 = (\circled{1} \quad 1)$$

$$B_2 = \begin{pmatrix} 1 \\ \circled{1} \end{pmatrix}$$

and

$$\Gamma(S) = \text{OPEN}(S, B_1) \vee \text{OPEN}(S, B_2)$$

Then Inv [Γ] consists of all images that can be formed by taking an EXTMAX (with possibly infinitely many inputs) of translations of B_1 and B_2. For instance,

$$T = \begin{pmatrix} 1 & 1 & 1 \\ 1 & 1 & * \\ \circled{1} & * & * \end{pmatrix}$$

$$= \text{TRAN}(B_1; 0, 2) \vee \text{TRAN}(B_1; 1, 2) \vee \text{TRAN}(B_1; 0, 1) \vee \text{TRAN}(B_2; 0, 0)$$

is invariant under Γ. Indeed, Γ removes a pixel from the domain of the input image if and only if that pixel has no activated strong neighbors. Hence, Inv [Γ] consists of all constant images for which each activated pixel possesses at least one activated strong neighbor. Given that characterization, the base $\{B_1, B_2\}$ can easily be found.

5.5 EUCLIDEAN GRANULOMETRIES

Whereas the Matheron theories concerning increasing τ-mappings and τ-openings have direct digital analogues, the granulometric theory is essentially Euclidean. However, it is very important from a modeling point of view and does have digital implications in terms of digitization.

The granulometries $t \to O(A, tB)$ were introduced in Section 3.5. The convexity of B plays a fundamental role in these granulometries because it results in decreasing monotonicity. (See Theorem 1.5.) In the present section we shall discuss a more general theory of granulometries developed by Matheron. The main result of the section will be Matheron's representation theorem for a class of granulometries known as Euclidean granulometries. Matheron has shown that the Euclidean granulometries, which model a certain class of well-behaved one-parameter families of filters, are always representable in terms of openings of the form $O(\cdot, tB)$. The representation is especially straightforward for convex structuring elements.

Because the particulars of the granulometric theory are mathematically abstract, this section will be somewhat brief. A more complete exposition is given in the appendix.

The notion of a granulometric mapping, as conceived by Matheron, has to do with the process of *sieving*. Imagine a wire mesh sieve into which particles of different diameters are tossed. Those with a diameter less than the mesh size, say t, will fall through, while those with a diameter larger than the mesh size will remain within the sieve. The action is like a filter: the smaller (diameter $< t$) particles are lost, while the larger (diameter $\geq t$) particles remain.

The situation is perhaps best illustrated by considering two sets, each a disjoint union of connected components. Accordingly, let

$$A = A_1 \cup A_2 \cup \cdots \cup A_m$$

and

$$B = B_1 \cup B_2 \cup \cdots \cup B_n$$

where $m \leq n$, and $A_k \subset B_k$ for $k = 1, 2, \ldots, m$. Moreover, suppose that the diameter of A_k is equal to r_k, where $r_1 \leq r_2 \leq \cdots \leq r_m$, and the diameter of B_k is equal to s_k, where $s_1 \leq s_2 \leq \cdots \leq s_n$. (See Figure 5.4.) Several sieving properties are evident. Let Ψ_t denote the function that outputs that part of the input set which remains in the sieve of mesh size t. Then $\Psi_t(A)$ is called the *t-oversize* of A. Also, if $r_k < t \leq r_{k+1}$, then

$$\Psi_t(A) = A_{k+1} \cup A_{k+2} \cup \cdots \cup A_m$$

Consequently, $\Psi_t(A) \subset A$; i.e., Ψ_t is antiextensive. Moreover, since $A_k \subset B_k$ for all k, $\Psi_t(A) \subset \Psi_t(B)$; i.e., Ψ_t is increasing. Finally, if t' is a different mesh size than t, then the effect of sieving by both mesh sizes is the same as sieving by only the greater of the two sizes—in other words,

$$\Psi_t[\Psi_{t'}(A)] = \Psi_{t'}[\Psi_t(A)] = \Psi_{\max(t, t')}(A)$$

Matheron has used these three properties of sieving to give the following axiomatic formulation of a granulometric process:

Sec. 5.5   Euclidean Granulometries   151

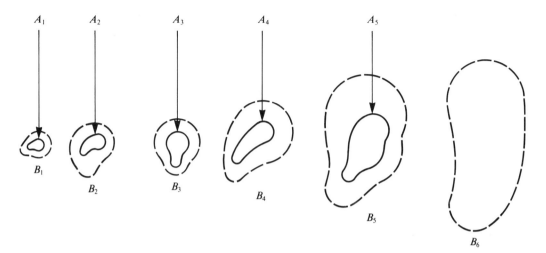

Figure 5.4 Particles to be sieved

A *granulometry* on $2^{R \times R}$, the collection of constant Euclidean images, is a family of mappings

$$\Psi_t \colon 2^{R \times R} \longrightarrow 2^{R \times R}, \, t > 0$$

satisfying the axioms

G-1: $\Psi_t(A) \subset A$ for any $t > 0$ (Ψ_t is antiextensive)
G-2: If $A \subset B$, then $\Psi_t(A) \subset \Psi_t(B)$ (Ψ_t is increasing)
G-3: $\Psi_t \circ \Psi_{t'} = \Psi_{t'} \circ \Psi_t = \Psi_{\max(t, t')}$, for all $t, t' > 0$

where "$\circ$" denotes function composition. Note that each mapping Ψ_t within the granulometry $\{\Psi_t\}$ satisfies the first two requirements of an algebraic opening, and also that the granulometries $\{O(\cdot, tB)\}$ of Section 3.5 satisfy the requirements of a general granulometry. For the special case $t = 0$, we define $\Psi_0(A) = A$.

Example 5.13:

   Let

$$H(t) = \{(x, y) \colon 0 \leq x \leq t \text{ and } y = 0\}$$

   and

$$V(t) = \{(x, y) \colon x = 0 \text{ and } 0 \leq y \leq t\}$$

Then $H(t)$ and $V(t)$ are, respectively, the horizontal and vertical line segments emanating from the origin. Now define

$$\Phi_t(A) = O[A, H(t)] \cup O[A, V(t)]$$

Then Φ_t satisfies the axioms for a granulometry. For fixed t, Φ_t filters by opening by a linear horizontal structuring element and by a linear vertical structuring element, and then by taking the union of the two openings.

The first proposition of this section, Theorem 5.11, states that, in terms of sieving, the greater the size of the mesh of the sieve, the less of the original set which remains within the sieve. While completely intuitive, this property is also a logical consequence of the granulometric axoims G1 through G3.

Theorem 5.11[15]. If $\{\Psi_t\}$ is a granulometry and $r \geq s$, then $\Psi_r(A) \subset \Psi_s(A)$.

Previously, it was noted that each Ψ_t within the granulometry $\{\Psi_t\}$ satisfies the first two requirements of an algebraic opening. In fact, by G-3, Ψ_t is idempotent; that is,

$$\Psi_t[\Psi_t(A)] = \Psi_{\max(t,t)}(A) = \Psi_t(A)$$

Hence, Ψ_t is an algebraic opening. In fact, we can go much further: as the next theorem states, general granulometries can be characterized in terms of algebraic openings and invariant classes. Note in the statement of the theorem the intuitive nature of G-2': for $r \geq s$, a set that is unaffected by the sieving function Ψ_r is also unaffected by the sieving function Ψ_s.

Theorem 5.12[16]. $\{\Psi_t\}$ is a granulometry if and only if

G-1': Ψ_t is an algebraic opening for all $t > 0$
G-2': If $r \geq s$, then Inv $[\Psi_r] \subset$ Inv $[\Psi_s]$

In terms of the granulometries $\{O(\cdot, tB)\}$ of Section 3.5, G-2' can be stated in the following form: if A is rB-open, then it is also sB-open. Put this way, the property is seen to be closely related to the expansivity of convex sets. (See Theorem 1.6.)

The following two axioms are added to G-1 through G-3 for a general granulometry to define a *Euclidean granulometry*:

G-4: For any $t > 0$, Ψ_t is a τ-mapping
G-5: For any $t > 0$ and for any image A, $\Psi_t(A) = t\Psi_1[(1/t)A]$

Together with Theorem 5.12, G-4 states that Ψ_t is a τ-opening. Intuitively, it says that the sieving operation is independent of the position of A in the plane R^2. Practically speaking, this means that the mesh of the sieve is uniform throughout its extent, i.e., it is space invariant. G-5 states that scaling an image by $1/t$, sieving by Ψ_1, and then rescaling by t is the same as sieving by Ψ_t. This identity appears most intuitive in the case of the usual mesh-type sieving. Furthermore, it is satisfied by the granulometries $\{O(\cdot, tB)\}$. Indeed, for $\Psi_t = O(\cdot, tB)$, properties M-12 and M-13 yield

$$t\Psi_1[(1/t)A] = tO[(1/t)A, B] = t[((1/t)A \ominus (-B)) \oplus B]$$
$$= [t((1/t)A \ominus (-B))] \oplus tB$$
$$= [A \ominus (-tB)] \oplus tB = O(A, tB) = \Psi_t(A)$$

It is the Euclidean granulometries that Matheron has characterized (Theorem 5.14), and G-5 makes it plain that the τ-opening Ψ_1 plays a special role. Hence, we

Sec. 5.5 Euclidean Granulometries

should expect that its invariant class is also of importance and that G-5 has a formulation in terms of invariant classes. Indeed, we have the following theorem:

Theorem 5.13[17]. Let $\{\Psi_t\}$ be a granulometry for which G-4 holds. Then G-5 is equivalent to Inv $[\Psi_t] = t$ Inv $[\Psi_1]$, where the latter equality means that $A \in$ Inv $[\Psi_t]$ if and only if $(1/t)A \in$ Inv $[\Psi_1]$.

Example 5.14:

Let $\{\Phi_t\}$ be the granulometry of Example 5.13. Because $O[\cdot, H(t)]$ and $O[\cdot, V(t)]$ are τ-mappings, so is Ψ_t. Now, each opening satisfies G-5, and furthermore, for any two sets A and B,

$$t[A \cup B] = (tA) \cup (tB)$$

Hence, Φ_t satisfies G-5, and $\{\Phi_t\}$ is a Euclidean granulometry. Indeed, it should be clear that the union of any collection of granulometries $\{O(\cdot, tB_k)\}$ is a Euclidean granulometry. Note, of course, that the convexity of B_k plays a crucial role.

It is no accident that we have concentrated on unions of openings by scalar multiples of convex structuring elements. The next theorem, due to Matheron, completely characterizes the Euclidean granulometries, and the role of convexity in this characterization will be seen shortly.

Theorem 5.14 (Matheron)[18]. A family of image-to-image mappings $\{\Psi_t\}$, $t > 0$, is a Euclidean granulometry if and only if there exists some class of images $\mathcal{B}_0$ such that

$$\Psi_t(A) = \bigcup_{B \in \mathcal{B}_0} \bigcup_{r \geq t} O(A, rB)$$

$\mathcal{B}_0$ is called a *generator* of the granulometry.

The double union in Theorem 5.14 makes it appear somewhat forbidding. However, consider the case where $\mathcal{B}_0$ consists of a single compact structuring element B. Then the theorem becomes

$$\Psi_t(A) = \bigcup_{r \geq t} O(A, rB)$$

In other words, each sieving function in the granulometry reduces to a union of openings. Furthermore, if B is convex, then Theorem 1.5 shows that $O(A, rB) \subset O(A, tB)$ for all $r \geq t$, and hence $\{\Psi_t\}$ reduces to the granulometry $\{O(\cdot, tB)\}$. On the other hand, if B is not convex, then, using Theorem 1.6, it can be shown that the union cannot be reduced to a single opening.

From the perspective of filtering, a Euclidean granulometry is a parametrized family of filters that provide increased filtering as $t \to \infty$. Since each Ψ_t is a τ-opening, each Ψ_t filters uniformly over R^2; that is, it filters in a manner compatible with translation, and its filtering does not disturb the order relation on the class of subsets of R^2. Moreover, axiom G-5 assures us that the family $\{\Psi_t\}$ behaves well with respect to scaling. Indeed, for positive r and t,

$$\Psi_t(rA) = t\Psi_1[(r/t)A] = [r(t/r)]\Psi_1[(r/t)A] = r\Psi_{t/r}(A)$$

In words, the output of any sieving function Ψ_t with input rA can be found by inputting A itself into the sieving function $\Psi_{t/r}$. What is important is that A determines the effect of filtering rA. In sum, the criteria for a Euclidean granulometry yield a class of filters that are highly congruous with human perception, at least insofar as that perception is considered to be Euclidean.

Given, then, the desirable properties of Euclidean granulometries, the importance of Theorem 5.14 is evident: if one wishes to construct a family of filters that satisfies G-1 through G-5, then the family must be given by a union of openings as described in Theorem 5.14. Moreover, should one desire to filter by employing a single generating structuring element, then the union will reduce to a single opening if and only if that structuring element is convex (assuming it is already compact). Further, should one wish to filter more finely by employing a generator with two structuring elements, say $\mathcal{B}_0 = \{B_1, B_2\}$, then the granulometry will reduce to

$$\Psi_t(\cdot) = O(\cdot, tB_1) \cup O(\cdot, tB_2)$$

if and only if both B_1 and B_2 are convex. This is exactly the situation considered in Example 5.13. Similar remarks hold for a generator consisting of some other finite number of structuring elements.

EXERCISES

5.1. For any constant digital image S, let the activated pixels of $\Psi(S)$ be all the activated pixels of S together with all pixels (i, j) for which $S(i, j - 1) = 1$. Find Ker $[\Psi]$.

5.2. For any constant digital image S, let $\mathbf{U}[S]$ denote the image whose domain consists of the set of all pixels (i, j) for which there exists an activated pixel (i, v) of S such that $j \leq v$. $\mathbf{U}[S]$ is called the *umbra* of S. Show that $\mathbf{U}$ is an increasing τ-mapping and find its kernel.

5.3. Find the dual mappings of the mappings Ψ and $\mathbf{U}$ of Exercises 5.1 and 5.2.

5.4. Find bases for the kernels of the mappings Ψ and $\mathbf{U}$ of Exercises 5.1 and 5.2.

5.5. Show that the mapping Ψ of Exercise 5.1 is 1-bounded, whereas the umbra $\mathbf{U}$ is not λ-bounded.

5.6. Let

$$E = \begin{pmatrix} * & * & 1 & 1 \\ * & * & * & * \\ * & * & * & * \\ \circledast & * & * & * \end{pmatrix}$$

The mapping DILATE(S, E) is λ-bounded. Find λ.

5.7. In the text it was stated that for a fixed structuring element E, the mappings DILATE$(\cdot, E)$, ERODE$(\cdot, E)$, OPEN$(\cdot, E)$, and CLOSE$(\cdot, E)$ are $[\kappa(E) - 1]$-bounded, $[\kappa(E) - \rho(E)]$-bounded, 0-bounded, and 0-bounded, respectively. Prove that these properties in fact hold.

5.8. For the mapping Ψ of Exercise 5.1 and the image

$$S = \begin{pmatrix} 1 & 1 & * & * & * \\ 1 & 1 & * & * & 1 \\ \circledast & 1 & 1 & 1 & 1 \end{pmatrix}$$

find $\Psi(S)$ by applying Theorem 5.8.

5.9. Apply Theorem 5.8 to find the topological closure of the image S of Exercise 5.8.

5.10. Apply Theorem 5.8 to find the erosion of the image S of Exercise 5.8 by the structuring element of Exercise 5.6.

5.11. Define the mapping Φ in the following way: pixel (i, j) in S is an element of $\Phi(S)$ if and only if there exists a pixel (u, v) in SQUARE(i, j), besides (i, j) itself, such that (u, v) is in the domain of S. SQUARE(i, j) consists of (i, j) and its neighbors.

   (a) Show that Φ is a τ-opening.

   (b) Find $\Phi(T)$ for the image

$$T = \begin{pmatrix} 1 & * & * & * & 1 & * \\ * & * & * & * & 1 & * \\ 1 & * & * & 1 & 1 & * \\ * & * & 1 & 1 & * & * \\ 1 & ① & 1 & * & * & * \\ * & * & * & * & * & 1 \end{pmatrix}$$

   (c) Find Inv $[\Phi]$.

   (d) Find a base for Φ.

   (e) Apply Theorem 5.10 to find $\Phi(T)$.

FOOTNOTES FOR CHAPTER 5

1. See Appendix: Proposition A.39.
2. See Appendix: Theorem A.5.
3. See Appendix: Proposition A.41.
4. See Appendix: Proposition A.41.
5. See Appendix: Theorem A.5.
6. See Appendix: Proposition A.34.
7. See Appendix: Proposition A.36.
8. E. R. Dougherty and C. R. Giardina, "A Digital Version of the Matheron Representation Theorem for Increasing τ-Mappings in Terms of a Basis for the Kernel," IEEE Computer Vision and Pattern Recognition (1986), 534–536.
9. Ibid., p. 534–536.
10. Ibid., p. 534–536.
11. Ibid., p. 534–536.
12. See Appendix: Proposition A.43.
13. See Appendix: Theorem A.6.
14. See Appendix: Proposition A.45.
15. See Appendix: Proposition A.46.
16. See Appendix: Proposition A.47.
17. See Appendix: Proposition A.48.
18. See Appendix: Theorem A.7.

6

Gray-scale Morphology

6.1 GRAY-SCALE MORPHOLOGICAL OPERATORS FOR EUCLIDEAN SIGNALS

We begin the presentation of gray-scale morphology in the context of one-dimensional Euclidean signals, since it is much easier to gain an intuitive understanding when working in one dimension, rather than two. Although we present rigorous definitions, important concepts are underscored by graphical illustrations. Those not concerned with the mathematical subtleties of the presentation should pay careful attention to the illustrations, for it is in these that the morphological geometry we attempt to faithfully capture in the digital setting is intuitively represented.

In dealing with Euclidean signals, which are real-valued functions defined on subsets of the real line R, we shall often employ specialized versions of the supremum (least upper bound) and the infimum (greatest lower bound). These operations are commonly employed in mathematical analysis and can be found in any introductory text. For those not familiar with the usual sup and inf operations, suffice it to say that they are variations of max and min, respectively, that apply to infinite sets. For finite sets, they reduce to their finite counterparts. In morphological analysis we employ variations of sup and inf that are analogous to EXTMAX and MIN, two of the digital morphological primitives. Moreover, when we proceed to the digital case, EXTMAX and MIN will once again play the key roles. Accordingly, if one has difficulty with

Sec. 6.1   Gray-Scale Morphology Operators for Euclidean Signals

the material in the present section, particular attention should be paid to the examples and the graphical illustrations. The theoretical peculiarities that arise in the Euclidean case, specifically in regard to sup and inf, do not arise in digital implementation.

Given a collection of signals $\{f_k\}$, possibly infinite, we define

$$[\text{EXTSUP}(f_k)](t) = \begin{cases} \sup [f_k(t)], & \text{if there exists at least one } k \text{ such that } f_k \text{ is defined} \\ & \text{at } t, \text{ and where the supremum is over all such } k \\ \text{undefined}, & \text{if } f_k(t) \text{ is undefined for all } k \end{cases}$$

and

$$[\text{INF}(f_k)](t) = \begin{cases} \inf [f_k(t)], & \text{if } f_k(t) \text{ is defined for all } k \\ \text{undefined}, & \text{otherwise} \end{cases}$$

In the event that the collection $\{f_k\}$ is finite, sup and inf reduce to max and min, respectively, and we will often write $\text{EXTSUP}(f_1, f_2, \ldots, f_m)$ and $\text{INF}(f_1, f_2, \ldots, f_m)$. As in the case of EXTMAX and MIN, the domain of $\text{EXTSUP}(f_k)$ is the union of the input domains, while the domain of $\text{INF}(f_k)$ is the intersection of the input domains.

Although the definitions appear rather cumbersome, EXTSUP and INF are rather straightforward from a graphical perspective, as the following examples show.

Example 6.1:

Let

$$f(t) = \begin{cases} 2t + 1, & \text{for } -1 \leq t < 0 \\ -2t + 1, & \text{for } 0 \leq t \leq 1 \end{cases}$$

and

$$g(t) = t, \quad \text{if } 0 \leq t \leq 2$$

Then

$$[\text{EXTSUP}(f, g)](t) = \begin{cases} 2t + 1, & \text{if } -1 \leq t \leq 0 \\ -2t + 1, & \text{if } 0 < t \leq \frac{1}{3} \\ t, & \text{if } \frac{1}{3} < t \leq 2 \end{cases}$$

and

$$[\text{INF}(f, g)](t) = \begin{cases} t, & \text{if } 0 \leq t \leq \frac{1}{3} \\ -2t + 1, & \text{if } \frac{1}{3} < t \leq 1 \end{cases}$$

(See Figure 6.1)

Example 6.2:

For $k = 1, 2, \ldots$, let $f_k(t) = f(t) - 1/k$, where f is the signal of Example 6.1. Then, since the domains of all the input signals are identical, we obtain

$$\text{EXTSUP}(f_k) = f$$

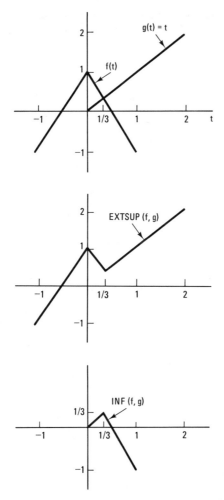

Figure 6.1 Illustration of Example 6.1

Note the role played by the supremum: for any fixed t in the domain of f,

$$\lim_{k \to \infty} f_k(t) = f(t)$$

and the supremum reflects the limit. In the finite case limits will not be required. (See Figure 6.2.)

Example 6.3:

For $k = 1, 2, \ldots$, let f_k be the upper semicircle of radius $1 - 2^{-k}$. Mathematically,

$$f_k(t) = \sqrt{(1 - 2^{-k})^2 - t^2}, \quad \text{for } 2^{-k} - 1 \leq t \leq 1 - 2^{-k}$$

The extended supremum of the f_k is the upper unit semicircle without the endpoints $t = 1$ and $t = -1$:

$$[\text{EXTSUP}(f_k)](t) = \sqrt{1 - t^2}, \quad \text{for } -1 < t < 1$$

Note that the domain of $\text{EXTSUP}(f_k)$ is the union of the domains of the f_k. (See Figure 6.3.) It is also clear from the figure that $[\text{INF}(f_k)](t) = \sqrt{\frac{1}{4} - t^2}$, for $-\frac{1}{2} \leq t \leq \frac{1}{2}$.

Sec. 6.1 Gray-Scale Morphology Operators for Euclidean Signals

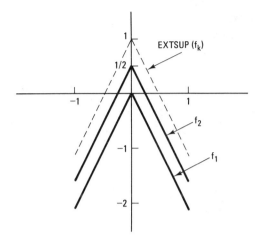

Figure 6.2 Illustration of Example 6.2

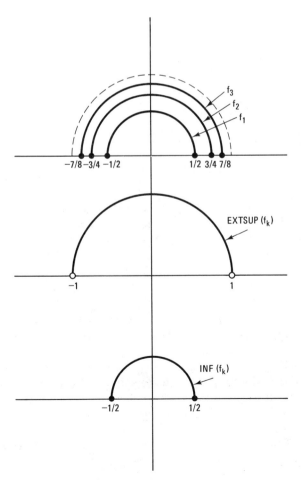

Figure 6.3 Illustration of Example 6.3

Example 6.4:

Let $f(t) = t$ for $0 \leq t \leq 2$. For any fixed x, the function $f(t - x)$ is simply f translated x units to the right, with domain $x \leq t \leq x + 2$. If we consider $\{x: 0 \leq x < 1\}$, a family of functions $\{f(t - x)\}$ is generated over the index set $[0, 1)$. By examining Figure 6.4, we see that

$$\text{EXTSUP}_{0 \leq x < 1}[f(t - x)] = \begin{cases} t, & \text{for } 0 \leq t \leq 2 \\ 2, & \text{for } 2 < t < 3 \\ \text{undefined}, & \text{elsewhere} \end{cases}$$

and

$$\text{INF}_{0 \leq x < 1}[f(t - x)] = \begin{cases} t - 1, & \text{for } 1 \leq t \leq 2 \\ \text{undefined}, & \text{elsewhere} \end{cases}$$

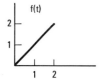

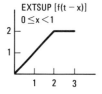

Figure 6.4 Illustration of Example 6.4

Normally, if we write $f \leq g$, we mean that f and g possess the same domain and $f(t) \leq g(t)$ for any t in that domain. Since our intention is to algebraically operate on signals that have different domains, however, we need to introduce an altered form of signal ordering. Accordingly, we define $f \ll g$ to mean that the domain of f is a subset of the domain of g and, for any t in the domain of f, $f(t) \leq g(t)$. Under this new order relation, $\text{INF}(f_k) \ll f_k$ for all k and $\text{EXTSUP}(f_k) \gg f_k$ for all k.

If $\{f_k\}$ is any collection of signals that contains a signal $f_{k'}$ such that $f_{k'} \gg f_k$ for all k, then $\text{EXTSUP}(f_k) = f_{k'}$. In a similar vein, if $\{f_k\}$ contains a signal $f_{k'}$ such that $f_{k'} \ll f_k$ for all k, then $\text{INF}(f_k) = f_{k'}$. In many morphological applications, the collection $\{f_k\}$ under consideration will contain either $\text{EXTSUP}(f_k)$ or $\text{INF}(f_k)$ within it.

Morphological algebra is extended to signals on subsets of R by employing the EXTSUP and INF operations. As in the Euclidean morphology of Chapter 1, it is the intuitive geometric notion of fitting, or probing, with structuring elements that underlies the methodology. Moreover, it is this Euclidean intuition that motivates the digital implementation to follow.

For any signal f, with domain D_f, and point x in R, we define the *translation* f_x by $f_x(t) = f(t - x)$. That is, f_x is f translated x units to the right. Hence, $D_{f_x} = D_f + x$.

Note that f_x is the usual translation that is employed in calculus. If, in addition, we desire a translation up y units, we simply consider $f(t) + y$, which possesses the same domain as f. Consequently, $f_x + y$ is a signal that is the same shape as f except that it has been translated over x and up y. (See Figure 6.5.)

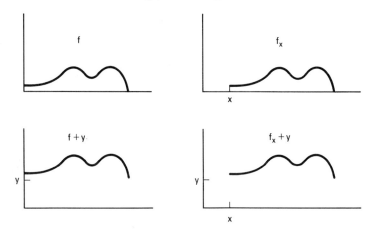

Figure 6.5 Translation of a signal

For signals f and g, with respective domains D_f and D_g, we define the *dilation* of f by g as

$$\mathcal{D}(f, g) = \text{EXTSUP}_{x \in D_f}[g_x + f(x)]$$

Geometrically, the dilation is obtained by taking an extended supremum of all copies of g that have been translated over x units and up $f(x)$ units. Calling the origin the *center* of the *structuring element* g, we construct the dilation by shifting g over x units and then translating the shifted copy upward so that its center coincides with $(x, f(x))$. Finally, EXTSUP is applied to the resulting copies of g. As in the two-valued Euclidean morphology, dilation is also called *Minkowski addition* and is denoted by $f \oplus g$.

The dilation methodology is illustrated in Figures 6.6 through 6.10. In each figure, a shows the structuring element g, b gives the signal f to be dilated, c portrays one or two copies of a translated structuring element $g_x + f(x)$, and d gives the dilation $\mathcal{D}(f, g)$. Note particularly the behavior of the domains, which have been emphasized in the drawings, and always keep in mind that $g_x + f(x)$ is a function of t with domain $D_g + x$. In terms of t,

$$[g_x + f(x)](t) = g(t - x) + f(x)$$

Observe the dilating (obviously) effect that dilation has on the graph of f in each case.

We now consider each figure individually. In Figure 6.6, $g(t) = 2$ for $-1 < t < 1$, and

$$f(t) = \begin{cases} 3, & \text{if } 2 < t < 6, t \neq 4 \\ 2, & \text{if } t = 4 \\ \text{undefined}, & \text{elsewhere} \end{cases}$$

The dilation is given by $\mathcal{D}(f, g) = 5$ for $1 < t < 7$.

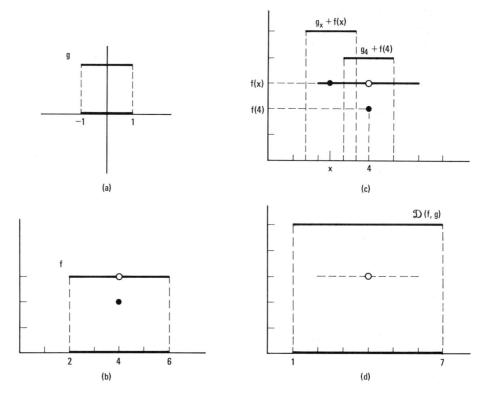

Figure 6.6 Euclidean gray-scale signal dilation I

In Figure 6.7, g is the same as in Figure 6.6 and so is f except at the point 4, where $f(4) = 4$. We have

$$\mathcal{D}(f, g) = \begin{cases} 6, & \text{if } 3 < t < 5 \\ 5, & \text{if } t \in (1, 3] \cup [5, 7) \\ \text{undefined}, & \text{elsewhere} \end{cases}$$

Notice how this output, compared with that of Figure 6.6, demonstrates the sensitivity of dilation to functional changes at a point of discontinuity.

In Figure 6.8, g is a semicircular signal and f is a signal consisting of two triangles. The dilation of f by g is found by tracing out the upper locus that results from placing the "center" of g along points on the graph of f. It is this type of geometric action that is generally associated with the heuristic interpretation of gray-scale dilation. As seen here, when the Euclidean signals involved are continuous, we generally obtain very intuitive dilations.

In Figure 6.9, g is a triangular waveform and f is a constant image with two discontinuities, one at $t = z$ and the other at $t = x$. Once again, note the manner in which the discontinuities affect the dilated output.

In Figure 6.10, the segment H of the graph of $\mathcal{D}(f, g)$ which runs one unit on either side of x is never attained by any translate; indeed, f is discontinuous at x, and $g_x + f(x)$ is one full unit below H. This shows the necessity of taking a supremum rather than a maximum in defining the gray-scale dilation in the Euclidean case.

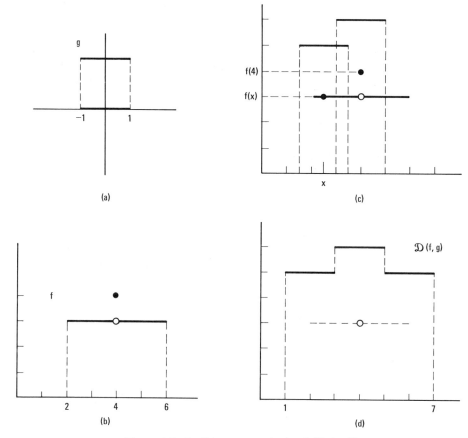

Figure 6.7 Euclidean gray-scale signal dilation II

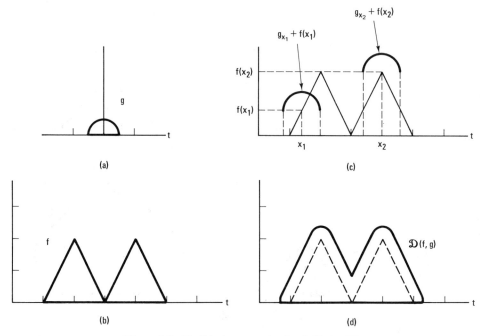

Figure 6.8 Euclidean gray-scale signal dilation III

163

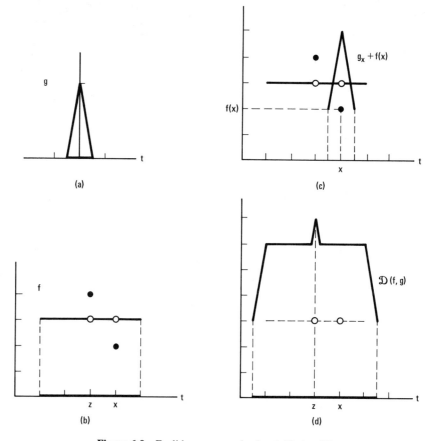

Figure 6.9 Euclidean gray-scale signal dilation IV

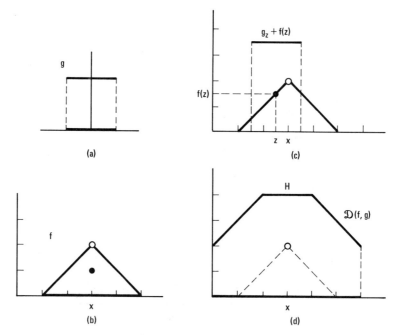

Figure 6.10 Euclidean gray-scale signal dilation V

Sec. 6.1 Gray-Scale Morphology Operators for Euclidean Signals

As in the case of two-valued dilation, both the commutative and associative laws hold. We state these as properties GM-1 and GM-2, respectively. Proofs that are valid in the digital setting will be presented in Section 6.5; nevertheless, both laws hold in the Euclidean case also.

Property GM-1. $f \oplus g = g \oplus f$

Property GM-2. $f \oplus (g \oplus h) = (f \oplus g) \oplus h$

GM-1 allows the interchange of the roles of f and g in *Minkowski addition*. Hence,

$$f \oplus g = \text{EXTSUP}_{x \in D_g}[f_x + g(x)]$$

Erosion, the dual of dilation, is defined pointwise for Euclidean signals by

$$[\mathcal{E}(f, g)](x) = \sup \{y: g_x + y \ll f\}$$

By stipulation, the erosion is undefined for any x for which the domain of g_x is not a subdomain of D_f. Note in the definition that the sup operation is the standard pointwise definition and that the ordering relation $\ll$ plays a role. Put crudely, to find the value of the erosion of f by g at the point x, we shift g so that it is centered at x and then we find the largest vertical translation y that will leave $g_x + y$ beneath f, remembering all the while that if $D_g + x$, the domain of g_x, is not a subset of the domain of f, the erosion is undefined at x. This "fitting" definition of erosion is analogous to the two-valued Euclidean definition, the only difference being the use of the $\ll$ ordering instead of subset ordering. Indeed, even this difference is merely formal. If we identify a subset A of R with the constant signal

$$f_A(t) = \begin{cases} 1, & \text{if } t \in A \\ \text{undefined}, & \text{otherwise} \end{cases}$$

then $A \subset B$ if and only if $f_A \ll f_B$. In fact, in the digital image case, it is precisely such an identification which allows us to view a constant image and its domain interchangeably.

Figures 6.11 through 6.14 employ the signals and structuring elements of Figures 6.6 through 6.9, respectively, to illustrate signal erosion. Notice the eroding (obviously) of the graphs of the signals and also the dimunition of the input domains.

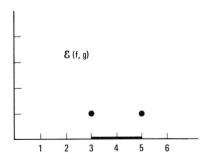

Figure 6.11 Euclidean gray-scale signal erosion I

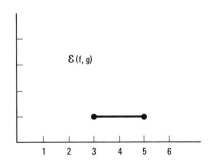

Figure 6.12 Euclidean gray-scale signal erosion II

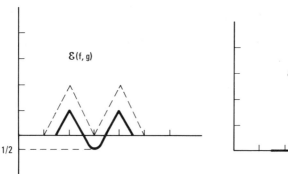

Figure 6.13 Euclidean gray-scale signal erosion III

Figure 6.14 Euclidean gray-scale signal erosion IV

Figure 6.15 illustrates the filtering effect of erosion. Note how the three points of discontinuity are smoothed in the output signal, which is simply the straight line b.

In order to maintain consistency with the two-valued morphology of Chapter 1, we introduce the *Minkowski subtraction* for gray-scale signals. For signals f and g, we have

$$f \ominus g = \text{INF}_{x \in D_g}[f_x + g(x)]$$

The duality in the definitions of $f \ominus g$ and $f \oplus g$ should be evident, at least insofar as symbolism is concerned.

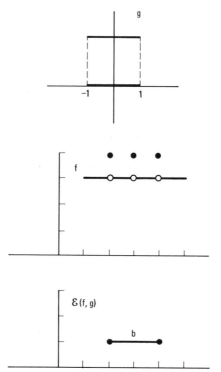

Figure 6.15 Euclidean gray-scale signal erosion V

Sec. 6.1   Gray-Scale Morphology Operators for Euclidean Signals   **167**

Of immediate consequence is Theorem 6.1, the gray-scale analogue of Theorem 1.1. Whereas in the preceding chapters we have cited either the appendix or references for the proofs of theorems, in the sequel the theory will be part of the exposition proper. The newness of the gray-scale theory necessitates this approach. For those not interested in the mathematical details, the examples and illustrations should provide a sufficient foundation for applications. The proof of Theorem 1.1 involves the theory of infima and suprema; however, a much easier proof can be provided in the fully digital environment. Before stating the theorem, we require a definition. For a signal h with domain D_h, the *reflection through the origin* of h, denoted $h^\wedge$, is defined by $h^\wedge(t) = -h(-t)$ for any $t \in D_{h^\wedge} = -D_h$. (See Figure 6.16.)

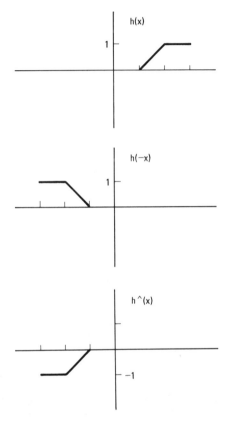

Figure 6.16 Reflection of signal through the origin

Theorem 6.1. For any two signals f and g, $\mathscr{E}(f, g) = f \ominus g^\wedge$.

Proof. We first show that $\mathscr{E}(f, g)$ is undefined at a point x if and only if $f \ominus g^\wedge$ is undefined at x. Indeed, $\mathscr{E}(f, g)$ is undefined at x if and only if $D_{g_x} \not\subset D_f$ if and only if there exists a $z \in D_g$ such that $z \notin D_f - x$ if and only if there exists a $z \in D_g$ such that $x \notin D_f - z$ if and only if

$$x \notin \bigcap_{z \in D_g} D_f - z$$

which is equivalent to $f \ominus g^\wedge$ being undefined at x.

Now suppose that both $f \ominus g^\wedge$ and $\mathcal{E}(f, g)$ are defined at x and that $g_x + y \ll f$. Then, for any $t \in D_{g_x}$,

$$g(t - x) + y \le f(t)$$

Letting $t' = t - x$, we have, for any $t' \in D_g$,

$$g(t') + y \le f(t' + x)$$

or, equivalently,

$$y \le f_{-t'}(x) - g(t')$$

Hence,

$$[\mathcal{E}(f, g)](x) \le [\text{INF}_{t' \in D_g}(f_{-t'} - g(t'))](x)$$
$$= [\text{INF}_{\tau \in D_{g^\wedge}}(f_\tau + g^\wedge(\tau))](x)$$
$$= [f \ominus g^\wedge](x)$$

To obtain the reverse inequality, suppose $[\mathcal{E}(f, g)](x) < w$. Then

$$g_x + w \not\ll f$$

Therefore, there exists a point $t \in D_{g_x}$ such that

$$g(t - x) + w > f(t)$$

Letting $t'' = t - x \in D_g$, we have

$$w > f_{-t''}(x) - g(t'') \ge [\text{INF}_{t'' \in D_g}(f_{-t''} - g(t''))](x) = [f \ominus g^\wedge](x)$$

Since w is arbitrary,

$$[f \ominus g^\wedge](x) \ll [\mathcal{E}(f, g)](x)$$

Because $g^{\wedge\wedge} = g$, we have an immediate corollary:

$$f \ominus g = f \ominus g^{\wedge\wedge} = \mathcal{E}(f, g^\wedge)$$

Moreover, in terms of the definition of Minkowski subtraction,

$$\mathcal{E}(f, g) = \text{INF}_{x \in D_g}[f_x + g^\wedge(x)] = \text{INF}_{x \in D_g}[f_{-x} - g(x)]$$

Figure 6.17 gives an illustration of erosion computed by using INF. The expression of erosion in terms of an infimum plays an important role in digital implementation, where the algorithmic relation between Minkowski subtraction and classical convolution will become clear.

Gray-scale *opening* can be defined analogously to its definition in the constant-image (1–*) case:

$$O(f, g) = \mathcal{D}[\mathcal{E}(f, g), g] = (f \ominus g^\wedge) \oplus g$$

As in Chapter 1, the intuitive meaning of the opening for morphological filtering is given in terms of a fitting criterion. Theorem 6.2, the gray-scale-signal version of Theorem 1.2, provides this criterion, which is crucial for morphological filtering.

Sec. 6.1 Gray-Scale Morphology Operators for Euclidean Signals

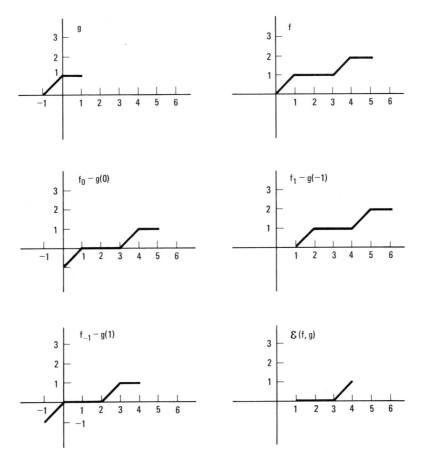

Figure 6.17 Erosion by using INF

Theorem 6.2. $O(f, g) = \text{EXTSUP}\{g_x + y: g_x + y \ll f\}$

Proof.

$$\mathcal{D}[\mathcal{E}(f, g), g] = \text{EXTSUP}_{x \in D_{\mathcal{E}(f,g)}}[g_x + (\mathcal{E}(f, g))(x)]$$
$$= \text{EXTSUP}_{x \in D_{\mathcal{E}(f,g)}}[g_x + \sup\{y: g_x + y \ll f\}]$$
$$= \text{EXTSUP}[g_x + y: g_x + y \ll f]$$

Figures 6.18 through 6.20 provide illustrations of the opening. Note that the domain of the opening is a subset of the input domain, but it is not necessarily a proper subset. Note also the filtering effect of the opening and how this filtering takes place underneath the graph of f.

Certain fundamental properties of the opening are apparent from the fitting criterion of Theorem 6.2. Of special importance to morphological filtering is Theorem 6.3, the gray-scale version of Theorem 1.3.

170 Gray-Scale Morphology   Chap. 6

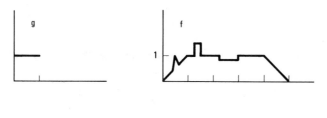

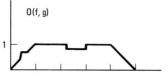

Figure 6.18 Euclidean gray-scale signal opening I

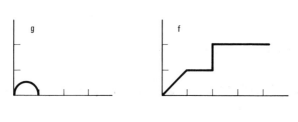

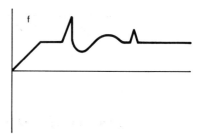

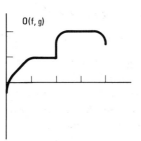

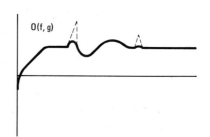

Figure 6.19 Euclidean gray-scale signal opening II

Figure 6.20 Euclidean gray-scale signal opening III

Sec. 6.1   Gray-Scale Morphology Operators for Euclidean Signals

Theorem 6.3. The gray-scale opening satisfies:

(i) $O(f, g) \ll f$ (antiextensivity)
(ii) If $f_1 \ll f_2$, then $O(f_1, g) \ll O(f_2, g)$ (increasing monotonicity)
(iii) $O[O(f, g), g] = O(f, g)$ (idempotence)

Many remarks concerning the two-valued Euclidean opening apply to the gray-scale situation, especially insofar as filtering is concerned. We specifically mention *translation invariance*:

$$O(f_x + y, g) = O(f, g)_x + y$$

Here one must be aware of the potential ambiguity of the terminology. A signal-to-signal operator Ψ will be called *translation invariant* if

$$\Psi(f_x + y) = \Psi(f)_x + y$$

The terminological difficulty is that whereas f_x is a horizontal translation by x and $f + y$ is a vertical translation by y, $f_x + y$ is a morphological translation in that its graph has been geometrically translated by the vector (x, y). Though new terminology could be introduced, we believe it is prudent to stay with the usual morphological expressions.

The gray-scale *closing* is defined by taking the dual of the opening:

$$C(f, g) = -O(-f, -g)$$

The closing is obtained by reflecting f and g through the x-axis, applying the opening, and then reflecting back again. Figures 6.21 and 6.22 illustrate this procedure.

Like the two-valued closing, the gray-scale closing is increasing and idempotent. That this is so is easy to deduce from the definition and the corresponding properties of the opening. For instance, idempotence is inferable as follows:

$$\begin{aligned} C[C(f, g), g] &= -O[-C(f, g), -g] \\ &= -O[O(-f, -g), -g] \\ &= -O(-f, -g) = C(f, g) \end{aligned}$$

The closing is also translation invariant:

$$C(f_x + y, g) = C(f, g)_x + y$$

The closing is not extensive: it may be that there exists a point x in the domain of f such that no translate of $-g$ fits beneath $-f$ in such a way as to contain x within its domain. In such a case, x would not be in the domain of $C(f, g)$ and hence, $C(f, g) \not\gg f$. Specifically, extensivity fails if the two-valued one-dimensional opening of D_f by D_g is a proper subset of D_f; otherwise, it holds. To describe this situation, we say that the closing is *pseudoextensive*. In the extended setting of Chapter 7, the closing will be fully extensive.

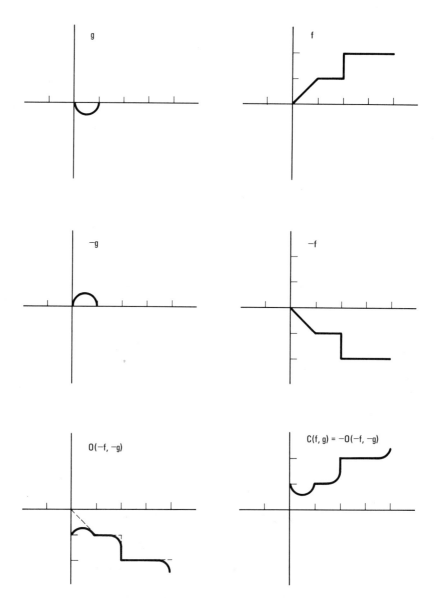

Figure 6.21 Euclidean gray-scale signal closing I

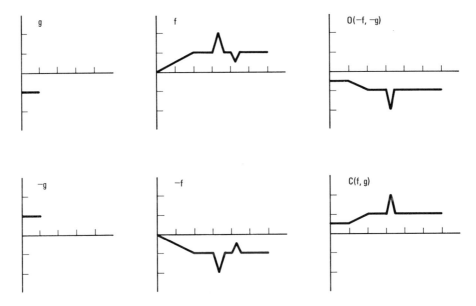

Figure 6.22 Euclidean gray-scale signal closing II

Note that we have *not* defined the closing in terms of dilation and erosion. This is because, as Figure 6.23 shows, the closing is not given by the expression

$$\mathcal{E}[\mathcal{D}(f, g^\wedge), g^\wedge]$$

which would be analogous to the two-valued situation. More will be said on this matter at the end of Section 6.5.

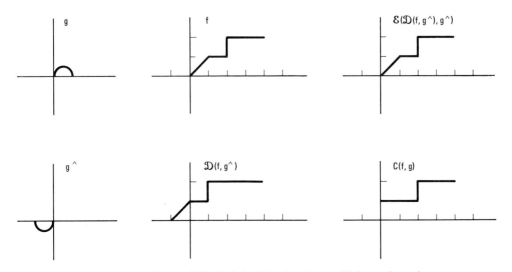

Figure 6.23 Relationship of closing to dilation and ersosion

6.2 UMBRA TRANSFORM

A very useful device for relating gray-scale morphology on signals to 1-* morphology on constant images is the *umbra transform* **U**. As with most transform techniques, the umbra transform is used to move the solution of problems from one setting to another.

Suppose $f(t)$ is a signal with domain D_f. Then the graph of f, as usually employed in calculus, is rigorously defined by

$$\mathbf{G}[f] = \{(t, f(t)): t \in D_f\}$$

In the terminology of image processing, **G**[f] is a constant Euclidean image; i.e., **G**[f] is a subset of the Euclidean plane R^2. Looking at it another way, as an operator, **G** maps gray-scale signals into constant images.

To facilitate further discussion, let $\mathcal{S}$ denote the collection of all Euclidean signals and $2^{R \times R}$ denote the collection of all subsets of the plane. Then

$$\mathbf{G}: \mathcal{S} \longrightarrow 2^{R \times R}$$

Moreover, since no two distinct signals possess the same graph, we can invert the graph operation. Indeed, we generally do not distinguish between a function and its graph.

Given the preceding definitions, the set of all points in the plane that lie below **G**[f], including **G**[f] itself, is called the *umbra* of f and is denoted by **U**[f]. Like **G**[f], **U**[f] is a Euclidean constant image (a subset of R^2) and is given by

$$\mathbf{U}[f] = \{(x, y): x \in D_f \text{ and } y \leq f(x)\}$$

(See Figure 6.24.)

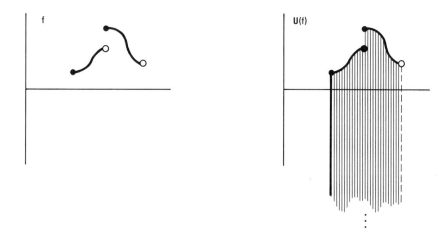

Figure 6.24 Umbra transform

Although this definition treats the umbra transform as if it were defined on signals, which is how we think about it intuitively, in fact it is a mapping between subsets of the plane. Specifically,

$$\mathbf{U}: 2^{R \times R} \longrightarrow 2^{R \times R}$$

by

$$\mathbf{U}[A] = \{(x, y): \text{there exists a } y' \text{ with } (x, y') \in A \text{ and } y \leq y'\}$$

In other words, for any set A, the umbra of A, denoted $\mathbf{U}[A]$, is the set of all points that lie beneath some point of A. If we now compare the definition of an umbra of a signal to an umbra of a set, we see that, for a signal f,

$$\mathbf{U}[f] = \mathbf{U}[\mathbf{G}[f]]$$

Figure 6.25 gives an illustration of several umbrae.

In general, a subset B of R^2 is an umbra if $\mathbf{U}[B] = B$, which says that the umbrae of B is once again B. To be an umbra, a set B must satisfy the property that if $(x, y') \in B$, then $(x, y) \in B$ for any $y \leq y'$. For a signal f of bounded domain, the umbra $\mathbf{U}[f]$ is a columnar set; that is, put crudely, umbrae possess vertical sides. This is not the case for signals over infinite domains.

The umbra in Figure 6.25(c) is an infinite vertical strip. For the moment, let us disregard such umbrae and consider only those for which each vertical line segment within the umbra possesses a finite supremum. Such umbrae will be referred to as *regular umbrae*. All the umbrae of Figure 6.25 are regular except the one in part (c). It is crucial to note that the umbrae of signals are regular.

We define the *domain* of an umbra to be the set of all points x on the horizontal axis for which there exists some point (x, y) in the umbra. For a signal f, the domain of $\mathbf{U}[f]$ is precisely the domain of f. The domains of the umbrae in Figure 6.25 are respectively $(-\infty, +\infty)$, $[0, 2] \cup [3, 4]$, $(1, 2)$, $[0, 1]$, and $\{3\}$.

Finally, and perhaps most importantly, for any regular umbra A, we define the *surface* of A, denoted $\mathbf{S}[A]$, to be the set of all points (x, y_x) such that x is in the domain of A and

$$y_x = \sup \{y: (x, y) \in A\}$$

Intuitively, if it lies in the umbra, (x, y_x) is the highest point in the umbra over x.

A regular umbra is said to be *closed* if it contains its surface. Note that an umbra may be closed as an umbra, but not closed topologically. (See Figure 6.25(d).) For a signal f, the surface of $\mathbf{U}[f]$ equals the graph of f; i.e.,

$$\mathbf{S}[\mathbf{U}[f]] = \mathbf{G}[f]$$

Identifying $\mathbf{G}[f]$ with f yields $f = \mathbf{S}[\mathbf{U}[f]]$.

For arbitrary sets A and B, the following basic properties hold:

U-1: $A \subset \mathbf{U}[A]$ (extensivity)
U-2: $\mathbf{U}[A] = \mathbf{U}[\mathbf{U}[A]]$ (idempotence)

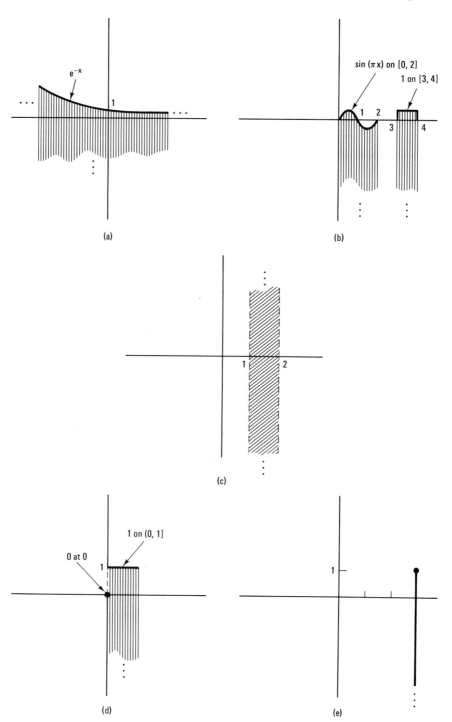

Figure 6.25 Several signal umbrae

Sec. 6.2 Umbra Transform

U-3: If $A \subset B$, then $U[A] \subset U[B]$ (increasing monotonicity)
U-4: $U[A + (x, y)] = U[A] + (x, y)$ (translation invariance)
U-5: $U[A] \subset U[S[U[A]]]$, assuming that $U[A]$ is regular
U-6: If $U[A]$ is closed, then $U[A] = U[S[U[A]]]$

The identity between the surface of the umbra of a signal f and the graph of f is of paramount importance from the transform point of view. Since a signal is uniquely defined by its graph and conversely, every regular umbra determines a unique signal, viz., that signal whose graph is the surface of the umbra in question. Thus, if $\mathcal{U}_r$ is the collection of all regular umbrae, $\mathcal{S}$ the collection of all signals, and $\mathbf{U}$ the umbra transform, then

$$\mathbf{U}: \mathcal{S} \longrightarrow \mathcal{U}_r$$

in a one-to-one manner. Moreover, if $\mathcal{U}_c$ is the collection of all closed umbrae, then the mapping

$$\mathbf{U}: \mathcal{S} \longrightarrow \mathcal{U}_c$$

is onto, and hence there is a unique inverse mapping

$$\mathbf{U}^{-1}: \mathcal{U}_c \longrightarrow \mathcal{S}$$

defined by $\mathbf{U}^{-1}[\mathbf{U}(f)] = f$, and concomitantly, $\mathbf{U}[\mathbf{U}^{-1}(F)] = F$. This inverse mapping is the surface operator $\mathbf{S}$. It takes the space $\mathcal{U}_c$ of closed regular umbrae as its input space and inverts the umbra transform $\mathbf{U}$ by pulling off the surface of F, that surface being the graph of the signal f. (See Figure 6.26.) As is customary, we then identify $G[f]$ with f.

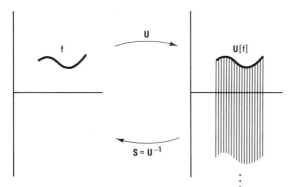

Figure 6.26 Umbra and surface transforms

The preceding inversion methodology is analogous to the inversion methodology for the Fourier transform. The fact that the Fourier transform can be inverted for certain classes of functions means precisely that it acts in a one-to-one and onto manner for those classes. Indeed, in general transform theory such inversion criteria are known as *uniqueness* conditions.

Since they are subsets in the plane, umbrae can be inputs to the morphological operations of Chapter 1. Furthermore, if A and B are umbrae, then so are the dilation,

$\mathcal{D}(A, B) = A \oplus B$, and the erosion, $\mathcal{E}(A, B) = A \ominus (-B)$, where $-B$ denotes the 180° rotation of B around the origin. Also, if A and B are regular umbrae, then so are $\mathcal{D}(A, B)$ and $\mathcal{E}(A, B)$. Note that for regular umbrae, the straight Minkowski subtraction $A \ominus B$ is empty.

The importance of the umbra transform is the manner in which it relates the gray-level morphology of signals to the two-valued morphology of the plane. In this regard, the following theorem is paramount.

Theorem 6.4. For any signals f and g,

$$\mathcal{D}(f, g) = S[\mathcal{D}(U[f], U[g])]$$

where the $\mathcal{D}$ on the left-hand side of the equation denotes gray-scale dilation, and the $\mathcal{D}$ on the right-hand side denotes Euclidean constant-image dilation.

Proof.

$$\mathcal{D}(U[f], U[g]) = \bigcup_{(x,y) \in U[f]} U[g] + (x, y) \tag{1}$$
$$= \bigcup_{(x,y) \in U[f]} U[g_x + y]$$

Now, for any (t, s) in the latter union, there exists an $(x, y) \in U[f]$ such that $(t, s) \in U[g_x + y]$; hence,

$$s \leq g_x(t) + y \leq [\mathcal{D}(f, g)](t)$$

where the second inequality follows from the fact that (x, y) lies in the umbra of f. Consequently, applying S to (1) yields

$$S[\mathcal{D}(U[f], U[g])] \leq \mathcal{D}(f, g)$$

Now suppose $[\mathcal{D}(f, g)](t) = y$. Then there exists a sequence $\{x_j\}$ such that

$$\lim_{j \to \infty} [g_{x_j} + f(x_j)](t) = y \tag{2}$$

But for all j,

$$(t, g(t - x_j) + f(x_j)) \in U[g_{x_j} + f(x_j)]$$
$$= U[g] + (x_j, f(x_j))$$
$$\subset \mathcal{D}(U[f], U[g])$$

Therefore, by (2),

$$y \leq S[\mathcal{D}(U[f], U[g])](t)$$

Example 6.5:

Let

$$g(t) = \begin{cases} 1 - t, & \text{if } 0 < t \leq 1 \\ 0, & \text{if } t = 0 \\ \text{undefined}, & \text{elsewhere} \end{cases}$$

and let $f(t) = 1$ for $0 \leq t \leq 1$. Then

$$\mathcal{D}(f, g) = \text{EXTSUP}_{x \in [0, 1]}[g_x + 1] = \begin{cases} 1, & \text{if } t = 0 \\ 2, & \text{if } t \in (0, 1] \\ 3 - t, & \text{if } t \in (1, 2] \\ \text{undefined}, & \text{elsewhere} \end{cases}$$

These signals are illustrated in Figure 6.27 along with the umbra transform methodology for obtaining the dilation.

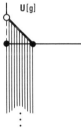
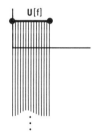

Figure 6.27 Dilation by means of umbra transform

Theorem 6.4 allows the transformation of gray-scale problems to constant-image problems. In terms of $\oplus$, it states that

$$f \oplus g = \mathbf{S}(\mathbf{U}[f] \oplus \mathbf{U}[g])$$

where once again the $\oplus$ is interpreted in the appropriate manner.

Suppose we apply the umbra transform to both sides of the equation in Theorem 6.4. We obtain

$$\mathbf{U}[f \oplus g] = \mathbf{U}[\mathbf{S}(\mathbf{U}[f] \oplus \mathbf{U}[g])]$$

If it turns out that $U[f] \oplus U[g]$ is closed, then, according to property U-6, $\mathbf{U}$ and $\mathbf{S}$ are inverses, and the preceding equality reduces to

$$U[f \oplus g] = U[f] \oplus U[g]$$

In the illustration of Figure 6.27, the Minkowski sum of the umbrae is not closed, so the preceding reduction does not apply. It is important to note, however, that for fully digital implementation, the reduction is always possible. (See Section 6.5.)

The following theorem, corresponding to Theorem 6.4, applies to erosion.

Theorem 6.5. $\mathcal{E}(f, g) = S[\mathcal{E}(U[f], U[g])]$

Proof. Suppose $g_x + y \ll f$. Then

$$U[g] + (x, y) \subset U[f]$$

and

$$(x, y) \in \mathcal{E}(U[f], U[g])$$

which implies that

$$y \leq S[\mathcal{E}(U[f], U[g])](x)$$

Hence, $\mathcal{E}(f, g) \ll S[\mathcal{E}(U[f], U[g])]$.

For the converse, suppose $S[\mathcal{E}(U[f], U[g])](x) = y$. Then there exists a sequence $\{(x, y_j)\}$ lying in $\mathcal{E}(U[f], U[g])$ such that y_j converges monotonically upward to y. In particular, for any j,

$$U[g] + (x, y_j) \subset U[f]$$

Hence, $g_x + y_j \ll f$, and it follows that $g_x + y \ll f$. Therefore, $y \leq [\mathcal{E}(f, g)](x)$.

In terms of Minkowski subtraction, Theorem 6.5 becomes

$$f \ominus g^\wedge = S(U[f] \ominus (-U[g]))$$

where $-U[g]$ is the 180° rotation of the constant image $U[g]$. Applying the umbra transform to both sides yields

$$U[f \ominus g^\wedge] = U[S(U[f] \ominus (-U[g]))]$$
$$= U[S(\mathcal{E}(U[f], U[g]))]$$

Once again, no further reduction is possible in the Euclidean case unless the erosion of the umbra $U[f]$ by $U[g]$ is closed. Moreover, in general, $U[g^\wedge] \neq -U[g]$. The methodology of Theorem 6.5 is illustrated in Figure 6.28.

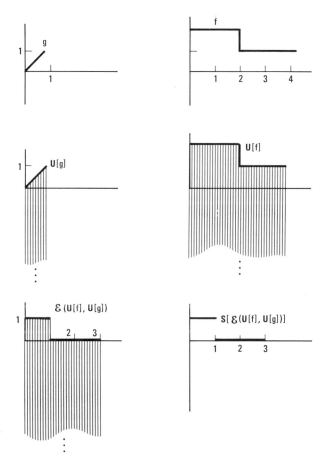

Figure 6.28 Erosion by means of umbra transform

6.3 GRAY-SCALE MORPHOLOGY FOR SAMPLED SIGNALS

Rather than proceed to the properties of the gray-scale Minkowski algebra, we shall first present the model for sampled signals. This will allow us to work out some actual digital examples and relate the somewhat abstract matter of the preceding two sections to a more computationally oriented setting.

Figure 6.29 shows a continuous signal and an equidistanced sampled version of that signal. Through a rescaling of the x-axis, we can assume that a sampled signal is a function from some subset of Z, the integers, into the real line R. Sampled signals are akin to digital images, which have domains in $Z \times Z$, except that sampled signals are simpler to work with. If D_f is the domain of f, then for each $i \in D_f \cap Z$, $f(i)$ denotes the intensity of the signal at i.

Now suppose f is a sampled signal with finite domain. Then f can be represented as a *bound row vector*

$$f = (a_1 \quad a_2 \quad \cdots \quad a_n)_r$$

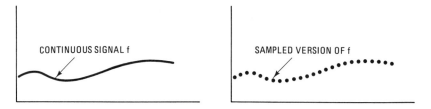

Figure 6.29 Sampled signal

where the subscript r denotes the leftmost element in the domain of f, $a_1 = f(r)$, and, in general, $a_k = f(r + k - 1)$. It should be obvious that a bound row vector is simply a special case of a bound matrix, except that there is only one row. Consequently, the operations discusssed in Chapter 1, which apply to bound matrices, have immediate analogues for bound vectors, the only differences being that TRAN requires only a single integer input, and that only NINETY^2, not NINETY, can be employed for rotation. Also, note that we speak of a *minimal bound vector* instead of a minimal bound matrix.

Since we need to work with signals of the form $f_i + y$, we must have an operator that "adds" a constant y to a signal. This operator, the *offset* operator, is defined by

$$[\mathrm{OFF}(f; y)](p) = f(p) + y$$

As will be seen in Section 6.5, OFF is actually a form of translation and, as such, is not viewed as a new operator in the morphological basis. For the moment, just note that the offsetting of a signal is equivalent to the upward translation of the signal's umbra.

We also require the negation of signals. For this purpose, we introduce the operator SUB defined by

$$[\mathrm{SUB}(f)](p) = -f(p)$$

SUB is actually a form of rotation and, as such, can be included in the augmented morphological basis.

Example 6.6:

Let
$$f = (2 \quad 1 \quad \tfrac{1}{2} \quad * \quad 1 \quad 0)_0$$

and
$$g = (1 \quad 8 \quad -1 \quad 2 \quad 0 \quad 0 \quad 2 \quad 1 \quad \tfrac{1}{3})_{-1}$$

Both signals are depicted in Figure 6.30. Operating in turn by EXTMAX, MIN, TRAN, OFF, SUB, and NINETY^2 yields

$$\mathrm{EXTMAX}(f, g) = (1 \quad 8 \quad 1 \quad 2 \quad 0 \quad 1 \quad 2 \quad 1 \quad \tfrac{1}{3})_{-1}$$

$$\mathrm{MIN}(f, g) = (2 \quad -1 \quad \tfrac{1}{2} \quad * \quad 0 \quad 0)_0$$

$$\mathrm{TRAN}(f; -3) = (2 \quad 1 \quad \tfrac{1}{2} \quad * \quad 1 \quad 0)_{-3}$$

$$\mathrm{OFF}(g; 1) = (2 \quad 9 \quad 0 \quad 3 \quad 1 \quad 1 \quad 3 \quad 2 \quad \tfrac{4}{3})_{-1}$$

$$\mathrm{SUB}(f) = (-2 \quad -1 \quad -\tfrac{1}{2} \quad * \quad -1 \quad 0)_0$$

Sec. 6.3 Gray-Scale Morphology for Sampled Signals

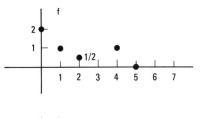

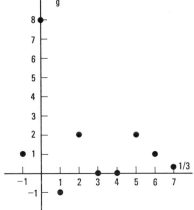

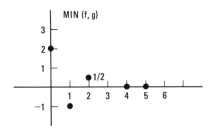

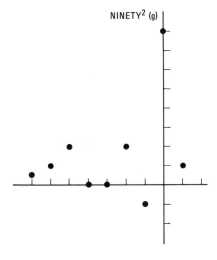

Figure 6.30 Signals of Example 6.6

and

$$\text{NINETY}^2(g) = (\tfrac{1}{3} \quad 1 \quad 2 \quad 0 \quad 0 \quad 2 \quad -1 \quad 8 \quad 1)_{-7}$$

Note that NINETY^2 utilizes the origin pixel as the pivot point.

It is now fairly straightforward to specify digital versions of the gray-scale morphological operations, where, for the sake of practicality, we assume that each signal possesses a finite domain. This assumption allows the valid use of EXTMAX and MIN, and the suppression of sup and inf. To begin, the *dilation*, or *Minkowski addition*, of two sampled signals, denoted DILATE(f, g), or $f \boxplus g$, is specified by the block diagram of Figure 6.31. Note how the flow of the diagram follows the original Euclidean definition, only in this instance digital operators are employed. The signal g is translated by all elements in the domain of f, and each resulting translate is offset by a constant that equals the appropriate value of f. The outputs of all the offsets are then input into an extended maximum in order to produce the desired dilation.

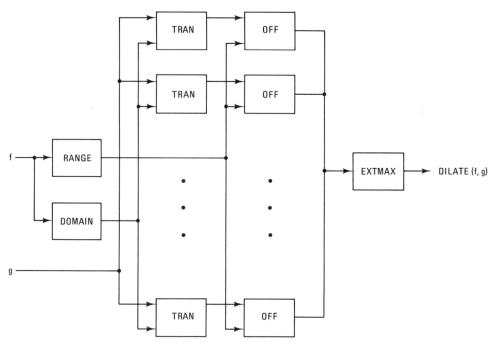

Figure 6.31 Block diagram for DILATE

Example 6.7:

Let

$$f = (2 \quad 2 \quad 8 \quad 4 \quad 2 \quad 1 \quad * \quad -1)_0$$

and

$$g = (6 \quad 4 \quad 6)_{-1}$$

Then
$$\text{RANGE}(f) = [2, 2, 8, 4, 2, 1, -1]$$
and
$$\text{DOMAIN}(f) = [0, 1, 2, 3, 4, 5, 7]$$

The inputs to the first OFF operator of Figure 6.31 are

$$(6 \quad 4 \quad 6)_{-1}$$

which is g translated by 0, and 2, which equals $f(0)$. Hence, the operator has output

$$g^0 = (8 \quad 6 \quad 8)_{-1}$$

where the superscript zero indicates that this output has resulted from the domain element 0 of f. The outputs of the subsequent OFF blocks are:

$$g^1 = (8 \quad 6 \quad 8)_0$$
$$g^2 = (14 \quad 12 \quad 14)_1$$
$$g^3 = (10 \quad 8 \quad 10)_2$$
$$g^4 = (8 \quad 6 \quad 8)_3$$
$$g^5 = (7 \quad 5 \quad 7)_4$$
$$g^7 = (5 \quad 3 \quad 5)_6$$

There is of course no g^6 since 6 is not in the domain of f.

Now, inputting all the g^k into EXTMAX yields

$$\text{DILATE}(f, g) = (8 \quad 8 \quad 14 \quad 12 \quad 14 \quad 10 \quad 8 \quad 7 \quad 3 \quad 5)_{-1}$$

Whereas f had a single point spike at $i = 2$, $f \boxplus g$ possesses a plateau three points wide and centered at $i = 2$.

Besides Minkowski addition, we can also generate definitions of Minkowski subtraction and erosion for sampled signals. Using the digital form of Theorem 6.1, Figure 6.32 gives a block diagram specification of erosion. Notice how it parallels the original, nondigital Euclidean specification. To obtain a block diagram of the Minkowski difference $f \boxminus g$, we simply forego the NINETY² and SUB operations, since these arise from the expression of erosion in terms of Minkowski subtraction. Specifically, for the digital case,

$$g^\wedge = \text{SUB}[\text{NINETY}^2(g)]$$

and

$$\text{ERODE}(f, g) = f \boxminus g^\wedge = f \boxminus \text{SUB}[\text{NINETY}^2(g)]$$

Example 6.8:

Let
$$f = (0 \quad 2 \quad 1 \quad 5 \quad 9 \quad 6 \quad 1 \quad 0)_2$$

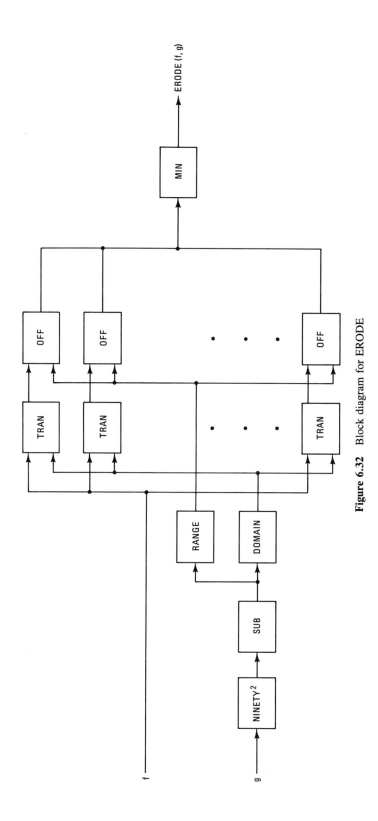

Figure 6.32 Block diagram for ERODE

Sec. 6.3   Gray-Scale Morphology for Sampled Signals   **187**

and
$$g = (5\ \ 5\ \ 4)_0$$
(See Figure 6.33.) Applying the schema of Figure 6.32, we first obtain
$$\text{RANGE}[\text{SUB}(\text{NINETY}^2(g))] = [-4, -5, -5]$$
and
$$\text{DOMAIN}[\text{SUB}(\text{NINETY}^2(g))] = [-2, -1, 0]$$

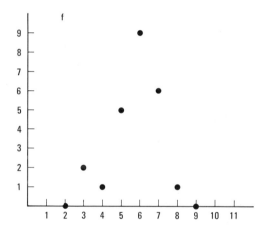

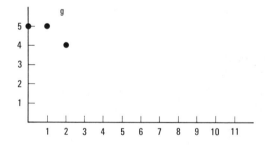

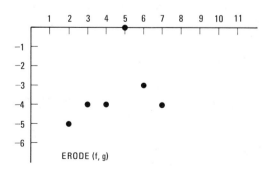

Figure 6.33 Digital erosion

Since there are only three points in the domain, we need consider only three translates of f, each of which must be offset by the appropriate constant. We obtain

$$f^{-2} = (-4 \quad -2 \quad -3 \quad 1 \quad 5 \quad 2 \quad -3 \quad -4)_0$$
$$f^{-1} = (-5 \quad -3 \quad -4 \quad 0 \quad 4 \quad 1 \quad -4 \quad -5)_1$$

and

$$f^0 = (-5 \quad -3 \quad -4 \quad 0 \quad 4 \quad 1 \quad -4 \quad -5)_2$$

where once again we have used the superscript notation to keep track of the translates. Applying MIN gives

$$\text{ERODE}(f, g) = (-5 \quad -4 \quad -4 \quad 0 \quad -3 \quad -4)_2$$

If, in the preceding example, we write the signals f^{-2}, f^{-1}, and f^0 in origin specification form, we can better see the manner in which the translations produce the erosion:

$$f^{-2} = (\boxed{-4} \quad -2 \quad -3 \quad 1 \quad 5 \quad 2 \quad -3 \quad -4 \quad * \quad *)$$
$$f^{-1} = (\boxed{*} \quad -5 \quad -3 \quad -4 \quad 0 \quad 4 \quad 1 \quad -4 \quad -5 \quad *)$$
$$f^0 = (\boxed{*} \quad * \quad -5 \quad -3 \quad -4 \quad 0 \quad 4 \quad 1 \quad -4 \quad -5)$$

Reading down the columns of the partitioned matrix

$$\begin{pmatrix} f^{-2} \\ \cdots \\ f^{-1} \\ \cdots \\ f^0 \end{pmatrix}$$

gives $\text{ERODE}(f, g)$. Note the correspondence between this method of computing the erosion of gray-scale signals and the method of serial products used for the convolution of digital signals. Because of the similarity, the partitioned matrix will be called the *serial matrix for erosion*. It is made into a bound matrix proper by giving it the subscripts r and t, where r is the leftmost activated point in the domain of the signal in row one, and t is one less than the number of rows, including rows of stars. The net effect is that the lower left pixel in the serial matrix is situated at $(r, 0)$. In Example 6.8, $r = 0$ and $t = 2$.

Notice the eroding effect in Example 6.8: there has been a diminution of the signal domain, and the original three-point plateau of increased amplitude has been reduced to a single point. Moreover, since the structuring element is not symmetric about the origin, the diminution is not symmetric; indeed, the point $i = 6$ was originally the center of the region of increased amplitude, whereas in the output there is increased amplitude only at $i = 5$.

As discussed in Section 6.1, the gray-scale erosion is defined directly in terms of fitting. In the case of sampled signals,

$$[\text{ERODE}(f, g)](i) = \max \{j: g_i + j \leq f\}$$
$$= \max \{j: \text{OFF}[\text{TRAN}(g; i); j] \leq f\}$$

Sec. 6.3   Gray-Scale Morphology for Sampled Signals   **189**

the latter expression being in terms of image operators. Though we could give a machine implementation in this form, we shall not do so. Suffice it to say that implementation in terms of Minkowski subtraction is of the same order of computational complexity as convolution.

At this point, a brief comment on notation is apropos. Although it would perhaps be best to be completely consistent and utilize procedural (acronymic) notation at all times, such a course would result in unwieldy expressions. Consequently, we shall mix notation. In those instances in which we are particularly concerned with algorithm specification, however, we shall employ procedural notation fully.

Example 6.9 illustrates the fitting method just discussed.

Example 6.9:

> Let f and g be as in Example 6.8. To find the erosion of f by g using the fitting criterion, we must obtain all translates of g to points in the set
>
> $$\{2, 3, 4, 5, 6, 7\}$$
>
> and then, for each translate, find the maximum j for which the sum of the translated image with j lies underneath f. Using Euclidean notation, the largest value of j for which $g_2 + j$ lies underneath f is -5, for which $g_3 + j$ lies underneath f is -4, for which $g_4 + j$ lies underneath f is -4, for which $g_5 + j$ lies underneath f is 0, for which $g_6 + j$ lies underneath f is -3, and for which $g_7 + j$ lies underneath f is -4. No other translates of g fit underneath f, and hence, employing the fitting criterion, we obtain the output of Example 6.8. (See Figure 6.33.)

Using the specifications for dilation and erosion, we can now implement the digital opening for sampled signals. We define the opening by exact analogy with the Euclidean case:

$$\text{OPEN}(f, g) = \text{DILATE}[\text{ERODE}(f, g), g] = (f \boxminus g^{\wedge}) \boxplus g$$

The geometric interpretation in terms of fitting is analogous to the Euclidean opening for gray-scale signals.

Example 6.10:

> Let f and g be as in Example 6.8. To compute the opening, we need only dilate the output of that example by g. Using the digital version of commutativity (GM-1), we can change the roles of the input signals in Figure 6.31. Accordingly, letting $h = \text{ERODE}(f, g)$, we obtain
>
> $$\text{RANGE}(g) = [5, 5, 4]$$
>
> and
>
> $$\text{DOMAIN}(g) = [0, 1, 2]$$
>
> We need consider only three translates of h. Upon applying OFF to these translates and the appropriate values of g, we get
>
> $$h^0 = (0 \quad 1 \quad 1 \quad 5 \quad 2 \quad 1)_2$$
>
> $$h^1 = (0 \quad 1 \quad 1 \quad 5 \quad 2 \quad 1)_3$$
>
> and
>
> $$h^2 = (-1 \quad 0 \quad 0 \quad 4 \quad 1 \quad 0)_4$$

Applying EXTMAX to h^0, h^1, and h^2 yields

$$\text{OPEN}(f, g) = (0 \quad 1 \quad 1 \quad 5 \quad 5 \quad 4 \quad 1 \quad 0)_2$$

The original domain of f is now recovered, and the spike of height 9 at $i = 6$ has been smoothed from beneath.

As a final illustration of operations on sampled signals, we present the gray-scale closing, which in the digital case is defined in a manner paralleling the Euclidean definition, viz.,

$$\text{CLOSE}(f, g) = \text{SUB}[\text{OPEN}(\text{SUB}(f), \text{SUB}(g))]$$

or, in mixed notation,

$$\text{CLOSE}(f, g) = -\text{OPEN}(-f, -g)$$

Example 6.11:

Let f and g be as in Example 6.8. Then

$$-f = (0 \quad -2 \quad -1 \quad -5 \quad -9 \quad -6 \quad -1 \quad 0)_2$$
$$-g = (-5 \quad -5 \quad -4)_0$$

and

$$h = \text{ERODE}(-f, -g) = (3 \quad -1 \quad -5 \quad -4 \quad -4 \quad -1)_2$$

In a manner similar to the computation of the erosion, using

$$\text{DOMAIN}(-g) = [0, 1, 2]$$

we can find $\text{DILATE}(h, -g)$ by reading down the columns of the partitioned matrix

$$\begin{pmatrix} h^0 \\ \cdots \\ h^1 \\ \cdots \\ h^2 \end{pmatrix}$$

In the case of dilation, we apply EXTMAX. The matrix, called the *serial matrix for dilation*, is given by

$$\begin{pmatrix} -2 & -6 & -10 & -9 & -9 & -6 & * & * \\ * & -2 & -6 & -10 & -9 & -9 & -6 & * \\ * & * & -1 & -5 & -9 & -8 & -8 & -5 \end{pmatrix}_{2,2}$$

Hence,

$$\text{DILATE}(h, -g) = (-2 \quad -2 \quad -1 \quad -5 \quad -9 \quad -6 \quad -6 \quad -5)_2$$

and

$$\text{CLOSE}(f, g) = \text{SUB}[\text{DILATE}(h, -g)]$$
$$= (2 \quad 2 \quad 1 \quad 5 \quad 9 \quad 6 \quad 6 \quad 5)_2$$

The next example uses the digital version of Theorem 6.2 to find an opening. Like the fitting specification of erosion, it, too, could be formalized into a machine algorithm for signals that are both sampled and quantized; however, we shall not pursue the matter.

Example 6.12:

Let
$$f = (3\ \ 5\ \ 3\ \ 4\ \ 9\ \ 9\ \ 8\ \ 10\ \ 3\ \ 3\ \ 2)_1$$
and
$$g = (2\ \ 2\ \ 2)_0$$

Then, according to Theorem 6.2,

$$\text{OPEN}(f, g) = \text{EXTMAX}\{g_i + j: g_i + j \ll f\}$$

The following images satisfy the relation $g_i + j \ll f$ and are maximal in that there are no "better fits":

$$g_1 + 1 = (3\ \ 3\ \ 3)_1$$
$$g_2 + 1 = (3\ \ 3\ \ 3)_2$$
$$g_3 + 1 = (3\ \ 3\ \ 3)_3$$
$$g_4 + 2 = (4\ \ 4\ \ 4)_4$$
$$g_5 + 6 = (8\ \ 8\ \ 8)_5$$
$$g_6 + 6 = (8\ \ 8\ \ 8)_6$$
$$g_7 + 1 = (3\ \ 3\ \ 3)_7$$
$$g_8 + 1 = (3\ \ 3\ \ 3)_8$$
$$g_9 + 0 = (2\ \ 2\ \ 2)_9$$

Applying EXTMAX to $g_1 + 1$ through $g_9 + 0$ yields

$$\text{OPEN}(f, g) = (3\ \ 3\ \ 3\ \ 4\ \ 8\ \ 8\ \ 8\ \ 8\ \ 3\ \ 3\ \ 2)_1$$

Thus, the input image f has been smoothed from beneath by the structuring element g.

6.4 UMBRA MATRIX

In digital problems not only do we sample the domain, but we also quantize the range. Upon a suitable rescaling, a fully digitized (i.e., quantized and sampled) signal has a domain and range that are both subsets of Z, the set of integers. Note that both the domain and range must be finite. All of the operations discussed in the previous section, including the machine implementations of erosion, opening, and closing by fitting, apply without alteration to fully digitized signals.

If $\mathcal{S}_d$ denotes the collection of all fully digitized signals, then the umbra transform $\mathbf{U}: \mathcal{S}_d \to 2^{Z \times Z}$, the collection of subsets of $Z \times Z$, which is identified with the collection of all constant (1–*) images. Specifically, for a fully digital signal f,

$$\mathbf{U}[f] = \{(i, j): i \in D_f \text{ and } j \leq f(i)\}$$

Analogously to the Euclidean case, $\mathbf{U}[f]$ consists of all pixels lying below $\mathbf{G}[f]$, which in this case is the digital graph of f.

We shall not repeat in detail the material of Section 6.2; we simply note that the theory goes through with essentially no change. The one noteworthy difference is that in the digital case the surface of a regular umbra is always part of the umbra. Consequently, all regular umbrae are closed, and we have the following theorem (see the remarks following Theorems 6.4 and 6.5):

Theorem 6.6. If f and g are fully digital signals, then

(*i*) $\mathbf{U}[\text{DILATE}(f, g)] = \text{DILATE}(\mathbf{U}[f], \mathbf{U}[g])$
(*ii*) $\mathbf{U}[\text{ERODE}(f, g)] = \text{ERODE}(\mathbf{U}[f], \mathbf{U}[g])$
(*iii*) For any regular umbra A, $A = \mathbf{U}[\mathbf{S}(A)]$

Properties (*i*) and (*ii*) have been termed the *umbra homomorphism theorems* by Haralick.[1] Their fundamental importance will become apparent in Section 6.5.

Given a digital signal

$$f = (b_1 \quad b_2 \quad \cdots \quad b_n)_r$$

in minimal form, where it is possible that some of the b_k, except for b_1 or b_n, are stars, we let

$$f_+ = \max [b_1, b_2, \ldots, b_n]$$

and

$$f_- = \min [b_1, b_2, \ldots, b_n]$$

where any b_k that is a star is ignored in the max and min operations. Then f_+ and f_- are the maximal and minimal amplitudes, respectively, of f. We define the *graph matrix* of f, denoted by $\overline{G}[f]$, to be the bound matrix $(a_{pq})_{r, f_+}$, of dimensions $(f_+ - f_- + 1)$ by n, where $a_{pq} = 1$ if, in placing the bound matrix over the grid so that the graph of the signal lies within the matrix frame, a_{pq} lies on b_q, and $a_{pq} = *$ otherwise. The *umbra matrix* of f, $\overline{U}[f]$, is then obtained from $\overline{G}[f]$ by activating all pixels that lie beneath an activated pixel of $\overline{G}[f]$.

Since every bound matrix represents a constant image, the umbra transform operates on umbra matrices. For any signal f,

$$\mathbf{U}[\overline{U}(f)] = \mathbf{U}[\overline{G}(f)] = \mathbf{U}[f]$$

and, for the surface operator,

$$\mathbf{S}[\overline{U}(f)] = \overline{G}[f]$$

where we have, as usual, identified the graph of f, which is a subset of $Z \times Z$, with its bound matrix representation $\overline{G}[f]$.

Example 6.13:

Let
$$f = (1 \quad 3 \quad 4 \quad * \quad -1)_2$$

Then
$$\overline{G}[f] = \begin{pmatrix} * & * & 1 & * & * \\ * & 1 & * & * & * \\ * & * & * & * & * \\ 1 & * & * & * & * \\ * & * & * & * & * \\ * & * & * & * & 1 \end{pmatrix}_{2,4}$$

and
$$\overline{U}[f] = \begin{pmatrix} * & * & 1 & * & * \\ * & 1 & 1 & * & * \\ * & 1 & 1 & * & * \\ 1 & 1 & 1 & * & * \\ 1 & 1 & 1 & * & * \\ 1 & 1 & 1 & * & 1 \end{pmatrix}_{2,4}$$

In general, a constant $(1-*)$ bound matrix $(a_{pq})_{rt}$ is called an *umbra matrix* if, for any $a_{p',q} = 1$, $a_{p,q} = 1$ for all $p \geq p'$. In words, once there is an activated pixel in a column, all pixels in the bound matrix frame beneath that activated pixel must also be activated. An umbra matrix is said to be *minimal* if there exists at least one column with exactly one activated pixel in it. Two umbra matrices are said to be *equivalent* if, when every column with a 1 in it is extended infinitely downward, both matrices result in the same umbra. Note that, by definition, the umbra matrix of f is given in minimal form. However, under the equivalence condition just given, it could be replaced by an equivalent umbra matrix. In what follows, we shall not differentiate between two equivalent umbra matrices, since they represent the same umbra. In simple terms, the umbra matrix is a device to represent an umbra, which is necessarily infinite, by a finite data structure.

Every digital signal with finite domain possesses a unique umbra matrix (where we take equivalent umbra matrices to be identical); and conversely, each umbra matrix corresponds to a unique digital signal. In terms of the umbra transform, $\mathbf{U}: f \to \overline{U}[f]$. Moreover, if F is an umbra matrix, $\mathbf{S}: F \to f$, where $\mathbf{S} = \mathbf{U}^{-1}$ is the surface operator and we identify f with its graph.

Example 6.14:

Consider the umbra matrix
$$F = \begin{pmatrix} * & * & 1 & 1 \\ * & * & 1 & 1 \\ 1 & * & 1 & 1 \\ 1 & * & 1 & 1 \end{pmatrix}_{2,8}$$

Then, in minimal form,
$$F = \begin{pmatrix} * & * & 1 & 1 \\ * & * & 1 & 1 \\ 1 & * & 1 & 1 \end{pmatrix}_{2,8}$$

and

$$S[F] = \begin{pmatrix} * & * & 1 & 1 \\ * & * & * & * \\ 1 & * & * & * \end{pmatrix}_{2,8} = \overline{G}[f]$$

where

$$f = (6 \ * \ 8 \ 8)_2$$

According to the digital version of Theorem 6.4,

$$f \boxplus g = S[U[f] \boxplus U[g]]$$

or, more precisely,

$$G[f \boxplus g] = S[U[f] \boxplus U[g]]$$

However, we do not distinguish between a signal and its graph. Moreover, according to Theorem 6.5,

$$\text{ERODE}(f, g) = f \boxminus g^\wedge = S[U[f] \boxminus (-U[g])]$$

The next example shows how the characterizations of signal dilation and erosion in terms of corresponding bound matrix operations on umbrae can be interpreted and implemented through the use of umbra matrices.

Example 6.15:

Let
$$f = (0 \ 2 \ 1 \ 5 \ 9 \ 6 \ 1 \ 0)_2$$

and
$$h = (2 \ 2 \ 1)_0$$

The umbra matrices of f and h are given in Figure 6.34, along with minimal umbra matrix representations of $\text{DILATE}(\overline{U}[f], \overline{U}[h])$, $\text{ERODE}(\overline{U}[f], \overline{U}[h])$, and $\text{OPEN}(\overline{U}[f], \overline{U}[h])$.

Since there are two rows in the minimal bound matrix for $\overline{U}[h]$, it is necessary to augment the minimal umbra matrix for $\overline{U}[f]$ by one row on the bottom before employing the fitting methodologies for erosion and opening of constant images. In general, if there are m rows in the minimal umbra matrix of the structuring element, then we must augment the minimal umbra matrix by $m - 1$ rows on the bottom before fitting the structuring element. This augmentation reflects the fact that we are really concerned with the umbra itself, not just a particular representation of the umbra.

Applying the surface operator, and identifying the graph matrix of a signal with the signal itself, we have

$$\text{DILATE}(f, h) = S[\text{DILATE}(\overline{U}[f], \overline{U}[h])]$$

$$= (2 \ 4 \ 4 \ 7 \ 11 \ 11 \ 10 \ 7 \ 2 \ 1)_2$$

Sec. 6.4 Umbra Matrix

$$\overline{U}[f] = \begin{pmatrix} * & * & * & * & * & * & 1 & * & * & * \\ * & * & * & * & * & * & 1 & * & * & * \\ * & * & * & * & * & * & 1 & * & * & * \\ * & * & * & * & * & * & 1 & 1 & * & * \\ * & * & * & * & * & 1 & 1 & 1 & * & * \\ * & * & * & * & * & 1 & 1 & 1 & * & * \\ * & * & * & * & * & 1 & 1 & 1 & * & * \\ * & * & * & 1 & * & 1 & 1 & 1 & * & * \\ * & * & * & 1 & 1 & 1 & 1 & 1 & * \\ \circledast & * & 1 & 1 & 1 & 1 & 1 & 1 & 1 & 1 \end{pmatrix}$$

$$\overline{U}[h] = \begin{pmatrix} 1 & 1 & * \\ 1 & 1 & 1 \\ \circledast & * & * \end{pmatrix}$$

$$\text{DILATE}(\overline{U}[f], \overline{U}[h]) = \begin{pmatrix} * & * & * & * & * & * & 1 & 1 & * & * & * & * \\ * & * & * & * & * & * & 1 & 1 & 1 & * & * & * \\ * & * & * & * & * & * & 1 & 1 & 1 & * & * & * \\ * & * & * & * & * & * & 1 & 1 & 1 & * & * & * \\ * & * & * & * & * & 1 & 1 & 1 & 1 & 1 & * & * \\ * & * & * & * & * & 1 & 1 & 1 & 1 & 1 & * & * \\ * & * & * & * & * & 1 & 1 & 1 & 1 & 1 & * & * \\ * & * & * & 1 & 1 & 1 & 1 & 1 & 1 & 1 & * & * \\ * & * & * & 1 & 1 & 1 & 1 & 1 & 1 & 1 & * & * \\ * & * & 1 & 1 & 1 & 1 & 1 & 1 & 1 & 1 & 1 & * \\ * & * & 1 & 1 & 1 & 1 & 1 & 1 & 1 & 1 & 1 & 1 \\ \circledast & * & * & * & * & * & * & * & * & * & * & * \end{pmatrix}$$

$$\text{ERODE}(\overline{U}[f], \overline{U}[h]) = \begin{pmatrix} * & * & * & * & * & 1 & * & * \\ * & * & * & * & * & 1 & * & * \\ * & * & * & * & * & 1 & * & * \\ \circledast & * & * & * & * & 1 & 1 & * \\ * & * & * & 1 & 1 & 1 & 1 & 1 \\ * & * & 1 & 1 & 1 & 1 & 1 & 1 \end{pmatrix}$$

$$\text{OPEN}(\overline{U}[f], \overline{U}[h]) = \begin{pmatrix} * & * & * & * & * & 1 & 1 & * & * & * \\ * & * & * & * & * & 1 & 1 & 1 & * & * \\ * & * & * & * & * & 1 & 1 & 1 & * & * \\ * & * & * & * & * & 1 & 1 & 1 & * & * \\ * & * & * & 1 & 1 & 1 & 1 & 1 & 1 & * \\ \circledast & * & 1 & 1 & 1 & 1 & 1 & 1 & 1 & 1 \end{pmatrix}$$

Figure 6.34 Umbra matrices for Example 6.15

$$\text{ERODE}(f, h) = \mathbf{S}[\text{ERODE}(\overline{U}[f], \overline{U}[h])]$$
$$= (-2 \quad -1 \quad -1 \quad 3 \quad 0 \quad -1)_2$$

and

$$\text{OPEN}(f, h) = \mathbf{S}[\text{OPEN}(\overline{U}[f], \overline{U}[h])]$$
$$= (0 \quad 1 \quad 1 \quad 5 \quad 5 \quad 4 \quad 1 \quad 0)_2$$

From Example 6.8, we have

$$\text{ERODE}(f, g) = (-5 \quad -4 \quad -4 \quad 0 \quad -3 \quad -4)_2$$

where

$$g = (5 \quad 5 \quad 4)_0$$

Hence,

$$\text{ERODE}(f, h) = \text{OFF}[\text{ERODE}(f, g); 3]$$

This agrees with property GM-6 of Section 6.5, which states that

$$\text{ERODE}[f, \text{OFF}(g; -t)] = \text{OFF}[\text{ERODE}(f, g); t]$$

In Example 6.10, we found the opening of f by g. The present result agrees with that earlier result. Indeed, it is an immediate consequence of Theorem 6.2 that

$$\text{OPEN}[f, \text{OFF}(g; t)] = \text{OPEN}[f, g]$$

If we attempt to apply the Minkowski subtraction form of Theorem 6.5 to umbra matrices using MIN instead of INF, we encounter a problem because $-\mathbf{U}[g]$ is not an umbra. Nevertheless, a MIN specification can still be given utilizing umbra matrices. Consider the MIN specification of signal erosion:

$$\text{ERODE}(f, g) = f \boxminus g^{\wedge} = \text{MIN}_{i \in D_g \wedge}[f_i + g^{\wedge}(i)]$$

From this expression, it can be seen that, in terms of umbrae, we need to translate the umbra of f by each element in the graph of $g^{\wedge}$ and then apply MIN. In terms of bound matrix operations, we have

$$\overline{U}[\text{ERODE}(f, g)] = \bigwedge_{(i,j) \in \text{DOMAIN}(\overline{G}[g^{\wedge}])} \text{TRAN}(\overline{U}[f]; i, j)$$

where the following stipulation is made regarding the particular representative of $\overline{U}[f]$ that is to be employed in a given term of the MIN operation: if j_0 is the minimum j for which there exists a pixel (i, j) in the domain of $\overline{G}[g^{\wedge}]$, then, for each translation (i, j), augment the minimal umbra matrix of $\text{TRAN}(\overline{U}[f]; i, j)$ by $j - j_0$ rows of ones. This stipulation is necessary because we want the MIN operation to take into account the fact that the actual umbra has all 1's beneath any bound matrix representing it.

Example 6.16:

Let

$$k = (-1 \quad 3 \quad 2 \quad 0 \quad 1)_1$$

Sec. 6.4 Umbra Matrix

and
$$h = (1 \quad 1 \quad -1)_0$$

Then
$$h^\wedge = (1 \quad -1 \quad -1)_{-2}$$

The umbra matrix $\overline{U}[k]$ and the graph matrix $\overline{G}[h^\wedge]$ are given in Figure 6.35. It is easily seen that

$$\text{DOMAIN}(\overline{G}[h^\wedge]) = [(-2, 1), (-1, -1), (0, -1)]$$

The minimal value of j in the domain is $j_0 = -1$. Hence, the translate $\text{TRAN}(\overline{U}[k]; -2, 1)$ is augmented by two rows of ones. The three translates required for MIN are shown in Figure 6.35, together with the umbra matrix for the erosion. Note that the final output, $\overline{U}[\text{ERODE}(k, h)]$, agrees with the output we would have obtained by eroding $\overline{U}[k]$ by $\overline{U}[h]$ using the fitting technique.

$$\overline{G}[h^\wedge] = \begin{pmatrix} 1 & * & * \\ * & * & \circledast \\ * & 1 & 1 \end{pmatrix}$$

$$\overline{U}[k] = \begin{pmatrix} * & * & 1 & * & * & * \\ * & * & 1 & 1 & * & * \\ * & * & 1 & 1 & * & 1 \\ \circledast & * & 1 & 1 & 1 & 1 \\ * & 1 & 1 & 1 & 1 & 1 \end{pmatrix}$$

$$\text{TRAN}(\overline{U}[k]; -2, 1) = \begin{pmatrix} * & 1 & * & * & * \\ * & 1 & 1 & * & * \\ * & 1 & 1 & * & 1 \\ * & 1 & 1 & 1 & 1 \\ 1 & \textcircled{1} & 1 & 1 & 1 \\ 1 & 1 & 1 & 1 & 1 \\ 1 & 1 & 1 & 1 & 1 \end{pmatrix}$$

$$\text{TRAN}(\overline{U}[k]; -1, -1) = \begin{pmatrix} * & * & 1 & * & * & * \\ * & * & 1 & 1 & * & * \\ * & \circledast & 1 & 1 & * & 1 \\ * & * & 1 & 1 & 1 & 1 \\ * & 1 & 1 & 1 & 1 & 1 \end{pmatrix}$$

$$\text{TRAN}(\overline{U}[k]; 0, -1) = \begin{pmatrix} * & * & * & 1 & * & * & * \\ * & * & * & 1 & 1 & * & * \\ * & \circledast & * & 1 & 1 & * & 1 \\ * & * & * & 1 & 1 & 1 & 1 \\ * & * & 1 & 1 & 1 & 1 & 1 \end{pmatrix}$$

$$\overline{U}[\text{ERODE}(k, h)] = \begin{pmatrix} * & * & 1 & * \\ \circledast & * & 1 & * \\ * & * & 1 & 1 \\ * & 1 & 1 & 1 \end{pmatrix}$$

Figure 6.35 Umbra matrices for Example 6.16

Although the notation in the preceding example is cumbersome due to the need to augment umbra matrices, the actual manipulations are quite straightforward once one has gained some experience working with them. Moreover, they help to make clear the more theoretical aspects of the umbra transform.

In terms of digital morphological algebra, umbra matrices demonstrate why the morphological basis is, in essence, not altered when we proceed to operate on gray-scale signals (or images): while it is true that we require other operators for direct gray-scale specification, the umbra transform makes it transparent that these operators merely correspond to TRAN, MIN, and EXTMAX (and perhaps some rotations or flips) in the umbra transform world, the world of constant images.

6.5 ALGEBRAIC PROPERTIES

In this section we present the fundamental algebraic properties relating to the gray-scale morphological operations on digital (quantized and sampled) signals. We restrict attention to the digital case for four reasons: (1) Computer implementation is the main goal of electronic image processing. (2) Theorem 6.6 applies. (3) There is no need to obscure the presentation with mathematical subtleties concerning suprema. (4) Umbra matrices can be employed to provide pragmatic and easily understood examples.

Many properties of this section correspond directly to properties given in Chapters 1 and 2; indeed, the proofs to be given demonstrate the manner in which the gray-scale properties are related to their 1–∗ counterparts by the umbra transform. Those with little interest in theory should at least skim the proofs in order to familiarize themselves with the use of the umbra transform. It will become obvious at once that most of the theory proceeds thus: the property to be proved is thrown back upon the constant-image case by use of the umbra transform, with the digital forms of Theorems 6.4 and 6.5 playing key roles. Ideas similar to some of those presented herein have been given by Haralick, Sternberg, and Zhuang.[2]

The first two properties have been stated in Section 6.1 for general Euclidean signals; we restate them here for completeness and also to give proofs that utilize the umbra transform. (Variables f, g, and h denote signals throughout this section.)

Property GM-1. $f \boxplus g = g \boxplus f$ (commutativity)

Proof. $f \boxplus g = S(U[f] \boxplus U[g]) = S(U[g] \boxplus U[f]) = f \boxplus g$

Property GM-2. $f \boxplus (g \boxplus h) = (f \boxplus g) \boxplus h$ (associativity)

Proof.
$f \boxplus (g \boxplus h) = S(U[f] \boxplus U[g \boxplus h])$
$= S(U[f] \boxplus (U[g] \boxplus U[h]))$
$= S((U[f] \boxplus U[g]) \boxplus U[h])$
$= S(U[f \boxplus g] \boxplus U[h])$
$= (f \boxplus g) \boxplus h$

Sec. 6.5 Algebraic Properties

To obtain the equalities in the proof of GM-2, we used, in order, Theorem 6.4, the umbra homomorphism, property M-2 from Table 2.1, the umbra homomorphism, and Theorem 6.4.

Property GM-3. $f \boxplus \text{TRAN}(g; i) = \text{TRAN}(f \boxplus g; i)$

Proof.
$$\begin{aligned}
f \boxplus \text{TRAN}(g; i) &= S(U[f] \boxplus U[\text{TRAN}(g; i)]) \\
&= S(U[f] \boxplus \text{TRAN}(U[g]; i, 0)) \\
&= S(\text{TRAN}(U[f] \boxplus U[g]; i, 0)) \\
&= S(\text{TRAN}(U[f \boxplus g]; i, 0)) \\
&= \text{TRAN}(S(U[f \boxplus g]); i, 0) \\
&= \text{TRAN}(f \boxplus g; i)
\end{aligned}$$

We have used the fact that a translation of a signal by i results in a translation of the umbra by $(i, 0)$ and conversely, the converse being applied in the last equality. Note the use of M-3 in the third equality.

Property GM-4. $f \boxplus \text{OFF}(g; j) = \text{OFF}(f \boxplus g; j)$

Proof.
$$\begin{aligned}
f \boxplus \text{OFF}(g; j) &= S(U[f] \boxplus U[\text{OFF}(g; j)]) \\
&= S(U[f] \boxplus \text{TRAN}(U[g]; 0, j)) \\
&= S(\text{TRAN}(U[f] \boxplus U[g]; 0, j)) \\
&= S(\text{TRAN}(U[f \boxplus g]; 0, j)) \\
&= \text{TRAN}(S(U[f \boxplus g]); 0, j) \\
&= \text{OFF}(f \boxplus g; j)
\end{aligned}$$

Note that offsetting a signal by j results in translating the umbra by $(0, j)$ and conversely. As in the proof of GM-3, M-3 was used in the third equality.

Example 6.17:

This example, which illustrates some of the notions used in the proof of Property GM-3, demonstrates how the translation of a signal results in a corresponding translation of the umbra matrix. It also helps elucidate the translation compatibility of GM-3. A similar example could be constructed for OFF in Property GM-4.

Let

$$f = (2 \quad 1 \quad 5 \quad 1 \quad 1)_1$$

and

$$g = (0 \quad 2 \quad 0)_0$$

Then

$$\text{TRAN}(g; -2) = (0 \ \ 2 \ \ 0)_{-2}$$

$$\overline{U}[f] = \begin{pmatrix} * & * & * & 1 & * & * \\ * & * & * & 1 & * & * \\ * & * & * & 1 & * & * \\ * & 1 & * & 1 & * & * \\ * & 1 & 1 & 1 & 1 & 1 \\ \circledast & * & * & * & * & * \end{pmatrix}$$

$$\overline{U}[g] = \begin{pmatrix} * & 1 & * \\ * & 1 & * \\ \textcircled{1} & 1 & 1 \end{pmatrix}$$

$$\text{TRAN}(\overline{U}[g]; -2, 0) = \begin{pmatrix} * & 1 & * \\ * & 1 & * \\ 1 & 1 & \textcircled{1} \end{pmatrix}$$

and

$$\overline{U}[f] \boxplus \text{TRAN}(\overline{U}[g]; -2, 0) = \begin{pmatrix} * & * & * & 1 & * & * & * \\ * & * & * & 1 & * & * & * \\ * & * & 1 & 1 & 1 & * & * \\ * & 1 & 1 & 1 & 1 & * & * \\ * & 1 & 1 & 1 & 1 & 1 & * \\ 1 & 1 & 1 & 1 & 1 & 1 & * \\ 1 & 1 & 1 & 1 & 1 & 1 & 1 \\ * & \circledast & * & * & * & * & * \end{pmatrix}$$

Applying the surface operator to the preceding umbra matrix yields the signal

$$h = (2 \ \ 4 \ \ 5 \ \ 7 \ \ 5 \ \ 3 \ \ 1)_{-1}$$

Finally, direct dilation of f by g yields

$$f \boxplus g = (2 \ \ 4 \ \ 5 \ \ 7 \ \ 5 \ \ 3 \ \ 1)_1$$

and $h = \text{TRAN}(f \boxplus g; -2)$.

The next two properties give translation invariance for erosion. Their proofs are similar to those for dilation; however, Theorem 6.5 is employed instead of Theorem 6.4, and we make use of the umbra homomorphism for digital erosion.

Property GM-5. $\text{ERODE}[f, \text{TRAN}(g; i)] = \text{TRAN}[\text{ERODE}(f, g); -i]$

Proof. $\text{ERODE}(f, \text{TRAN}(g; i))$

$$= \mathbf{S}(\text{ERODE}(\mathbf{U}[f], \mathbf{U}[\text{TRAN}(g; i)]))$$

$$= \mathbf{S}(\text{ERODE}(\mathbf{U}[f], \text{TRAN}(\mathbf{U}[g]; i, 0))$$

$$= \mathbf{S}(\text{TRAN}(\text{ERODE}(\mathbf{U}[f], \mathbf{U}[g]); -i, 0))$$

Sec. 6.5 Algebraic Properties

$$= S(TRAN(U[ERODE(f, g)]; -i, 0))$$
$$= TRAN(S(U[ERODE(f, g)]); -i, 0)$$
$$= TRAN(ERODE(f, g); -i)$$

The proof is similar to that for GM-4 except for the sign change in the third equality, which uses the identity

$$ERODE[A, TRAN(B; i, j)] = TRAN[ERODE(A, B); -i -j]$$

shown for constant Euclidean images in Section 1.3.

Example 6.18:

Let f and g be as in Exercise 6.17. Then

$$TRAN(g; 2) = (0 \quad 2 \quad 0)_2 = (\circledast \quad * \quad 0 \quad 2 \quad 0)$$

and

$$ERODE[f, TRAN(g; 2)] = (-1 \quad 1 \quad -1)_{-1}$$

while

$$ERODE(f, g) = (-1 \quad 1 \quad -1)_1$$

Property GM-6. $ERODE[f, OFF(g; j)] = OFF[ERODE(f, g); -j]$

Proof. The proof, which is similar to that of GM-5, is left to the interested reader.

Example 6.19.

Again, let f and g be as in Example 6.17. Then

$$OFF(g; -3) = (-3 \quad -1 \quad -3)_0$$

and

$$ERODE[f; OFF(g; -3)] = (2 \quad 4 \quad 2)_0$$
$$= OFF[ERODE(f, g); 3]$$

The next three properties concern monotonicity and follow directly from the definitions of dilation and erosion given in Section 6.1. Nevertheless, for pedagogic purposes, we shall prove GM-7 by using the umbra transform.

Property GM-7. If $f_1 \ll f_2$, then $DILATE(f_1, g) \ll DILATE(f_2, g)$

Proof. Since $f_1 \ll f_2$, the domain of f_1 is a subset of the domain of f_2. Moreover, on the intersection of their domains, $f_1 \leq f_2$. Consequently, $U[f_1]$ is a subset of $U[f_2]$. Applying M-9 from Table 2.1 yields

$$DILATE(f_1, g) = S(DILATE(U[f_1], U[g])) \ll S(DILATE(U[f_2], U[g]))$$
$$= DILATE(f_2, g)$$

Property GM-8. If $f_1 \ll f_2$, then ERODE$(f_1, g) \ll$ ERODE(f_2, g)

Property GM-9. If $g_1 \ll g_2$, then ERODE$(f, g_2) \ll$ ERODE(f, g_1)

The proof of the next property is somewhat tricky. It uses M-18 of Table 2.1 together with Theorem 6.6(*iii*).

Property GM-10. ERODE[ERODE$(f, g), h$] = ERODE[f, DILATE(g, h)]

Proof. ERODE(ERODE$(f, g), h$)

$\quad$ = ERODE(**S**(ERODE(**U**[f], **U**[g])), h)

$\quad$ = **S**(ERODE(**U**[**S**(ERODE(**U**[f], **U**[g]))], **U**[h]))

$\quad$ = **S**(ERODE(ERODE(**U**[f], **U**[g]), **U**[h]))

$\quad$ = **S**(ERODE(**U**[f], DILATE(**U**[g], **U**[h])))

$\quad$ = **S**(ERODE(**U**[f], **U**[**S**(DILATE(**U**[g], **U**[h]))]))

$\quad$ = **S**(ERODE(**U**[f], **U**[DILATE(g, h)]))

$\quad$ = ERODE$(f$, DILATE$(g, h))$

The next property is the duality between gray-scale Minkowski addition and gray-scale Minkowski subtraction. The proof utilizes the familiar relationship between the maximum and minimum of a finite collection of functions that possess the same domain:

$$\max [f_k] = -\min [-f_k]$$

In the morphological situation, however, a bit of subtlety is introduced due to the use of EXTMAX and MIN on functions that do not possess the same domain. The effect of these two operators, which take into account functions with varying domains, can be seen quite clearly by examining the relevant serial matrices. This is done in Example 6.20.

Property GM-11. On the domain of $f \boxminus g$,

$$f \boxplus g = -[(-f) \boxminus (-g)]$$

Proof. $f \boxplus g = \text{EXTMAX}_{i \in D_g}[f_i + g(i)]$

$\quad = -\text{MIN}_{i \in D_g}[-f_i - g(i)] \quad$ (on $D_{(-f) \boxminus (-g)} = D_{f \boxminus g}$)

$\quad = -\text{MIN}_{i \in D_{-g}}[-f_i + (-g(i))]$

$\quad = -[(-f) \boxminus (-g)]$

Example 6.20:

Let

$$f = (2 \ 3 \ 2 \ 2 \ 8 \ 9 \ 8 \ 9)_0$$

Sec. 6.5 Algebraic Properties

and

$$g = (0 \quad 5 \quad 2 \quad 2)_{-1}$$

Then

$$f \boxed{+} g = (2 \quad 7 \quad 8 \quad 7 \quad 8 \quad 13 \quad 14 \quad 13 \quad 14 \quad 11 \quad 11)_{-1}$$
$$-g = (0 \quad -5 \quad -2 \quad -2)_{-1}$$
$$-f = (-2 \quad -3 \quad -2 \quad -2 \quad -8 \quad -9 \quad -8 \quad -9)_0$$

and

$$(-f) \boxed{-} (-g) = (-7 \quad -8 \quad -13 \quad -14 \quad -13)_2$$

But the last is precisely $-(f \boxed{+} g)$ restricted to the domain of $(-f) \boxed{-} (-g)$, which domain is exactly the same as the domain of $f \boxed{-} g$.

The essence of the duality relation can be seen by examining the serial matrices for $f \boxed{+} g$ and $(-f) \boxed{-} (-g)$. These are respectively given by

$$H = \begin{pmatrix} 2 & 3 & 2 & 2 & 8 & 9 & 8 & 9 & * & * & * \\ * & 7 & 8 & 7 & 7 & 13 & 14 & 13 & 14 & * & * \\ * & * & 4 & 5 & 4 & 4 & 10 & 11 & 10 & 11 & * \\ * & * & * & 4 & 5 & 4 & 4 & 10 & 11 & 10 & 11 \end{pmatrix}_{-1,3}$$

and $-H$. Letting $H = (c_{pq})$ gives $-H = (-c_{pq})$. The points of the domain of $f \boxed{-} g$ are precisely the i values of those columns of H which do not contain stars, that is, $2 \leq i \leq 6$. For any such i,

$$[f \boxed{+} g](i) = \max [c_{pq}: 1 \leq p \leq 4]$$
$$= -\min [-c_{pq}: 1 \leq p \leq 4]$$
$$= -[(-f) \boxed{-} (-g)](i)$$

In terms of dilation and erosion, GM-11 takes the form

$$\text{DILATE}(f, g) = \text{SUB}[\text{ERODE}(-f, -g^{\wedge})]$$

on the domain of $\text{ERODE}(f, g^{\wedge})$. Figure 6.36 gives the expression for DILATE in terms of ERODE in full operator form. Note that $-g^{\wedge} = \text{NINETY}^2(g)$

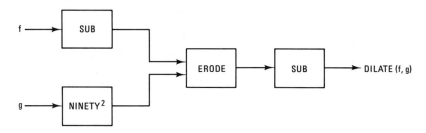

Figure 6.36 DILATE in terms of ERODE

The next three properties are the gray-scale versions of M-15, M-16, and M-17, respectively. The proofs make use of the following elementary properties of the umbra and surface transforms, insofar as these transforms relate to EXTMAX and MIN:

$$U[\text{EXTMAX}(f, g)] = \text{EXTMAX}(U[f], U[g]) \qquad (3)$$

$$U[\text{MIN}(f, g)] = \text{MIN}(U[f], U[g]) \qquad (4)$$

$$S[\text{EXTMAX}(U[f], U[g])] = \text{EXTMAX}(f, g) \qquad (5)$$

$$S[\text{MIN}(U[f], U[g])] = \text{MIN}(f, g) \qquad (6)$$

We employ the infix notation $\vee$ and $\wedge$ in the proofs for notational simplification.

Property GM-12. $f \boxplus \text{EXTMAX}(g, h) = \text{EXTMAX}(f \boxplus g, f \boxplus h)$

Proof.
$$\begin{aligned}
f \boxplus \text{EXTMAX}(g, h) &= S(U[f] \boxplus U[g \vee h]) \\
&= S(U[f] \boxplus (U[g] \vee U[h])) \\
&= S((U[f] \boxplus U[g]) \vee (U[f] \boxplus U[h])) \\
&= S(U[f \boxplus g] \vee U[f \boxplus h]) \\
&= S(U[f \boxplus g]) \vee S(U[f \boxplus h]) \\
&= \text{EXTMAX}(f \boxplus g, f \boxplus h)
\end{aligned}$$

Property GM-13.

$$\text{ERODE}[f, \text{EXTMAX}(g, h)] = \text{MIN}[\text{ERODE}(f, g), \text{ERODE}(f, h)]$$

Proof.

$$\begin{aligned}
\text{ERODE}(f, g \vee h) &= S(\text{ERODE}(U[f], U[g \vee h])) \\
&= S(\text{ERODE}(U[f], U[g] \vee U[h])) \\
&= S(\text{ERODE}(U[f], U[g]) \wedge \text{ERODE}(U[f], U[h])) \\
&= S(U[\text{ERODE}(f, g)] \wedge U[\text{ERODE}(f, h)]) \\
&= \text{MIN}(\text{ERODE}(f, g), \text{ERODE}(f, h))
\end{aligned}$$

Property GM-14.

$$\text{ERODE}[\text{MIN}(f, g), h] = \text{MIN}[\text{ERODE}(f, h), \text{ERODE}(g, h)]$$

Proof.

$$\begin{aligned}
\text{ERODE}(f \wedge g, h) &= S(\text{ERODE}(U[f \wedge g], U[h])) \\
&= S(\text{ERODE}(U[f] \wedge U[g], U[h])) \\
&= S(\text{ERODE}(U[f], U[g]) \wedge \text{ERODE}(U[f], U[h])) \\
&= S(U[\text{ERODE}(f, g)] \wedge U[\text{ERODE}(f, h)]) \\
&= \text{MIN}(\text{ERODE}(f, g), \text{ERODE}(f, h))
\end{aligned}$$

Example 6.21:

We illustrate properties (3) through (6) stated prior to GM-12 using

$$f = (2 \quad 3 \quad -1 \quad 1)_1$$

and

$$g = (2 \quad 1 \quad 1 \quad 1)_2$$

and their respective umbra matrices

$$\overline{U}[f] = \begin{pmatrix} * & * & 1 & * & * \\ * & 1 & 1 & * & * \\ * & 1 & 1 & * & 1 \\ \circledast & 1 & 1 & * & 1 \\ * & 1 & 1 & 1 & 1 \end{pmatrix}$$

and

$$\overline{U}[g] = \begin{pmatrix} * & * & 1 & * & * & * \\ * & * & 1 & 1 & 1 & 1 \\ \circledast & * & 1 & 1 & 1 & 1 \\ * & * & 1 & 1 & 1 & 1 \end{pmatrix}$$

(Notice how $\overline{U}[g]$ has been augmented so that it possesses the same number of rows beneath the origin as does $\overline{U}[f]$.) We obtain

$$\text{EXTMAX}(\overline{U}[f], \overline{U}[g]) = \begin{pmatrix} * & * & 1 & * & * & * \\ * & 1 & 1 & * & * & * \\ * & 1 & 1 & 1 & 1 & 1 \\ \circledast & 1 & 1 & 1 & 1 & 1 \\ * & 1 & 1 & 1 & 1 & 1 \end{pmatrix}$$

and

$$\text{MIN}(\overline{U}[f], \overline{U}[g]) = \begin{pmatrix} * & * & 1 & * & * \\ * & * & 1 & * & 1 \\ \circledast & * & 1 & * & 1 \\ * & * & 1 & 1 & 1 \end{pmatrix}$$

Also,

$$\text{EXTMAX}(f, g) = (2 \quad 3 \quad 1 \quad 1 \quad 1)_1$$

and

$$\text{MIN}(f, g) = (2 \quad -1 \quad 1)_2$$

The umbra matrices of these two signals are identical to the respective umbra matrices already computed. Moreover, the surfaces of the latter umbra matrices are identical to EXTMAX(f, g) and MIN(f, g), respectively.

Example 6.22:

Let

$$f = (4 \quad 3 \quad 4 \quad 8 \quad 1 \quad 0 \quad 1)_0$$
$$g = (1 \quad 1 \quad 1 \quad 8 \quad 9 \quad 8)_0$$

and
$$k = (0\ \ 4\ \ 4)_1$$
Then
$$\text{MIN}(f, g) = (1\ \ 1\ \ 1\ \ 8\ \ 1\ \ 0)_0$$
$$\text{ERODE}[\text{MIN}(f, g), k] = (-3\ \ -3\ \ -3\ \ -4)_{-1}$$
$$\text{ERODE}(f, k) = (-1\ \ 0\ \ -3\ \ -4\ \ -4)_{-1}$$
$$\text{ERODE}(g, k) = (-3\ \ -3\ \ 1\ \ 4)_{-1}$$
and GM-14 holds since
$$\text{MIN}[\text{ERODE}(f, k), \text{ERODE}(g, k)] = (-3\ \ -3\ \ -3\ \ -4)_{-1}$$
$$= \text{ERODE}[\text{MIN} f, g), k]$$

We conclude this section with some manipulations involving the digital gray-scale opening and closing.

Example 6.23:

Let
$$f = (2\ \ 2\ \ 9\ \ 9\ \ 9\ \ 2\ \ 2\ \ 2)_0$$
and
$$g = (0\ \ 1\ \ 2\ \ 1\ \ 0)_0$$
Using the fitting criterion for the opening, we apply EXTMAX to
$$g_0 + 1 = (1\ \ 2\ \ 3\ \ 2\ \ 1)_0$$
$$g_1 + 2 = (2\ \ 3\ \ 4\ \ 3\ \ 2)_1$$
$$g_2 + 1 = (1\ \ 2\ \ 3\ \ 2\ \ 1)_2$$
and
$$g_3 + 0 = (0\ \ 1\ \ 2\ \ 1\ \ 0)_3$$
to obtain
$$\text{OPEN}(f, g) = (1\ \ 2\ \ 3\ \ 4\ \ 3\ \ 2\ \ 1\ \ 0)_0$$
On the other hand,
$$\text{ERODE}(f, g) = (1\ \ 2\ \ 1\ \ 0)_0$$
Hence, the serial matrix of the dilation of $\text{ERODE}(f, g)$ by g is given by

$$\begin{pmatrix} 1 & 2 & 1 & 0 & * & * & * & * \\ * & 2 & 3 & 2 & 1 & * & * & * \\ * & * & 3 & 4 & 3 & 2 & * & * \\ * & * & * & 2 & 3 & 2 & 1 & * \\ * & * & * & * & 1 & 2 & 1 & 0 \end{pmatrix}_{0,4}$$

Applying EXTMAX down the columns gives $\text{OPEN}(f, g)$, and the two methods are in agreement.

Sec. 6.5 Algebraic Properties

By now it should be clear to those who have experience with digital convolution that we could have proceeded to develop the digital gray-scale signal algebra through the use of serial matrices; after all, the computation of the opening also involves a type of serial matrix. However, the umbra transform provides easy proofs in the digital case, and facility with it will provide other benefits, especially in the extension of the Matheron representation theorems to gray-scale morphological filters. For image (not signal) processing, the umbra has a serious drawback because its matrices are three-dimensional. But so, too, are the serial matrices.

Theorem 6.6(*i*) and (*ii*) set forth two umbra homomorphism properties. A similar property holds for the opening, viz.,

$$U[OPEN(f, g)] = OPEN(U[f], U[g])$$

This property is easily deduced from Theorem 6.6:

$$U[OPEN(f, g)] = U[(f \boxminus g^\wedge) \boxplus g]$$
$$= U[f \boxminus g^\wedge] \boxplus U[g]$$
$$= (U[f] \boxminus (-U[g])) \boxplus U[g]$$
$$= OPEN(U[f], U[g])$$

Applying the surface operator yields

$$OPEN(f, g) = S(OPEN(U[f], U[g]))$$

Let us now apply the duality property, GM-11, twice to the expression for the opening in terms of dilation and erosion in a formal manner, i.e., without concerning ourselves with domain restrictions. We obtain

$$CLOSE(f, g) = -OPEN(-f, -g)$$
$$= -[((-f) \boxminus (-g^\wedge)) \boxplus (-g)]$$
$$= [-((-f) \boxminus (-g^\wedge))] \boxminus g$$
$$= (f \boxplus g^\wedge) \boxminus g$$
$$= ERODE[DILATE(f, g^\wedge), g^\wedge]$$

which is analogous to the expression for the closing in terms of erosion and dilation that holds in the constant-image case. However, since GM-11 was applied without regard to the domains involved, we must proceed with caution. Indeed, the following example shows that the preceding equality does not necessarily hold on the entire domain of $CLOSE(f, g)$.

Example 6.24:

Let

$$f = (8 \quad 8 \quad 2 \quad 6 \quad 6)_0$$

and

$$g = (3 \quad 4 \quad 5)_0$$

Then

$$-f = (-8 \quad -8 \quad -2 \quad -6 \quad -6)_0$$
$$-g = (-3 \quad -4 \quad -5)_0$$

and OPEN$(-f, -g)$ is given by taking extended maxima over the columns of the following serial-type matrix which results from fittings of $-g$ underneath $-f$:

$$\begin{pmatrix} -8 & -9 & -10 & * & * \\ * & -8 & -9 & -10 & * \\ * & * & -5 & -6 & -7 \end{pmatrix}_{0,2}$$

Hence,

$$\text{CLOSE}(f, g) = -\text{OPEN}(-f, -g) = (8 \quad 8 \quad 5 \quad 6 \quad 7)_0$$

On the other hand,

$$g^{\wedge} = (-5 \quad -4 \quad -3)_{-2}$$

and DILATE$(f, g^{\wedge})$ is obtained from the serial matrix

$$\begin{pmatrix} 3 & 3 & -3 & 1 & 1 & * & * \\ * & 4 & 4 & -2 & 2 & 2 & * \\ * & * & 5 & 5 & -1 & 3 & 3 \end{pmatrix}_{-2,2}$$

Consequently,

$$\text{DILATE}(f, g^{\wedge}) = (3 \quad 4 \quad 5 \quad 5 \quad 2 \quad 3 \quad 3)_{-2}$$

Fitting $g^{\wedge}$ gives

$$\text{ERODE}[\text{DILATE}(f, g^{\wedge}), g^{\wedge}] = (8 \quad 8 \quad 5 \quad 6 \quad 6)_0$$

which is not identical to CLOSE(f, g).

As Example 6.24 illustrates, one must be careful in attempting to find direct relationships between the algebraic properties of the gray-scale morphological algebra and those of the constant-image morphological algebra. As we have just seen, formal manipulations can result in incorrect deductions, even in the finite digital case.

6.6 GRAY-SCALE MORPHOLOGY FOR EUCLIDEAN IMAGES

Gray-scale morphology for images is essentially a generalization of gray-scale morphology for signals. The major differences are that the graph of a signal is a curve whereas the graph of an image is a surface, and digital signals are bound vectors whereas digital images are bound matrices. Because of the similarity of the fundamentals, Sections 6.6 and 6.7 will simply provide a synopsis of the gray-scale image morphology as it extends the signal morphology. Properties GM-1 through GM-14 apply, as do Theorems 6.1 through 6.6, modified of course to handle images.

A Euclidean image $f(x, y)$ is a function defined on some subset of the plane R^2. In the terminology of calculus, it is a real-valued function of two real variables. Intuitively, at each point, $f(x, y)$ denotes the gray-level of the image at that point, and the graph of f represents the gray-level image that we perceive.

Sec. 6.6 Gray-Scale Morphology for Euclidean Images

The operations EXTSUP and INF are defined as for signals, except now we must write $[\text{EXTSUP}(f_k)](x, y)$ and $[\text{INF}(f_k)](x, y)$, where (x, y) denotes a point in the plane. EXTSUP gives the supremum over the union of the domains, and INF gives the infimum over the intersection of the domains. These are the extensions of EXTMAX and MIN of Chapter 2 to the case of gray-scale Euclidean images. Again, the relation $f \ll g$ will mean that $D_f \subset D_g$ and $f \leq g$ on D_f.

The *translation* of an image f by the point (x, y) will be denoted by $f_{x,y}$ and defined by

$$f_{x,y}(s, t) = f(s - x, t - y)$$

Consequently, $D_{f_{x,y}} = D_f + (x, y)$. A translation upward, which we shall call an *offset* of the image, is defined by $f(x, y) + z$. Accordingly, the image $f_{x,y} + z$ is simply the image f moved x and y units in the domain space and up z units in the range space. Geometrically, the graph of $f_{x,y} + z$ is the graph of f translated in three-dimensional space by the vector (x, y, z).

The *dilation* of the gray-scale images f and g, with respective domains D_f and D_g, is defined by

$$\mathcal{D}(f, g) = \text{EXTSUP}_{(x,y) \in D_f}[g_{x,y} + f(x, y)]$$

Geometrically, the interpretation is analogous to the case of signals, except here we translate copies of the structuring element across the plane x and y units and then up $f(x, y)$ units. In other words, the center (origin) of the structuring element is shifted to the spatial point $(x, y, f(x, y))$. As usual, we also call $\mathcal{D}(f, g)$ the *Minkowski sum* and write $f \oplus g$.

Figures 6.37 and 6.38 illustrate gray-scale image dilation, the former using a cone-shaped structuring element and the latter using a hemisphere. As with signals, note the expansional effect on both the image and its domain.

As noted, GM-1 holds and dilation is commutative; i.e., $f \oplus g = g \oplus f$. Hence, as in the case of signals, we can shift f instead of g:

$$f \oplus g = \text{EXTSUP}_{(x,y) \in D_g}[f_{x,y} + g(x, y)]$$

Erosion of images is defined pixelwise by

$$[\mathcal{E}(f, g)](x, y) = \sup \{z: g_{x,y} + z \ll f\}$$

This fitting-type definition is exactly analogous to the corresponding one for signals. Figures 6.39 and 6.40 utilize the images of Figures 6.37 and 6.38, respectively, to illustrate image erosion.

Minkowski subtraction is defined by

$$f \ominus g = \text{INF}_{(x,y) \in D_g}[f_{x,y} + g(x, y)]$$

Theorem 6.1 applies, except that the reflection of g through the origin is now defined by $g^{\wedge}(x, y) = -g(-x, -y)$, where $D_{g^{\wedge}} = -D_g$, the 180° rotation of D_g in the plane. (See Figure 6.41.) Using the Minkowski subtraction formulation, we have

$$\mathcal{E}(f, g) = \text{INF}_{(x,y) \in D_g}[f_{-x,-y} - g(x, y)]$$

The gray-scale *opening* of an image is again a dilation of an erosion:

$$O(f, g) = \mathcal{D}[\mathcal{E}(f, g), g]$$

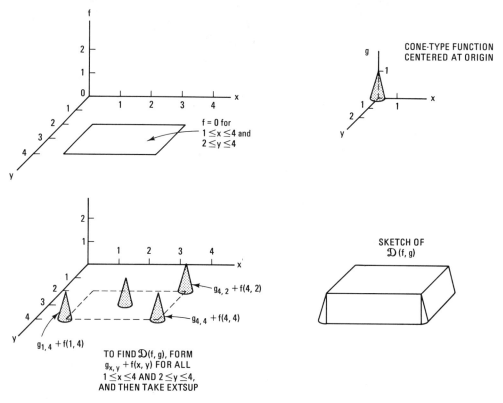

Figure 6.37 Euclidean gray-scale image dilation I

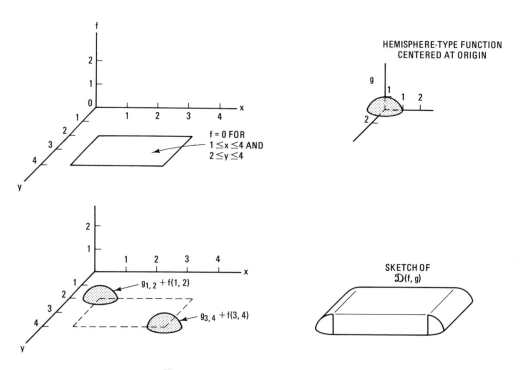

Figure 6.38 Euclidean gray-scale image dilation II

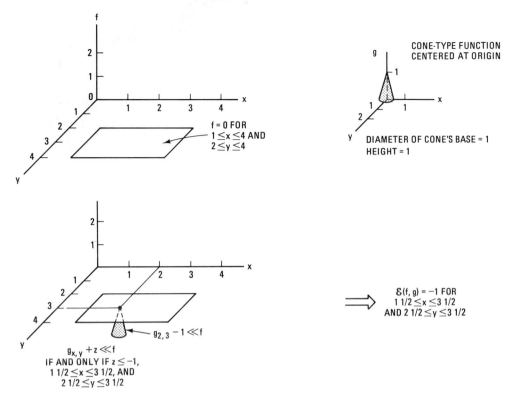

Figure 6.39 Euclidean gray-scale image erosion I

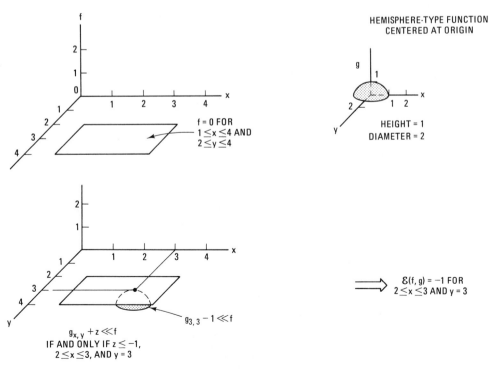

Figure 6.40 Euclidean gray-scale image erosion II

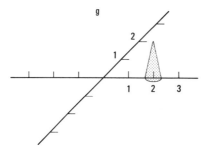

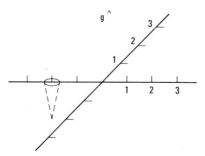

Figure 6.41 Reflection of image through the origin

Theorems 6.2 and 6.3 apply directly. Moreover, the opening is separately invariant under both horizontal translations and offsettings of the structuring element. Figures 6.42 and 6.43 show the openings corresponding to the images and structuring elements of Figures 6.37 and 6.38, respectively.

The gray-scale *closing* of an image is defined as the dual of the opening:

$$C(f, g) = -O(-f, -g)$$

Geometrically, the closing is obtained by reflecting f and g through the xy-plane, opening, and then reflecting back again. The closing is pseudoextensive, increasing, idempotent, and invariant under both horizontal translations and offsettings of the structuring element. Note that whereas the opening filters from beneath, the closing filters from above. Figures 6.44 and 6.45 employ the images of Figures 6.37 and 6.38, respectively, to illustrate the closing.

Given a set A in three-dimensional space, we define the *umbra* of A by

$$\mathbf{U}[A] = \{(x, y, z): \text{ there exists a } z' \text{ such that } (x, y, z') \in A \text{ and } z \leq z'\}$$

$\mathbf{U}[A]$ is the set of all points in R^3 that lie underneath some point in A. A subset B of R^3 is called an umbra if $\mathbf{U}[B] = B$. A *regular* umbra is one for which all vertical line segments in the umbra possess finite suprema. The domain of an umbra is the set of all points (x, y) in the plane for which there exists some point (x, y, z) in the umbra. The *surface*, $\mathbf{S}[A]$, of a regular umbra A is the set of all points (x, y, z') such that (x, y) is in the domain of A and

$$z' = \sup \{z: (x, y, z) \in A\}$$

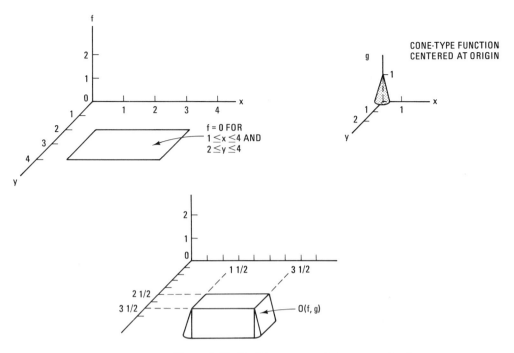

Figure 6.42 Euclidean gray-scale image opening I

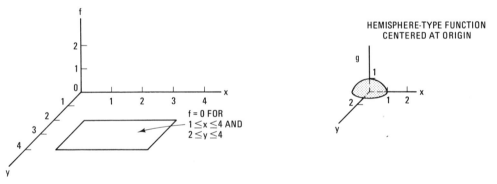

O(f, g) HAS MAXIMUM HEIGHT OF 0 ATTAINED FOR $2 \leq x \leq 3$ AND $y = 3$

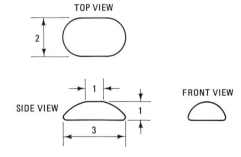

Figure 6.43 Euclidean gray-scale image opening II

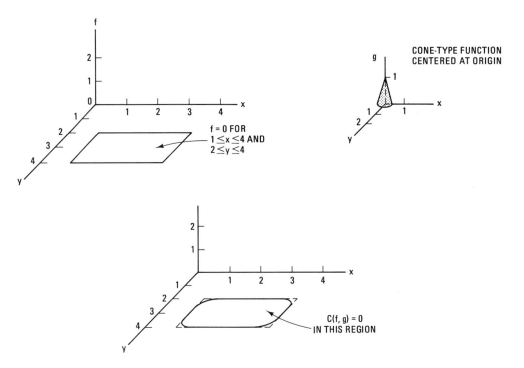

Figure 6.44 Euclidean gray-scale image closing I

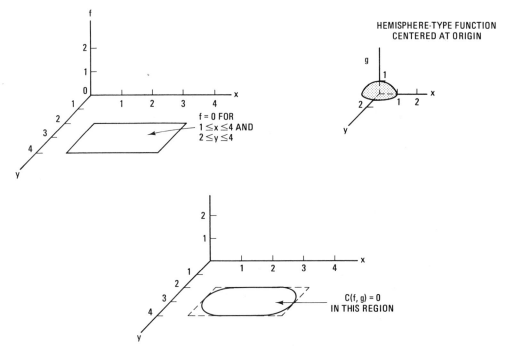

Figure 6.45 Euclidean gray-scale image closing II

An umbra is said to be *closed* if it contains its surface. Properties U-1 through U-6 hold for three-dimensional umbrae, where of course U-4 obtains with respect to translation invariance by triples (x, y, z). Figure 6.46 shows several umbrae.

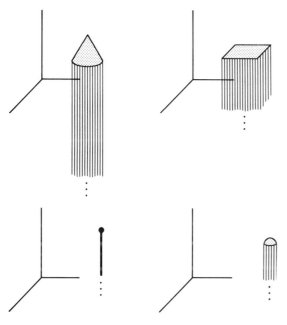

Figure 6.46 Several image umbrae

Now let f be an image with domain D_f. Then the graph of f is defined by

$$\mathbf{G}[f] = \{(x, y, f(x, y)): (x, y) \in D_f\}$$

$\mathbf{G}[f]$ is the usual graph considered in the calculus of two variables. From the perspective of image processing, it is a constant Euclidean three-dimensional image, a subset of R^3. The set of all points in R^3 that lie underneath $\mathbf{G}[f]$ is called the *umbra* of f. Denoted by $\mathbf{U}[f]$, the umbra of f is a three-dimensional constant Euclidean image (subset of R^3) and is given by

$$\mathbf{U}[f] = \mathbf{U}[\mathbf{G}[f]] = \{(x, y, z): (x, y) \in D_f \text{ and } z \leq f(x, y)\}$$

Every image f determines a unique closed regular umbra $\mathbf{U}[f]$, and conversely, every closed regular umbra determines a unique image f. Specifically, for any closed regular umbra F, $\mathbf{S}[F]$ is the graph of the unique image f it determines. As usual, we identify the graph of an image with the image itself, i.e., $\mathbf{S}[F] = \mathbf{G}[f] = f$.

Before proceeding, note that the Euclidean morphology of Chapter 1 can be generalized to three dimensions without difficulty. Indeed, in his original formulation of mathematical morphology, Matheron considered two-valued morphology in n-dimensional Euclidean space R^n. Although we shall not pursue the issue in detail, it is important to recognize that the theory of Chapter 1 is not dimensionally dependent. The interested reader should scan that chapter with an eye toward R^3. He or she will see at once that all results depend solely on the properties of set theory and vector translation, and those properties are the same in R^3 (or, in general, R^n) as they are in R^2.

The fundamental relationship between signal dilation and umbra dilation is expressed in Theorem 6.4. The theorem goes over without change to the dilation of images and their umbrae. The same holds true for gray-scale image erosion (Theorem 6.5). Thus, as in the case of signals, the umbra transform provides a link between the morphology of gray-scale images and that of three-dimensional constant images.

6.7 GRAY-SCALE MORPHOLOGY FOR DIGITAL IMAGES

According to the definition given in Chapter 2, a digital image is a mapping from a subset of $Z \times Z$ into R. In the terminology of Section 6.3, we would refer to these as *sampled* images. A fully digital image would be one in which the range is *quantized*. We shall adhere to the convention of saying that only a quantized digital image is fully digital in order to distinguish such images from those which are merely digital (i.e., sampled).

Digital images with finite domains are represented as bound matrices. Just as in the case of bound row vectors (finite sampled signals), we define the operators SUB and OFF by

$$[\text{SUB}(f)](i, j) = -f(i, j)$$

and

$$[\text{OFF}(f; y)](i, j) = f(i, j) + y$$

The specifications for the digital dilation and erosion of images are given by Figures 6.31 and 6.32, respectively, except that the blocks now define operators on bound matrices. We employ the same notation for images as for signals. Also, as in the earlier case, the specification of the Minkowski difference leaves off the NINETY^2 and SUB blocks in Figure 6.32.

Example 6.25:

Let

$$f = \begin{pmatrix} 0 & 2 & 2 & 2 & 1 \\ 1 & 2 & 6 & 2 & 1 \\ 0 & 6 & 7 & 2 & 1 \\ 1 & 1 & 6 & 1 & * \\ 1 & 0 & 2 & 2 & 1 \end{pmatrix}_{1,5}$$

and

$$g = \begin{pmatrix} 0 & 3 \\ 3 & 4 \end{pmatrix}_{0,1}$$

Since dilation is commutative, we can interchange the roles of f and g in Figure 6.31 so that only four translations need to be effected, one for each element in the domain of g.

Proceeding, we have

$$\text{DOMAIN}(g) = [(0, 1), (0, 0), (1, 1), (1, 0)]$$

and

$$\text{RANGE}(g) = [0, 3, 3, 4]$$

Sec. 6.7 Gray-Scale Morphology for Digital Images

Note the lexicographical manner in which the DOMAIN and RANGE stacks are pushed. There are four inputs to EXTMAX:

$$f_{0,1} + 0 = \begin{pmatrix} * & 0 & 2 & 2 & 2 & 1 \\ * & 1 & 2 & 6 & 2 & 1 \\ * & 0 & 6 & 7 & 2 & 1 \\ * & 1 & 1 & 6 & 1 & * \\ * & 1 & 0 & 2 & 2 & 1 \\ * & * & * & * & * & * \\ \circledast & * & * & * & * & * \end{pmatrix}$$

$$f_{0,0} + 3 = \begin{pmatrix} * & 3 & 5 & 5 & 5 & 4 \\ * & 4 & 5 & 9 & 5 & 4 \\ * & 3 & 9 & 10 & 5 & 4 \\ * & 4 & 4 & 9 & 4 & * \\ * & 4 & 3 & 5 & 5 & 4 \\ \circledast & * & * & * & * & * \end{pmatrix}$$

$$f_{1,1} + 3 = \begin{pmatrix} * & * & 3 & 5 & 5 & 5 & 4 \\ * & * & 4 & 5 & 9 & 5 & 4 \\ * & * & 3 & 9 & 10 & 5 & 4 \\ * & * & 4 & 4 & 9 & 4 & * \\ * & * & 4 & 3 & 5 & 5 & 4 \\ * & * & * & * & * & * & * \\ \circledast & * & * & * & * & * & * \end{pmatrix}$$

and

$$f_{1,0} + 4 = \begin{pmatrix} * & * & 4 & 6 & 6 & 6 & 5 \\ * & * & 5 & 6 & 10 & 6 & 5 \\ * & * & 4 & 10 & 11 & 6 & 5 \\ * & * & 5 & 5 & 10 & 5 & * \\ * & * & 5 & 4 & 6 & 6 & 5 \\ \circledast & * & * & * & * & * & * \end{pmatrix}$$

Next, we compute

$$\text{EXTMAX}[f_{0,1} + 0, f_{0,0} + 3, f_{1,1} + 3, f_{1,0} + 4]$$

We simply read "down" each pixel and take the extended maximum of the four values taken on by the pixel. For instance, the four values of pixel (1, 3) are 1, 3, *, and *. Upon the completion of EXTMAX, the output image is

$$\text{DILATE}(f, g) = \begin{pmatrix} * & 0 & 3 & 5 & 5 & 5 & 4 \\ * & 3 & 5 & 6 & 9 & 6 & 5 \\ * & 4 & 6 & 9 & 10 & 6 & 5 \\ * & 3 & 9 & 10 & 11 & 6 & 5 \\ * & 4 & 5 & 9 & 10 & 5 & 4 \\ * & 4 & 5 & 5 & 6 & 6 & 5 \\ \circledast & * & * & * & * & * & * \end{pmatrix}$$

In the original image f, there was a darkened triangle of three 6's and one 7. Dilation by g has resulted in a darkened region of eight pixels, each with gray value at least equal to 9.

In Example 6.25 there were four images $f_{i,j} + g(i, j)$, one for each pixel in the domain of g. Imagine that we were to place these images horizontally on a set of shelves, so that from top down the ordering on the shelves would be identical to the ordering within the stack DOMAIN(g). The result would be a three-dimensional bound matrix called the *serial matrix* for the dilation. Each one of the images $f_{i,j} + g(i, j)$ is a *horizontal section* of the serial matrix, and DILATE(f, g) is computed pixelwise by taking the extended maximum of all values over a given pixel. For instance, looking at the serial matrix for Example 6.25, there is a column vector of values over pixel (1, 3) given by

$$\begin{pmatrix} 1 \\ 3 \\ * \\ * \end{pmatrix}$$

Reading down this column vector and applying EXTMAX, we obtain

$$[\text{DILATE}(f, g)](1, 3) = 3$$

Although it might be difficult to write a three-dimensional data structure on paper, such a structure can certainly be represented within a machine.

Since the Minkowski subtraction utilizes the same images $f_{i,j} + g(i, j)$ as the Minkowski sum, it possesses the same serial matrix. The difference in the output results from the application of MIN instead of EXTMAX.

Example 6.26:

Let f and g be the images of Example 6.25. Applying MIN to $f_{0,1} + 0$, $f_{0,0} + 3$, $f_{1,1} + 3$, and $f_{1,0} + 4$ yields

$$f \boxminus g = \begin{pmatrix} * & * & 2 & 5 & 2 & 1 \\ * & * & 3 & 6 & 2 & 1 \\ * & * & 1 & 4 & 1 & * \\ * & * & 0 & 2 & 2 & * \\ * & * & * & * & * & * \\ \circledast & * & * & * & * & * \end{pmatrix}$$

Example 6.27:

We find ERODE(f, g) for the images f and g of Example 6.25. To begin,

$$g^{\wedge} = \begin{pmatrix} -4 & -3 \\ -3 & 0 \end{pmatrix}_{-1, 0}$$

$$\text{DOMAIN}(g^{\wedge}) = [(-1, 0), (-1, -1), (0, 0), (0, -1)]$$

and

$$\text{RANGE}(g) = [-4, -3, -3, 0]$$

Sec. 6.7 Gray-Scale Morphology for Digital Images

Hence, the four inputs to MIN in Figure 6.32 are

$$f_{-1,0} - 4 = \begin{pmatrix} -4 & -2 & -2 & -2 & -3 \\ -3 & -2 & 2 & -2 & -3 \\ -4 & 2 & 3 & -2 & -3 \\ -3 & -3 & 2 & -3 & * \\ -3 & -4 & -2 & -2 & -3 \\ \circledast & * & * & * & * \end{pmatrix}$$

$$f_{-1,-1} - 3 = \begin{pmatrix} -3 & -1 & -1 & -1 & -2 \\ -2 & -1 & 3 & -1 & -2 \\ -3 & 3 & 4 & -1 & -2 \\ -2 & -2 & 3 & -2 & * \\ \circled{-2} & -3 & -1 & -1 & -2 \end{pmatrix}$$

$$f_{0,0} - 3 = \begin{pmatrix} * & -3 & -1 & -1 & -1 & -2 \\ * & -2 & -1 & 3 & -1 & -2 \\ * & -3 & 3 & 4 & -1 & -2 \\ * & -2 & -2 & 3 & -2 & * \\ * & -2 & -3 & -1 & -1 & -2 \\ \circledast & * & * & * & * & * \end{pmatrix}$$

and

$$f_{0,-1} + 0 = \begin{pmatrix} * & 0 & 2 & 2 & 2 & 1 \\ * & 1 & 2 & 6 & 2 & 1 \\ * & 0 & 6 & 7 & 2 & 1 \\ * & 1 & 1 & 6 & 1 & * \\ \circledast & 1 & 0 & 2 & 2 & 1 \end{pmatrix}$$

Applying MIN yields

$$\text{ERODE}(f, g) = \begin{pmatrix} * & -2 & -1 & -2 & -3 \\ * & -3 & 2 & -2 & -3 \\ * & -3 & -2 & -3 & * \\ * & -4 & -3 & -2 & * \\ \circledast & * & * & * & * \end{pmatrix}$$

Aside from noting the obvious diminution of the domain, if we were to imagine fitting the structuring element g underneath the surface of the image f, we would see that it would fit the protrusion of four darkened pixels in f in only one way: it would have to be situated beneath pixels (2, 3), (3, 3), and (3, 4). This would place the center of the structuring element over pixel (2, 3), and the greatest offset of g that would still leave a fit would be 2. Indeed, we see that

$$[\text{ERODE}(f, g)](2, 3) = 2$$

and no other pixel in the erosion has a nonnegative gray value.

In this example, the serial matrix for the erosion is obtained by horizontally stacking the four inputs

$$f_{-i,-j} - g(i, j)$$

where (i, j) varies over the domain of g. The output can then be obtained pixelwise by taking the minimum down the column of values, including stars, over each pixel.

The preceding example mentioned "fitting" without being precise as to the order relation involved. In fact, in that example we were employing the image version of the relation ≪ that was previously defined for digital signals. Thus, as in that case, we can rigorously define the order relation ≪ on gray-scale digital images, namely: $f \ll g$ if and only if $D_f \subset D_g$ and $f(i, j) \leq g(i, j)$ for any $(i, j) \in D_f$. This definition agrees with the definition of ≪ as it was applied to constant digital images; indeed, if S and T are constant digital images, then $S \ll T$ means that $D_S \subset D_T$ and $S(i, j) = T(i, j)$ on D_S. Note also that in the case of gray-scale Euclidean images the order relation ≪ reduces to set inclusion ($\subset$) when applied to constant images. (See Section 6.1 for an analogous statement regarding constant signals.)

As usual, the opening is computed by taking the dilation of an erosion. Its interpretation in terms of fitting is that we take the extended maximum of all copies of the structuring element that fit underneath the image. As with digital erosion, we will be fitting bound matrices.

Now, for OFF$[g_{i,j}; y]$ to fit underneath f, we must have

$$g_{i,j} + y \ll f$$

But this implies that the domain of $g_{i,j} + y$ is a subdomain of the domain of f. Consequently, as long as the domains in question are finite, the domain of the opening is the opening of the domains, where we treat the domains as constant (1–∗) images.

Example 6.28:

Letting f and g be the images of Example 6.25, and letting $k = $ ERODE(f, g) be the erosion found in Example 6.27, we compute

$$\text{OPEN}(f, g) = \text{DILATE}(k, g)$$

Using the domain and range information for g that was found in Example 6.25, we obtain four inputs to the EXTMAX of Figure 6.31:

$$k_{0,1} + 0 = \begin{pmatrix} * & -2 & -1 & -2 & -3 \\ * & -3 & 2 & -2 & -3 \\ * & -3 & -2 & -3 & * \\ * & -4 & -3 & -2 & * \\ * & * & * & * & * \\ \circledast & * & * & * & * \end{pmatrix}$$

$$k_{0,0} + 3 = \begin{pmatrix} * & 1 & 2 & 1 & 0 \\ * & 0 & 5 & 1 & 0 \\ * & 0 & 1 & 0 & * \\ * & -1 & 0 & 1 & * \\ \circledast & * & * & * & * \end{pmatrix}$$

$$k_{1,1} + 3 = \begin{pmatrix} * & * & 1 & 2 & 1 & 0 \\ * & * & 0 & 5 & 1 & 0 \\ * & * & 0 & 1 & 0 & * \\ * & * & -1 & 0 & 1 & * \\ * & * & * & * & * & * \\ \circledast & * & * & * & * & * \end{pmatrix}$$

Sec. 6.7   Gray-Scale Morphology for Digital Images

$$k_{1,0} + 4 = \begin{pmatrix} * & * & 2 & 3 & 2 & 1 \\ * & * & 1 & 6 & 2 & 1 \\ * & * & 1 & 2 & 1 & * \\ * & * & 0 & 1 & 2 & * \\ \circledast & * & * & * & * & * \end{pmatrix}$$

Applying EXTMAX to the four inputs $k_{i,j} + g(i, j)$ yields

$$\text{OPEN}(f, g) = \begin{pmatrix} * & -2 & 1 & 2 & 1 & 0 \\ * & 1 & 2 & 5 & 2 & 1 \\ * & 0 & 5 & 6 & 2 & 1 \\ * & 0 & 1 & 2 & 1 & * \\ * & -1 & 0 & 1 & 2 & * \\ \circledast & * & * & * & * & * \end{pmatrix}$$

Note how the domain of OPEN(f, g) is a proper subset of the domain of f; indeed, pixel (5, 1) has been removed since the domain of g does not fit into the domain of f in such a way as to include (5, 1). To be precise, in constant-image form, the domains of f and g are

$$D_f = \begin{pmatrix} 1 & 1 & 1 & 1 & 1 \\ 1 & 1 & 1 & 1 & 1 \\ 1 & 1 & 1 & 1 & 1 \\ 1 & 1 & 1 & 1 & * \\ 1 & 1 & 1 & 1 & 1 \end{pmatrix}_{1,5}$$

and

$$D_g = \begin{pmatrix} 1 & 1 \\ \circled{1} & 1 \end{pmatrix}$$

Hence,

$$\text{OPEN}(D_f, D_g) = \begin{pmatrix} 1 & 1 & 1 & 1 & 1 \\ 1 & 1 & 1 & 1 & 1 \\ 1 & 1 & 1 & 1 & 1 \\ 1 & 1 & 1 & 1 & * \\ 1 & 1 & 1 & 1 & * \end{pmatrix}_{1,5}$$

Observe how the darkened triangle of four pixels in the original image has been reduced to three pixels: the structuring element cannot fit beneath the image tightly and still remain beneath pixel (3, 2), at which f has gray value 6.

Finally, notice the dampening effect of the gray value 0 at pixel (0, 1) in the structuring element: instead of having the triangular image

$$\begin{pmatrix} * & 6 \\ 6 & 7 \end{pmatrix}_{2,4}$$

as a subimage of the opening, we have the image

$$\begin{pmatrix} * & 5 \\ 5 & 6 \end{pmatrix}_{2,4}$$

Unless there is a particular reason for keeping $g(0, 1) = 0$ in the structuring element, for tight filtering with essentially the same effect it would be better to use the structuring element

$$\begin{pmatrix} * & 3 \\ 3 & 4 \end{pmatrix}_{0, 1}.$$

Example 6.29:

It was mentioned just before Example 6.28 that the domain of the opening of a digital image is the opening of the domain by the domain of the structuring element, provided these domains are finite. Such a restriction plays no role in practice; however, it is required in theory. For example, consider the image f whose domain is the set of pixels $(i, 0)$ such that i is a positive integer, and which possesses gray values defined by $f(i, 0) = -i$. If g is the image with the same domain, but with constant gray value 0, then no image of the form $g_{i,j} + y$ fits beneath f. Hence,

$$\text{OPEN}(f, g) = \varnothing$$

the null image, while

$$\text{OPEN}(D_f, D_g) = D_f$$

As in the case of signals, the closing for digital images is defined as the dual of the opening:

$$\text{CLOSE}(f, g) = \text{SUB}[\text{OPEN}(\text{SUB}(f), \text{SUB}(g))]$$

CLOSE filters the image from the top by fitting the structuring element from above. Since the closing is specified in terms of the opening, we can forego an example in order to avoid redundancy.

All that was discussed regarding fully digital signals applies to fully digital images. For example, we can construct the umbra matrix of a fully digital image, which turns out to be a three-dimensional constant bound matrix. Because the theory is completely parallel, we can be brief.

First of all, our comments regarding morphology on subsets of R^3 apply to the morphology of constant images in $Z \times Z \times Z$. Second, the construction of the umbra matrix and the concomitant definitions are analogous, except that here one must go through the messy details involved in three-dimensional bound matrix manipulation, which we shall not do. Of special note, however, is the transform pair

$$\mathbf{U}: f \longrightarrow \mathbf{U}[f]$$

and

$$\mathbf{S}: \mathbf{U}[f] \longrightarrow f$$

The transform and inversion techniques that worked for signals are directly applicable to gray-scale images. Moreover, Theorems 6.4, 6.5, and 6.6 all apply. Hence, the theory of Section 6.5 goes through without alteration, and properties GM-1 through GM-14 hold.

In sum, if one has a full understanding of the material covered in Sections 6.4 and 6.5, there should be no difficulty in seeing the generalizations of that material to the image setting.

EXERCISES

6.1. Let $f(t) = t^3$ for $-2 \leq t \leq 1$, and $g(t) = t^2$ for $-4 \leq t \leq 0$. Find EXTSUP(f, g) and INF(f, g). Draw a graph to illustrate the results.

6.2. For $k = 1, 2, 3, \ldots$, let
$$f_k(t) = \begin{cases} 1, & \text{if } 0 \leq t \leq 2 - 1/k \\ \text{undefined}, & \text{otherwise} \end{cases}$$
Find EXTSUP(f_k) and INF(f_k).

6.3. Let
$$f_1(t) = \begin{cases} 1, & \text{for } -2 \leq t \leq 0 \\ 1 + t, & \text{for } 0 \leq t \leq 2 \end{cases}$$
$$f_2(t) = \begin{cases} 9, & \text{for } -0.1 \leq t \leq 0.1 \\ 0, & \text{for } -2 \leq t < -0.1 \text{ or } 0.1 < t \leq 2 \end{cases}$$
$$f_3(t) = -f_2(t)$$
and
$$g(t) = -|t|, \text{ for } -1 \leq t \leq 1$$

Find, for $i = 1, 2, 3$:
- (a) $\mathcal{D}(f_i, g)$
- (b) $\mathcal{E}(f_i, g)$
- (c) $f_i \ominus g$
- (d) $O(f_i, g)$
- (e) $C(f_i, g)$
- (f) $C(f_i, -g)$
- (g) $\mathcal{E}(f_i, g^\wedge)$
- (h) $\mathcal{D}[\mathcal{E}(f_i, g), g]$
- (i) $\mathcal{E}[\mathcal{D}(f_i, g^\wedge), g^\wedge]$
- (j) $O[C(f_i, -g), g]$
- (k) $C[O(f_i, g), -g]$

6.4. Repeat Exercise 6.3; however, this time use
- (a) $g(t) = -1$, for $0 \leq t \leq 1$
- (b) $g(t) = 0$, for $2 \leq t \leq 3$
- (c) $g(t) = -|t - 1|$, for $0 \leq t \leq 2$
- (d) $g(t) = \begin{cases} -|t|, & \text{for } -1 \leq t \leq 1 \text{ and } t \neq 0 \\ 4, & \text{for } t = 0 \end{cases}$
- (e) $g(t) = \begin{cases} -|t|, & \text{for } -1 \leq t \leq 1 \text{ and } t \neq 0 \\ -2, & \text{for } t = 0 \end{cases}$
- (f) $g(t) = \begin{cases} 1, & \text{for } -\frac{1}{2} \leq t \leq \frac{1}{2} \\ 0, & \text{for } -1 \leq t < -\frac{1}{2} \text{ or } \frac{1}{2} < t \leq 1 \end{cases}$

6.5. Prove Theorem 6.3 by using Theorem 6.2.

6.6. Use Theorem 6.3 together with the definition of the closing to prove that the closing is pseudoextensive and increasing.

6.7. Let
$$A = \{(x, y): 0 \leq y < x^2 \text{ and } 0 \leq x \leq 2\}$$
Find U[A] and S[A]. Note that U[S[A]] $\neq$ U[A].

6.8. Prove properties U-1 through U-6.
6.9. Let
$$f_1 = (1 \quad 2 \quad 1 \quad 1 \quad 8 \quad 8 \quad 1 \quad 0 \quad 1)_3$$
$$f_2 = (1 \quad 2 \quad 3 \quad * \quad 2 \quad 1 \quad 2 \quad -7 \quad -7)_{-1}$$
$$f_3 = (1 \quad 9 \quad 1 \quad 9 \quad 1 \quad 9 \quad 1 \quad 9)_0$$
and
$$g = (1 \quad 1 \quad 2)_{-5}$$

For $i = 1, 2, 3$, find:
- **(a)** DILATE(f_i, g)
- **(b)** ERODE(f_i, g)
- **(c)** OPEN(f_i, g)
- **(d)** CLOSE(f_i, g)
- **(e)** CLOSE$(f_i, -g)$
- **(f)** ERODE$(f_i, g^\wedge)$
- **(g)** ERODE[DILATE$(f_i, g^\wedge), g^\wedge$]

In parts (b) and (f), use both the fitting and the Minkowski subtraction methodologies.

6.10. Repeat Exercise 6.9, except use
- **(a)** $g = (0 \quad 9)_0$
- **(b)** $g = (0 \quad 0 \quad 9 \quad 0)_0$

6.11. Find the minimal umbra matrices for the signals f_1, f_2, and f_3 of Exercise 6.9.

6.12. Let
$$f = (2 \quad 0 \quad 5 \quad 2 \quad 1)_0$$
and
$$g = (2 \quad 3 \quad 4)_{-1}$$

Using umbra matrices, find ERODE$(f, -g)$ and OPEN(f, g) by fitting. Then find ERODE(f, g) by employing the Minkowski subtraction formulation.

6.13. Let f and g be as in Exercise 6.12, and let
$$h = (0 \quad 1 \quad 0)_0$$

Using $f, g, h, i = 3$, and $j = -2$, illustrate properties GM-1 through GM-6, and GM-10 through GM-13.

6.14. Let
$$f = \begin{pmatrix} 1 & 0 & 1 & 0 & 1 & 0 \\ 1 & 7 & 7 & 2 & 0 & 0 \\ 2 & 6 & 7 & 0 & 1 & 0 \\ 1 & 1 & 0 & 1 & -6 & 1 \\ * & 1 & 0 & 1 & 1 & 1 \end{pmatrix}_{0,0}$$
and
$$g = \begin{pmatrix} * & 1 \\ * & 1 \end{pmatrix}$$

Find DILATE(f, g), ERODE(f, g), OPEN(f, g), and CLOSE(f, g).

FOOTNOTES FOR CHAPTER 6

1. R. M. Haralick, S. R. Sternberg, and X. Zhuang, "Grayscale Morphology," IEEE Computer Vision and Pattern Recognition, (1986), p. 543–550.
2. Ibid., p. 543–550.

7

Gray-scale Morphological Filters: Theory

7.1 EXTENDED SIGNALS

Because of the close relation between a signal and its umbra, we need to consider signals whose umbrae are not regular. To do this, we simply allow a signal to take on the value of $+\infty$. Rigorously, an *extended* Euclidean signal is a function whose domain is a subset of R and whose codomain is $R \cup \{+\infty\}$. We shall write $\mathscr{S}^{\#}$ to denote the class of extended signals.

Example 7.1:

Let

$$f(t) = \begin{cases} \dfrac{1}{t^2}, & \text{for } -1 \leq t < 0 \\ +\infty, & \text{for } t = 0 \\ \dfrac{1}{t^2}, & \text{for } 0 < t < 1 \\ +\infty, & \text{for } 1 \leq t \leq 2 \end{cases}$$

Then the domain of f is the closed interval $[-1, 2]$. Note that $f(0) = f(\frac{4}{3}) = +\infty$. The graph of f is shown in Figure 7.1, where the double line ($=$) is used to indicate the value at $+\infty$.

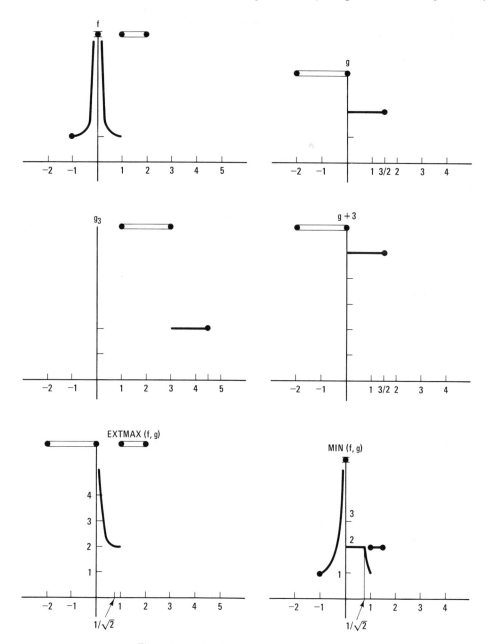

Figure 7.1 Signals for Examples 7.1 and 7.2

Since a new value has been introduced, we need to adjust the operational definitions that are in use. No difficulties arise with translation. However, we must be careful with the extended supremum (EXTSUP), the infimum (INF), and the offset (OFF). EXTSUP is handled by defining

$$[\text{EXTSUP}(f_k)](t) = +\infty$$

Sec. 7.1 Extended Signals

if there exists a k' such that $f_{k'}(t) = +\infty$. In other words, max$(y, +\infty) = +\infty$ for any real number y. As for INF, the only time

$$[\text{INF}(f_k)](t) = +\infty$$

is when $f_k(t) = +\infty$ for all k. In other words, min$(y, +\infty) = y$ for any real number y. Finally, insofar as offsetting is concerned, if $f(t) = +\infty$, then $f(t) + y = +\infty$ for any real number y. That is, $y + +\infty = +\infty$ for any $y \in R$.

Example 7.2:

Let f be the signal given in Example 7.1, and let

$$g(t) = \begin{cases} +\infty, & \text{for } -2 \leq t \leq 0 \\ 2, & \text{for } 0 < t \leq \frac{3}{2} \end{cases}$$

Then

$$g_3(t) = \begin{cases} +\infty, & \text{for } 1 \leq t \leq 3 \\ 2, & \text{for } 3 < t \leq \frac{9}{2} \end{cases}$$

$$g(t) + 3 = \begin{cases} +\infty, & \text{for } -2 \leq t \leq 0 \\ 5, & \text{for } 0 < t \leq \frac{3}{2} \end{cases}$$

$$\text{EXTMAX}(f, g) = \begin{cases} +\infty, & \text{for } -2 \leq t \leq 0 \\ 1/t^2, & \text{for } 0 < t \leq 1/\sqrt{2} \\ 2, & \text{for } 1/\sqrt{2} < t < 1 \\ +\infty, & \text{for } 1 \leq t \leq 2 \end{cases}$$

and

$$\text{MIN}(f, g) = \begin{cases} 1/t^2, & \text{for } -1 \leq t < 0 \\ +\infty, & \text{for } t = 0 \\ 2, & \text{for } 0 < t \leq 1/\sqrt{2} \\ 1/t^2, & \text{for } 1/\sqrt{2} < t < 1 \\ 2, & \text{for } 1 \leq t \leq \frac{3}{2} \end{cases}$$

Since only two signals are involved, we have used EXTMAX and MIN. The graphs of g and the four outputs are given in Figure 7.1.

When we write $f \ll g$ for extended signals, we mean that D_f, the domain of f, is a subset of D_g, the domain of g, and that $f(t) \leq g(t)$ on D_f. However, we must keep in mind the relations $+\infty \leq +\infty$ and $y \leq +\infty$, where y is a real number.

The umbra transform of an extended signal is defined in the same manner as for a finite signal; however, we use the fact that $+\infty$ is greater than any real number. Consequently, if $f(t) = +\infty$, then $(t, y) \in \mathbf{U}[f]$ for any $y \in R$. What is of interest is that the surface of any umbra now turns out to be a signal, whether or not the umbra is regular. Restricted to finite signals, the surface operator **S** could only be applied to regular umbrae. For extended signals, however, **S** can be applied to any umbra, and hence $\mathbf{S}: \mathcal{U} \to \mathcal{S}^*$, where $\mathcal{U}$ denotes the set of all umbrae. This latter point is crucial to the theoretical development of increasing τ-mappings and τ-openings in the context of signals. Figure 7.2 gives the umbra of the signal g depicted Figure 7.1.

In the nonextended setting we said that a regular umbra was closed if it contained its surface. In the extended case, where the surface signal can take on the value $+\infty$, the surface cannot be mathematically interpreted as a subset of R^2 (although one might do so for the sake of intuition). In the present context, an umbra A is closed if for any x in its domain, one of two conditions is satisfied:

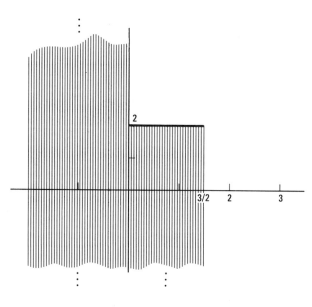

Figure 7.2 Umbra of an extended signal

1. $S[A](x)$ is finite and $(x, S[A](x)) \in A$.
2. $S[A](x) = +\infty$.

As a transform, **U** maps $\mathscr{S}^{\#}$ into $\mathscr{U}$ in a one-to-one manner. Hence, if we let $\mathscr{U}_c$ denote the collection of closed umbrae, then

$$\mathbf{U}: \mathscr{S}^{\#} \longrightarrow \mathscr{U}_c$$

and **U** is both one-to-one and onto. Consequently,

$$\mathbf{S}: \mathscr{U}_c \longrightarrow \mathscr{S}^{\#}$$

in both a one-to-one and onto manner also.

All of the preceding remarks have immediate analogues in the digital signal, Euclidean image, and digital image settings. In the digital settings, the bound matrix structure is particularly well suited to the inclusion of the value $+\infty$: together with the value $*$, which acts like $-\infty$ in the $\leqslant$ relation, it provides an important symmetry. (Of course, we could have introduced a star into Euclidean signal theory; but our purpose throughout has been to emphasize digital processing, and it is in the bound matrix context that the star is most natural.) Given these analogues, we can avoid the redundant restatement of the theory regarding $+\infty$ in all the other settings, and concentrate instead on two examples, one for digital signals and the other for digital images.

Example 7.3:

Consider the extended fully digital signals

$$f = (2 \quad -1 \quad 3 \quad +\infty \quad +\infty \quad 2 \quad 1 \quad * \quad 2)_0$$

and

$$g = (2 \quad -2 \quad +\infty \quad 5 \quad * \quad 1 \quad 1)_1$$

Sec. 7.1 Extended Signals

Some values of f and g are $f(1) = -1, f(3) = +\infty, f(17) = *$, and $g(5) = *$. We have

$$\text{TRAN}(g; 2) = (2 \quad -2 \quad +\infty \quad 5 \quad * \quad 1 \quad 1)_3$$
$$\text{OFF}(g; 4) = (6 \quad 2 \quad +\infty \quad 9 \quad * \quad 5 \quad 5)_1$$
$$\text{EXTMAX}(f, g) = (2 \quad 2 \quad 3 \quad +\infty \quad +\infty \quad 2 \quad 1 \quad 1 \quad 2)_0$$
$$\text{MIN}(f, g) = (-1 \quad -2 \quad +\infty \quad 5 \quad * \quad 1)_1$$

The umbra of f is depicted in Figure 7.3.

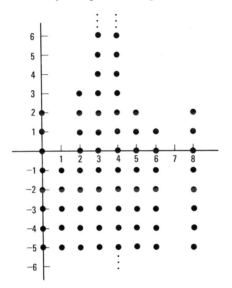

Figure 7.3 Umbra of signal from Example 7.3

Example 7.4:

Consider the extended images

$$f = \begin{pmatrix} 1 & * & 2 \\ +\infty & 1 & 0 \end{pmatrix}_{-2, 0}$$

and

$$g = \begin{pmatrix} +\infty & +\infty & 7 \\ 2 & * & -9 \end{pmatrix}_{-2, 0}$$

We have

$$\text{TRAN}(g; 1, -8) = \begin{pmatrix} +\infty & +\infty & 7 \\ 2 & * & -9 \end{pmatrix}_{-1, -8}$$

$$\text{OFF}(g; 2) = \begin{pmatrix} +\infty & +\infty & 9 \\ 4 & * & -7 \end{pmatrix}_{-2, 0}$$

$$\text{EXTMAX}(f, g) = \begin{pmatrix} +\infty & +\infty & 7 \\ +\infty & 1 & 0 \end{pmatrix}_{-2, 0}$$

$$\text{MIN}(f, g) = \begin{pmatrix} 1 & * & 2 \\ 2 & * & -9 \end{pmatrix}_{-2, 0}$$

The definitions of erosion, dilation, and opening may be readily extended to extended signals; we need only take into account the adjustments to the definitions of EXTSUP (EXTMAX), INF (MIN), TRAN, and OFF. The following example illustrates the methodology in the case of digital signals.

Example 7.5:

Let

$$f = (2 \quad 4 \quad 4 \quad +\infty \quad 5 \quad 6 \quad 4)_0$$

and

$$g = (1 \quad 2 \quad 2)_0$$

Then taking extended maximums down the columns of the *extended* serial matrix

$$\begin{pmatrix} 3 & 5 & 5 & +\infty & 6 & 7 & 5 & * & * \\ * & 4 & 6 & 6 & +\infty & 7 & 8 & 6 & * \\ * & * & 4 & 6 & 6 & +\infty & 7 & 8 & 6 \end{pmatrix}_{0,2}$$

yields

$$\text{DILATE}(f, g) = (3 \quad 5 \quad 6 \quad +\infty \quad +\infty \quad +\infty \quad 8 \quad 8 \quad 6)_0$$

Also, using fitting, we obtain

$$\text{ERODE}(f, g) = (1 \quad 2 \quad 3 \quad 3 \quad 2)_0$$

The opening is obtained through the use of a serial matrix of the maximum fits of signals of the form $g_i + j$:

$$\begin{pmatrix} 2 & 3 & 3 & * & * & * & * \\ * & 3 & 4 & 4 & * & * & * \\ * & * & 4 & 5 & 5 & * & * \\ * & * & * & 4 & 5 & 5 & * \\ * & * & * & * & 3 & 4 & 4 \end{pmatrix}_{0,4}$$

Taking extended maximums down the columns yields

$$\text{OPEN}(f, g) = (2 \quad 3 \quad 4 \quad 5 \quad 5 \quad 5 \quad 4)_0$$

Notice how the "infinite spike" at $i = 3$ has been filtered out.

Let us now consider two signals f' and f'' that are the same as f in Example 7.5 except that $f'(3) = *$ and $f''(3) = -99$, whereas $f(3) = +\infty$. Then

$$\text{OPEN}(f', g) = (2 \quad 3 \quad 3 \quad * \quad 3 \quad 4 \quad 4)_0$$

and

$$\text{OPEN}(f'', g) = (2 \quad 3 \quad 3 \quad -99 \quad 3 \quad 4 \quad 4)_0$$

The value -99 in f'' represents a long negative spike, and as such, it is left unaltered by the opening. But the $*$, or undefined value, in f' behaves in exactly the same way. Thus, just as $+\infty$ behaved as an infinite spike in Example 7.5, the value $*$ behaves, relative to the opening, as an infinite spike in the negative direction. Indeed, all points outside the domain of any signal act like $-\infty$ relative to both the usual and the extended opening.

Sec. 7.1   Extended Signals

Taking the preceding remarks as representing our intuitions, we can now proceed to extend the definition of the closing to the extended signal context. The approach is quite clear: we must define our operations in such a manner as to preserve the duality between OPEN and CLOSE. Accordingly, if $f(t) = +\infty$, we define

$$(-f)(t) = [\text{SUB}(f)](t) = *$$

On the other hand, if $f(t) = *$, which means that f is undefined at t, we define

$$(-f)(t) = [\text{SUB}(f)](t) = +\infty$$

In other words, we let $*$ act as though it were $-\infty$. From a purely mathematical viewpoint, what we are doing is changing the domain by taking points outside D_f and giving them the value $+\infty$ on D_{-f}, and taking points with value $+\infty$ in D_f and not including them in the domain of D_{-f}. However, it should be remembered that in computing CLOSE, we apply SUB twice. In the end, the definition of SUB has been motivated by modeling considerations: we want a value of $*$ to behave like an infinite negative spike, we want a value of $+\infty$ to behave like an infinite positive spike, and we want to preserve the duality between OPEN and CLOSE.

With the foregoing definition of SUB, it is easy to specify CLOSE: we simply employ the usual definition and adopt the convention that $*$ acts like $-\infty$ in the ordering scheme. This method of treating $*$ defines the manner in which fitting will take place. Indeed, the manner in which EXTMAX and MIN have been defined throughout is consistent with this new technique. (In general image algebra, the interpretation of $*$ as $-\infty$ does not work; therefore, we only operate in this manner in the special case of the morphological closing of extended signals (and extended images).)

Before proceeding to an example, one more convention must be adopted. Since, in the treatment just outlined, we assume that all values outside the frame of a bound matrix are $*$, and the operation SUB makes them $+\infty$, we shall employ the superscript s to indicate that the bound matrix is serving the special purpose at hand—i.e., that all values outside the frame are to be treated as $+\infty$. The situation is analogous to that of complementary bound matrices, where we use the superscript c to indicate that all values outside the $1-*$ bound matrix are 1's.

Example 7.6:

Suppose we wish to smooth the signal

$$f = (1 \quad 2 \quad 1 \quad +\infty \quad 2 \quad 3 \quad * \quad 3 \quad 2)_0$$

by flattening it from above. Moreover, suppose we wish the filtering to be very "fine," with the structuring element only two points wide. It would seem that we should then apply CLOSE with the structuring element

$$g = (0 \quad 0)_0$$

Proceeding in the extended sense,

$$\text{SUB}(f) = (-1 \quad -2 \quad -1 \quad * \quad -2 \quad -3 \quad +\infty \quad -3 \quad -2)_0^s$$

and

$$\text{SUB}(g) = (0 \quad 0)_0^s$$

Then, under our conventions regarding SUB, we get

$$[\text{SUB}(g)](-1) = [\text{SUB}(g)](2) = +\infty$$

Next, using a straight fitting criterion, and recalling the $+\infty$ outside the frames of the bound vectors, we obtain OPEN$(-f, -g)$ to be the null signal $(*)_0$. Indeed, because of the values of $+\infty$ outside the frame of the bound vector representing SUB(g), no translate of the form $-g_i + j$ fits under $-f$. Finally, applying SUB to OPEN$(-f, -g)$ gives the identically infinite signal. Specifically,

$$\text{CLOSE}(f, g) = (+\infty)_0^s$$

which certainly is not the desired smoothing.

Although the approach we have taken would have worked perfectly in the finite, nonextended situation, we can see from the preceding example that it does not work in the extended setting. There are two reasons for this: (1) SUB has taken $*$, which from a modeling point of view means undefined, and turned it into $+\infty$; and (2) related to this, CLOSE(f, g) is specified using SUB(g). Consequently, whereas dilation, erosion, and opening behave in complete accordance with our intuitions in the extended setting, closing requires a bit more care to be taken. Nevertheless, we can still accomplish the desired ends.

Instead of using the signal g, let us employ the extended signal

$$h = (0 \quad 0)_0^s$$

as a structuring element. Then

$$\text{SUB}(h) = (0 \quad 0)_0$$

$$\text{OPEN}(-f, -h) = (-1 \quad -2 \quad -2 \quad * \quad -3 \quad -3 \quad -3 \quad -3 \quad -2)_0^s$$

and

$$\text{CLOSE}(f, h) = (1 \quad 2 \quad 2 \quad +\infty \quad 3 \quad 3 \quad 3 \quad 3 \quad 2)_0$$

Looking at the original signal f, this output is exactly in line with our intuitions: it has smoothed from the top, left the infinite positive spike at $i = 3$ unaltered, and has filtered out the infinite negative spike at $i = 6$. Whereas we might view the domain of SUB(h) as consisting of just two points, in any fitting or order-relation approach all extended signals have all of Z as their domain, with some values being $+\infty$ and some being $*$, which acts like $-\infty$.

It is evident that in the extended-signal world the closing behaves in a manner that at first appears odd. In fact, just the opposite is true: it behaves exactly as we should expect it to behave. It is a dual operation, and duality is closely tied to the order relation. Although we shall not specifically discuss gray-scale duality, as we did in Chapter 5 for the two-valued case, the behavior of the closing is totally consistent with duality theory. Even more—its behavior is completely intuitive: simply think of infinite spikes protruding from backgrounds of infinite extent.

While many applications of gray-scale morphology are not concerned with extended signals, the behavior of such signals is fundamental to the theory of gray-scale morphological filter representation. In subsequent sections we shall develop gray-scale versions of the Matheron representation theorems for increasing τ-mappings and τ-openings. In addition, we shall discuss the problem of finding bases that will allow finite (and in many instances, trivial) expansions for increasing τ-mappings in terms of gray-scale erosions. Not only do the main theorems depend upon extended signals for their proofs, but practical filters must be defined carefully with respect to their behavior at undefined points so that they will possess workable bases.

7.2 GRAY-SCALE MORPHOLOGICAL FILTERS FOR SIGNALS

In Chapter 5 we discussed the Matheron theory concerning the representation of morphological filters. We now present a corresponding theory applicable to gray-scale morphological filtering. We begin with the class $\mathcal{S}^*$ of extended signals defined on subsets of the real line.

Consider a mapping of the form $\Psi: \mathcal{S}^* \to \mathcal{S}^*$, a so-called signal-to-signal mapping. Ψ is said to be *translation invariant*, or a τ-*mapping*, if, for any $f \in \mathcal{S}^*$,

$$\Psi(f_x + y) = [\Psi(f)]_x + y$$

In other words, if an extended signal f is translated by x and then offset by y, Ψ of the resulting signal is equal to the translation by x and the offsetting by y of $\Psi(f)$. Furthermore, Ψ is *increasing* if, whenever $f \ll g$, then $\Psi(f) \ll \Psi(g)$. As in the two-valued case, an increasing τ-mapping is called a *morphological filter*. (Recall that in Section 6.1 we defined translation invariance in terms of mappings on nonextended signals. Whenever we say that Ψ is a τ-mapping, we implicitly mean that it is translation invariant on $\mathcal{S}^*$.)

In Section 6.2 we defined the surface of an umbra. We now extend the surface transform to all subsets of R^2 by defining $\mathbf{S}[A] = \mathbf{S}[\mathbf{U}[A]]$ for any $A \subset R^2$. Note that if, for some $x_0 \in R$,

$$\sup \{y: (x_0, y) \in A\} = +\infty$$

then $\mathbf{U}[A]$ contains the infinite vertical line $x = x_0$, and $\mathbf{S}[A](x_0) = +\infty$, which is now an acceptable value for a signal.

For any mapping Ψ defined on $\mathcal{S}^*$, there exists an induced mapping

$$\hat{\Psi}: 2^{R \times R} \longrightarrow \mathcal{U}$$

where $\mathcal{U}$ is the collection of all umbrae, and where $\hat{\Psi}$ is defined by

$$\hat{\Psi}(A) = \mathbf{U}[\Psi(\mathbf{S}[A])]$$

In words, to find $\hat{\Psi}(A)$, take the surface of A, operate on it by Ψ, and then form the umbra. (See Figure 7.4.)

On the other hand, suppose

$$\Phi: 2^{R \times R} \longrightarrow 2^{R \times R}$$

Then there is an induced mapping

$$\check{\Phi}: \mathcal{S}^* \longrightarrow \mathcal{S}^*$$

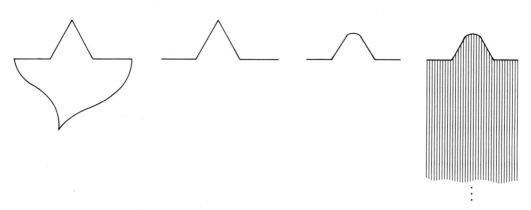

Figure 7.4 Action of induced mapping

defined by

$$\check{\Phi}(f) = S[\Phi(U[f])]$$

In words, to find $\check{\Phi}(f)$, take the umbra transform of f, operate on it by Φ, and then take the surface transform.

We next present some basic properties concerning the mappings Ψ, $\hat{\Psi}$, Φ, and $\check{\Phi}$. Some of the proofs involve delicate arguments concerning suprema; these will be annotated with an asterisk. For those readers interested in applications, the illustrations and subsequent examples should provide sufficient understanding.

For any extended signal f,

$$f_x + y = S[U[f_x + y]] = S[U[f] + (x, y)]$$

To simplify notation, we shall often let $\mathbf{z} = (x, y)$ and write $f_\mathbf{z}$ in lieu of $f_x + y$. In this notation, the preceding identity becomes

$$f_\mathbf{z} = S[U[f] + \mathbf{z}]$$

The definitions of translation invariance, increasing monotonicity, the surface operator, and the induced mappings $\hat{\Psi}$ and $\check{\Phi}$ have immediate analogues in the digital setting. No separate attention will be paid to the Euclidean and digital cases except when there is a material difference.

The proofs of several important properties follow.

Property GF-1. $\hat{\Psi} \circ U = U \circ \Psi$ on $\mathcal{S}^\#$.

Proof. Since, $S[U[f]] = f$,

$$\hat{\Psi}(U[f]) = U[\Psi(S[U[f]])] = U[\Psi(f)]$$

which, in function composition notation, is GF-1. (See Figure 7.5.)

Sec. 7.2   Gray-Scale Morphological Filters for Signals   **235**

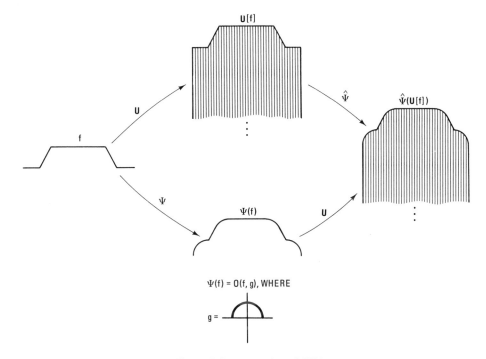

Figure 7.5   Illustration of GF-1

An immediate consequence of GF-1 is the fundamental identity

$$\mathbf{S} \circ \hat{\Psi} \circ \mathbf{U} = \mathbf{S} \circ \mathbf{U} \circ \Psi = \Psi$$

Property GF-2. If Ψ is increasing, then so is $\hat{\Psi}$.

Proof. If $A \subset B$, then $\mathbf{S}[A] \ll \mathbf{S}[B]$, and hence,

$$\hat{\Psi}(A) = \mathbf{U}[\Psi(\mathbf{S}[A])] \subset \mathbf{U}[\Psi(\mathbf{S}[B])] = \hat{\Psi}(B)$$

Property GF-3: If Ψ is a τ-mapping, then so is $\hat{\Psi}$.

Proof.
$$\begin{aligned}\hat{\Psi}(A + \mathbf{z}) &= \mathbf{U}[\Psi(\mathbf{S}[A + \mathbf{z}])] = \mathbf{U}[\Psi(\mathbf{S}[A]_{\mathbf{z}})] \\ &= \mathbf{U}[\Psi(\mathbf{S}[A])_{\mathbf{z}}] = \mathbf{U}[\Psi(\mathbf{S}[A])] + \mathbf{z} \\ &= \hat{\Psi}(A) + \mathbf{z}\end{aligned}$$

Example 7.7:

In this example we consider a morphological filter that belongs to the class of order-statistic filters, which we shall study in greater detail in Section 8.3. Given a digital signal f, and a point i in the domain of f, list in order of magnitude the values $f(i - 1)$, $f(i)$, $f(i + 1)$, and $f(i + 2)$. Define the value of $\Upsilon(f)$ at the point i to be the second-largest value among the four signal values. Should any of the four values be a $*$, we define $[\Upsilon(f)](i) = *$. For instance, if $f(i - 1) = 2$, $f(i) = 8$, $f(i + 1) = -3$, and

$f(i + 2) = 3$, then $[Y(f)](i) = 3$. If $f(i - 1) = +\infty$, $f(i) = 0$, $f(i + 1) = 2$, and $f(i + 2) = 2$, then $[Y(f)](i) = 2$.

Suppose
$$f = (3\ 2\ 5\ 5\ 2\ 6\ 7\ 3\ 5)_0$$
Then
$$Y(f) = (5\ 5\ 5\ 6\ 6\ 6)_1$$

Notice the smoothing effect of Y.

Now consider the constant image

$$T = \begin{pmatrix} * & * & * & * & * & * & 1 & * & * \\ * & * & * & * & * & 1 & 1 & * & * \\ * & * & 1 & 1 & * & 1 & * & * & 1 \\ * & * & * & * & * & 1 & 1 & * & 1 \\ 1 & * & 1 & 1 & * & 1 & 1 & 1 & 1 \\ 1 & 1 & 1 & * & 1 & 1 & 1 & * & 1 \end{pmatrix}_{0,7}$$

Since $S[T] = f$,
$$\hat{Y}(T) = U[Y(S[T])] = U[Y(f)]$$

which is an umbra. We represent it by the umbra matrix

$$\hat{Y}(T) = \begin{pmatrix} * & * & * & 1 & 1 & 1 \\ 1 & 1 & 1 & 1 & 1 & 1 \end{pmatrix}_{1,6} = \bar{U}\,[Y(f)]$$

The next two properties deal with monotonicity and τ-mapping relations between Φ and $\check{\Phi}$.

Property GF-4. If Φ is increasing, then so is $\check{\Phi}$.

Proof. $f \ll g$ implies $U[f] \subset U[g]$; hence, since Φ is increasing,
$$\check{\Phi}(f) = S[\Phi(U[f])] \ll S[\Phi(U[g])] = \check{\Phi}(g)$$

Property GF-5. If Φ is a τ-mapping, then so is $\check{\Phi}$.

Proof.
$$\check{\Phi}(f_z) = S[\Phi(U[f_z])] = S[\Phi(U[f] + z)]$$
$$= S[\Phi(U[f]) + z] = S[\Phi(U[f])]_z$$
$$= \check{\Phi}(f)_z$$

The following three properties are straightforward in the digital setting, $Z \times Z \to Z$; however, proving them in the Euclidean setting requires arguments concerning the supremum that invoke limits.

Property GF-6. $S[\mathscr{E}(U[A], U[B])] = S[\mathscr{E}(U[S[A]], U[B])]$

Proof.* $U[A] \subset U[S[A]]$. Therefore, by M-10,
$$\mathscr{E}(U[A], U[B]) \subset \mathscr{E}(U[S[A]], U[B])$$

and hence,
$$S[\mathscr{E}(U[A], U[B])] \ll S[\mathscr{E}(U[S[A]], U[B])]$$

Conversely, suppose $S[\mathscr{E}(U[S[A]], U[B])](x) = y$. Then there exist points (x, y_j) such that (x, y_j) is in the erosion and $\lim_{j \to \infty} y_j = y$. Now, since (x, y_j) is in the erosion, it follows that
$$U[B] + (x, y_j) \subset U[S[A]]$$
Hence, for $j = 1, 2, 3, \ldots,$
$$U[B] + (x, y_j - \tfrac{1}{j}) \subset U[S[A]] + (0, \tfrac{-1}{j}) \subset U[A]$$
Therefore,
$$(x, y_j - \tfrac{1}{j}) \in \mathscr{E}(U[A], U[B])$$
Letting $j \to \infty$, we have
$$y \leq S[\mathscr{E}(U[A], U[B])](x)$$

Property GF-7. $S[\mathscr{E}(U[A], U[B])] = S[\mathscr{E}(U[A], U[S[B]])]$

*Proof.** By M-11,
$$\mathscr{E}(U[A], U[S[B]]) \subset \mathscr{E}(U[A], U[B])$$
and hence,
$$S[\mathscr{E}(U[A], U[S[B]])] \ll S[\mathscr{E}(U[A], U[B])]$$

Conversely, suppose $S[\mathscr{E}(U[A], U[B])](x) = y$. Then there exist points (x, y_j) such that $\lim_{j \to \infty} y_j = y$ and
$$U[B] + (x, y_j) \subset U[A]$$
Hence, for any positive integer j,
$$U[S[U[B]]] + (x, y_j - \tfrac{1}{j}) \subset U[A]$$
Therefore,
$$(x, y_j - \tfrac{1}{j}) \in \mathscr{E}(U[A], U[S[U[B]]])$$
Letting $j \to \infty$, and noting that $S[U[B]] = S[B]$, we have
$$y \leq S[\mathscr{E}(U[A], U[S[B]])](x)$$

Property GF-8. For any collection $\{B_k\}$ of subsets of R^2,
$$S\left[\bigcup_{k=1}^{\infty} B_k\right] = \underset{k}{\text{EXTSUP}}(S[B_k])$$

Proof.* Suppose

$$S\left[\bigcup_{k=1}^{\infty} B_k\right](x) = y$$

Then there exist points (x, y_j) such that $\lim_{j \to \infty} y_j = y$ and (x, y_j) lies in at least one of the B_k. The latter condition implies that $y_j \leq S[B_k](x)$ for at least one k. Hence,

$$y \leq \text{EXTSUP}(S[B_k])(x)$$

Conversely, suppose $w = \text{EXTSUP}(S[B_k])(x)$. Let $\epsilon > 0$. Then there exists a k' such that

$$|w - S[B_{k'}](x)| < \frac{\epsilon}{2}$$

Moreover, there exits a w_0 such that $(x, w_0) \in B_{k'}$ and

$$|w_0 - S[B_{k'}](x)| < \frac{\epsilon}{2}$$

By the triangle inequality,

$$|w_0 - w| \leq |w_0 - S[B_{k'}](x)| + |w - S[B_{k'}](x)| < \frac{\epsilon}{2} + \frac{\epsilon}{2} = \epsilon$$

Hence, since (x, w_0) lies in the union of the B_k,

$$S\left[\bigcup_{k=1}^{\infty} B_k\right](x) > w - \epsilon$$

and, since ϵ was arbitrary,

$$S\left[\bigcup_{k=1}^{\infty} B_k\right](x) \geq w$$

Example 7.8:

In the digital setting, we can replace union and EXTSUP by EXTMAX in GF-8. Let

$$B_1 = \begin{pmatrix} * & 1 & 1 & * \\ * & 1 & * & 1 \\ \textcircled{1} & * & * & 1 \end{pmatrix}$$

$$B_2 = \begin{pmatrix} 1 & * & * & * \\ * & 1 & * & 1 \\ \textcircled{1} & 1 & * & 1 \end{pmatrix}$$

and

$$B_3 = \begin{pmatrix} * & * & * & * & 1 \\ * & * & * & * & 1 \\ \textcircled{*} & * & * & 1 & * \end{pmatrix}$$

Sec. 7.2 Gray-Scale Morphological Filters for Signals

Then

$$\text{EXTMAX}(B_1, B_2, B_3) = \begin{pmatrix} 1 & 1 & 1 & * & 1 \\ * & 1 & * & 1 & 1 \\ \textcircled{1} & 1 & * & 1 & * \end{pmatrix}$$

and, in bound matrix format,

$$S[\text{EXTMAX}(B_1, B_2, B_3)] = \begin{pmatrix} 1 & 1 & 1 & * & 1 \\ * & * & * & 1 & * \\ \textcircled{*} & * & * & * & * \end{pmatrix}$$

On the other hand, in signal form,

$$S[B_1] = (0 \quad 2 \quad 2 \quad 1)_0$$
$$S[B_2] = (2 \quad 1 \quad * \quad 1)_0$$

and

$$S[B_3] = (* \quad * \quad * \quad 0 \quad 2)_0$$

so that

$$\text{EXTMAX}(S[B_1], S[B_2], S[B_3]) = (2 \quad 2 \quad 2 \quad 1 \quad 2)_0$$

GF-6 and GF-7 together yield the fundamental property GF-9, which is a generalization of Theorem 6.5 that applies to sets in R^2.

Property GF-9. $\mathcal{E}(S[A], S[B]) = S[\mathcal{E}(U[A], U[B])]$

Proof.

$$\begin{aligned}\mathcal{E}(S[A], S[B]) &= S[\mathcal{E}(U[S[A]], U[S[B]])] && \text{(Theorem 6.5)} \\ &= S[\mathcal{E}(U[A], U[S[B]])] && \text{(GF-6)} \\ &= S[\mathcal{E}(U[A], U[B])] && \text{(GF-7)}\end{aligned}$$

Note that the use of Theorem 6.5 in the proof of GF-9 is legitimate since the proof given originally for the theorem carries over without change to the extended setting.

Whereas in the Euclidean case GF-9 requires GF-6 and GF-7 in its derivation, in the digital setting it follows easily from the umbra homomorphism for erosion, Theorem 6.6(ii). Beginning with the umbra homomorphism, and treating $S[A]$ as a signal, we have

$$\begin{aligned}U[\text{ERODE}(S[A], S[B])] &= \text{ERODE}(U[S[A]], U[S[B]]) \\ &= \text{ERODE}(U[A], U[B])\end{aligned}$$

where we have used the fact that in the digital setting $U[S[A]] = U[A]$. Now apply the surface operator to both sides of the preceding equation to obtain

$$S[U[\text{ERODE}(S[A], S[B])]] = S[\text{ERODE}(U[A], U[B])]$$

But for any signal f, $S[U[f]] = f$, and hence the last equation reduces to

$$\text{ERODE}(S[A], S[B]) = S[\text{ERODE}(U[A], U[B])]$$

which is precisely GF-9 in digital form.

Although we shall not go through the details, a similar result, which is a generalization of Theorem 6.4, applies to dilation:

Property GF-10. $\mathcal{D}(S[A], S[B]) = S[\mathcal{D}(U[A], U[B])]$

In order to arrive at a gray-scale version of Theorem 5.2, it is necessary to define the *kernel* of a signal-to-signal increasing τ-mapping. For such a mapping Ψ,

$$\text{Ker}[\Psi] = \{f\colon [\Psi(f)](0) \geq 0\}$$

Thus, the kernel consists of all those signals for which the output signal is nonnegative at the origin.

Example 7.9:

Let Ψ be defined by $[\Psi(f)](t) = f(t) + r$, where r is a constant. Then Ψ simply offsets f by r. The kernel of Ψ is the collection of all signals f such that

$$f(0) + r \geq 0$$

In other words, $f \in \text{Ker}[\Psi]$ if and only if $f(0) \geq -r$. In this instance, implicit in f being in the kernel of Ψ is that zero lies in the domain of f.

Example 7.10:

Let $\Psi(f) = O(f, g)$, where

$$g(t) = \begin{cases} 1, & \text{if } -1 \leq x \leq 1 \\ \text{undefined}, & \text{elsewhere} \end{cases}$$

Then $f \in \text{Ker}[\Psi]$ if and only if $[O(f, g)](0) \geq 0$. Figure 7.6 shows four signals, two of which are in the kernel, and two of which are not in the kernel, of Ψ. Note that in each case $f = O(f, g)$.

Property GF-11. Suppose Ψ is an increasing τ-mapping on $\mathcal{S}^{\#}$. Then $A \in \text{Ker}[\hat{\Psi}]$ if and only if $\Psi(S[A])(0) \geq 0$

Proof. $A \in \text{Ker}[\hat{\Psi}]$ if and only if

$$(0, 0) \in \hat{\Psi}(A) = U[\Psi(S[A])]$$

if and only if $\Psi(S[A])(0) \geq 0$.

The next theorem gives the desired representation of gray-scale increasing τ-mappings. Just as Theorem 5.2 showed that all increasing τ-mappings of constant images are expressable in terms of constant-image erosions, Theorem 7.1 shows that all increasing τ-mappings of signals can be represented in terms of gray-scale erosions.

Sec. 7.2 Gray-Scale Morphological Filters for Signals

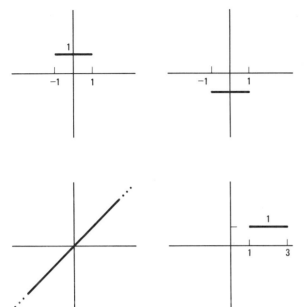

Figure 7.6 Kernel and nonkernel signals

Theorem 7.1. If Ψ is an increasing τ-mapping, then

$$\Psi(f) = \text{EXTSUP}_{g \in \text{Ker}[\Psi]} \mathcal{E}(f, g)$$

Proof.

$$\Psi(f) = S[\hat{\Psi}(U[f])] \qquad \text{(GF-1)}$$

$$= S\left[\bigcup_{A \in \text{Ker}[\hat{\Psi}]} \mathcal{E}(U[f], A)\right] \qquad \text{(Theorem 5.2)}$$

$$= \text{EXTSUP}_{A \in \text{Ker}[\hat{\Psi}]} S[\mathcal{E}(U[f], A)] \qquad \text{(GF-8)}$$

$$= \text{EXTSUP}_{A \in \text{Ker}[\hat{\Psi}]} S[\mathcal{E}(U[f], U[A])]$$

$$= \text{EXTSUP}_{A \in \text{Ker}[\hat{\Psi}]} S[\mathcal{E}(U[U[f]], U[A])] \qquad \text{(idempotence of U)}$$

$$= \text{EXTSUP}_{A \in \text{Ker}[\hat{\Psi}]} \mathcal{E}(S[U[f]], S[A]) \qquad \text{(GF-9)}$$

$$= \text{EXTSUP}_{A \in \text{Ker}[\hat{\Psi}]} \mathcal{E}(f, S[A]) \qquad (f = S[U[f]])$$

$$= \text{EXTSUP}_{\{A: \Psi(S[A])(0) \geq 0\}} \mathcal{E}(f, S[A]) \qquad \text{(GF-11)}$$

$$= \text{EXTSUP}_{\{g: [\Psi(g)](0) \geq 0\}} \mathcal{E}(f, g)$$

$$= \text{EXTSUP}_{g \in \text{Ker}[\Psi]} \mathcal{E}(f, g)$$

The next-to-last equality follows from the fact that there is a one-to-one correspondence between surfaces of sets and signals, and the last follows from the definition of the kernel. The fourth equality results from the obvious fact that eroding an umbra by a set gives the same output as eroding by the umbra of that set.

Note the use of extended signals in the proof of Theorem 7.1: we need $\hat{\Psi}$ to be defined on all subsets of the plane in order to apply Theorem 5.2.

Just as the Euclidean version of Theorem 5.2 becomes concrete in the digital setting, so will Theorem 7.1 become less abstract when applied to $Z \times Z$.

Example 7.11:

Consider the mapping Y of Example 7.7. Then $[Y(f)](0) \geq 0$ if f is defined at -1, 0, 1, and 2, and if at least two of the values $f(-1)$, $f(0)$, $f(1)$, and $f(2)$ are nonnegative. Precisely such signals comprise the kernel of Y. For instance,

$$f = (-1 \quad -3 \quad -2 \quad 2 \quad 1 \quad -4)_{-2}$$

lies in Ker [Y] since $[Y(f)](0) = 1$, whereas

$$g = (2 \quad 2 \quad 2 \quad * \quad 2 \quad 1)_{-1}$$

and

$$h = (0 \quad 2 \quad -6 \quad -3 \quad -1 \quad 0)_{-2}$$

do not lie in the kernel since $[Y(g)](0) = *$ and $[Y(h)](0) = -1$.

Applying Theorem 7.1 to Y, we obtain

$$Y(f) = \text{EXTSUP}_{g \in \text{Ker}[Y]} \text{ERODE}(f, g)$$

where we have kept the EXTSUP operation since there are infinitely many elements in the kernel.

7.3 BASIS FOR THE KERNEL

In Section 5.3 we used property M-11 as they key to the construction of a basis for the kernel of a constant-image increasing τ-mapping. Now using property GM-9, the gray-scale signal version of M-11, we shall present a somewhat parallel development for signal-to-signal increasing τ-mappings. Since out goal is a digitally implementable version of Theorem 7.1, we shall restrict our attention to the digital setting; however, those familiar with the theory of partially ordered sets will no doubt recognize that we could have proceeded in the Euclidean framework insofar as the general theory is concerned.

Let Ψ be an increasing τ-mapping. Then a subcollection of elements in Ker $[\Psi]$ is a *basis* $\mathcal{M}$ for the kernel if

(1) No two elements of $\mathcal{M}$ are properly related by the order relation $\ll$.
(2) For any signal $f \in \text{Ker } [\Psi]$, there exists a signal $f' \in \mathcal{M}$ such that $f' \ll f$.

Condition 1 means that if $g, h \in \mathcal{M}$ and $g \ll h$, then $g = h$.

Using the preceding definition, we shall develop the basis theory for signals that corresponds to the theory of Section 5.3.

Theorem 7.2. If there exists a basis for the kernel of an increasing τ-mapping on signals, then that basis is unique.

Sec. 7.3 Basis for the Kernel

Proof. Suppose $\mathcal{M}$ and $\mathcal{N}$ are bases of Ker $[\Psi]$ and suppose $f \in \mathcal{M}$. Then, since $f \in$ Ker $[\Psi]$, there exists an $f' \in \mathcal{N}$ such that $f' \ll f$. Similarly, there exists an $f'' \in \mathcal{M}$ such that $f'' \ll f'$. Hence, $f'' \ll f' \ll f$. But by condition 1 of the definition of a basis, $f'' = f$, and hence $f = f'$. Consequently, $f \in \mathcal{N}$, and $\mathcal{N} \supset \mathcal{M}$. And by a similar argument, $\mathcal{M} \supset \mathcal{N}$.

Given this uniqueness of the basis for the kernel, we shall henceforth write Bas $[\Psi]$ to denote $\mathcal{M}$.

A signal-to-signal mapping Ψ is said to be *continuous from above* if, whenever

$$f_1 \gg f_2 \gg f_3 \gg \cdots$$

then

$$\Psi[\text{INF}(f_k)] = \text{INF}[\Psi(f_k)]$$

Continuity from above relative to signals is analogous to continuity from above relative to constant-image mappings. Hence, we should expect the next theorem to hold, the proof of which should be skipped by those not familiar with Zorn's lemma.

Theorem 7.3. If Ψ is an increasing τ-mapping on signals that is continuous from above, then there exists a basis for the kernel.

Proof. Define

$$\mathcal{M} = \{f \in \text{Ker }[\Psi] : \text{there does not exist an } f' \neq f \text{ in Ker }[\Psi] \text{ with } f' \ll f\}$$

Then condition 1 for the basis is clearly fulfilled. To show condition 2, let f be any signal in Ker $[\Psi]$, and let $\mathcal{P}$ be the class of all f' in Ker $[\Psi]$ such that $f' \ll f$. Then $\mathcal{P}$ is partially ordered by $\ll$. Now, suppose $\{f_1, f_2, \ldots\}$ is a totally ordered subclass of $\mathcal{P}$, and let

$$\bar{f} = \text{INF}_k(f_k)$$

Then, since Ψ is continuous from above and $f_k \in$ Ker $[\Psi]$ for any k, we have

$$0 \leq [\text{INF}(\Psi(f_k))](0) = [\Psi(\text{INF}(f_k))](0) = [\Psi(\bar{f})](0)$$

and hence $\bar{f} \in$ Ker $[\Psi]$. Moreover, $\bar{f}$ is a lower bound for $\{f_k\}$. Consequently, by Zorn's lemma, $\mathcal{P}$ possesses a minimal element, say f^*. But $f^* \in \mathcal{M}$, or else there would exist some signal f^{**} in $\mathcal{P}$ such that $f^{**} \ll f^*$ properly, and this would contradict the minimality of f^*. But then, the signal f^* is the element demanded by condition 2, and it follows that $\mathcal{M}$ is a basis for Ker $[\Psi]$.

Example 7.12:

Consider the mapping Y whose kernel was found in Example 7.11. We show that Y is not continuous from above. For $k = 1, 2, 3, \ldots$, let

$$f_k = (0 \quad 0 \quad 0 \quad -k)_{-1}$$

Then for any k,

$$Y(f_k) = (0)_0$$

and hence
$$\text{INF}[Y(f_k)] = (0)_0$$
However,
$$\text{INF}(f_k) = (0 \ \ 0 \ \ 0)_{-1}$$
where we have used the fact that, in the extended setting, inf $(-k) = *$. Hence, $Y[\text{INF}(f_k)]$ is the null signal, and Y is not continuous from above.

Now, since Theorem 7.3 gives a sufficient, but not necessary, condition for the existence of a basis, one might still hope that Y possesses a basis. However, it is a fairly straightforward task to show that it does not. Indeed, if it did, and if f were an element of the basis, then f would have to be activated at precisely the points $-1, 0, 1,$ and 2, or else its domain could be trivially reduced. Moreover, at least two of its values would have to be nonnegative. Thus, whatever the other two values might be, we could always reduce them to produce a signal f' such that $f' \ll f$ properly and $f' \in \text{Ker}\,[Y]$. But since $f' \in \text{Ker}\,[Y]$, there exists an f'' in the basis such that $f'' \ll f'$. But this means that $f'' \ll f$, and hence the first requirement of a basis would be invalidated.

To illustrate the foregoing remarks, suppose
$$f = (0 \ \ -2 \ \ 0 \ \ -9)_{-1}$$
lies in the basis. Then, since
$$f' = (0 \ \ -3 \ \ 0 \ \ -9)_{-1}$$
lies in Ker $[Y]$, there exists an $f'' \ll f'$ in the basis. But this implies that $f'' \ll f$.

The preceding example at first appears to throw a shadow on our ability to obtain bases for increasing τ-mappings that are both simple and useful. In fact, however, it merely shows us that we have not been sufficiently careful in defining the mapping Y insofar as its "boundary conditions" are concerned. Instead of defining $Y(f) = *$ whenever one of the values $f(-1), f(0), f(1),$ or $f(2)$ happens to be $*$, we should have proceeded to consider $*$ as part of the ordering. Indeed, in the order relation on extended signals, $*$ acts like $-\infty$. The next example remedies the situation and illustrates a fundamental point: even though we are only concerned in practice with the behavior of a signal-to-signal mapping as that behavior manifests itself on finite signals, we must pay attention to the specification of the mapping on extended signals also.

Example 7.13:

Redefine the mapping Y of Example 7.7 so that the value $*$ is treated as $-\infty$ in the ordering. For instance, if $f(i - 1) = 2, f(i) = *, f(i + 1) = 8$, and $f(i + 2) = +\infty$, then $[Y(f)](i) = 8$; and if $f(i - 1) = 2, f(i) = *, f(i + 1) = *$, and $f(i + 2) = *$, then $[Y(f)](i) = *$. Once again, Y is an increasing τ-mapping. Its kernel is given by all signals that possess at least two nonnegative signal values among the values $f(-1), f(0), f(1),$ and $f(2)$. Moreover, a signal is not rejected for the kernel if one or two of those values happen to be stars. Thus, whereas the signal g of Example 7.11 did not lie in the kernel according to the original definition of Y, it does lie in the kernel according to the new definition.

Sec. 7.3   Basis for the Kernel

Most importantly, with the new definition, there exists a basis for Ker [Y]. Indeed, it contains only six elements:

$$f_1 = (0 \quad 0)_{-1}$$
$$f_2 = (0 \quad * \quad 0)_{-1}$$
$$f_3 = (0 \quad * \quad * \quad 0)_{-1}$$
$$f_4 = (0 \quad 0)_{0}$$
$$f_5 = (0 \quad * \quad 0)_{0}$$
$$f_6 = (0 \quad 0)_{1}$$

According to the property GM-9, if g_1 and g_2 are two signals such that $g_1 \ll g_2$, then

$$\text{ERODE}(h, g_2) \ll \text{ERODE}(h, g_1)$$

and we can state the extended-signal version of Theorem 5.5, which follows from Theorem 7.1 just as Theorem 5.5 followed from Theorem 5.2.

Theorem 7.4.   If Ψ is an increasing τ-mapping that possesses a basis for the kernel, then

$$\Psi(f) = \text{EXTSUP}_{g \in \text{Bas}[\Psi]} \text{ERODE}(f, g)$$

Example 7.14:

Let Y be defined as in Example 7.13, and let $f_1, f_2, \ldots, f_6$ be the basis elements given in that example. By Theorem 7.4, for any signal h,

$$Y(h) = \text{EXTMAX}_{k=1,2,\ldots,6} \text{ERODE}(h, f_k)$$

For instance, let

$$h = (1 \quad 8 \quad 9 \quad 1 \quad 9 \quad 1 \quad 9)_0$$

Then

$$Y(h) = (1 \quad 8 \quad 8 \quad 9 \quad 9 \quad 9 \quad 9 \quad 1)_{-1} \qquad (1)$$

To use the basis expansion, we find the six erosions $\text{ERODE}(h, f_k)$ and then apply EXTMAX. Stacked in order, the six erosions yield a serial matrix, and the output is obtained by taking extended maximums down the columns. The six erosions are

$$\text{ERODE}(h, f_1) = (1 \quad 8 \quad 1 \quad 1 \quad 1 \quad 1)_1$$
$$\text{ERODE}(h, f_2) = (1 \quad 1 \quad 9 \quad 1 \quad 9)_1$$
$$\text{ERODE}(h, f_3) = (1 \quad 8 \quad 1 \quad 1)_1$$
$$\text{ERODE}(h, f_4) = (1 \quad 8 \quad 1 \quad 1 \quad 1 \quad 1)_0$$
$$\text{ERODE}(h, f_5) = (1 \quad 1 \quad 9 \quad 1 \quad 9)_0$$
$$\text{ERODE}(h, f_6) = (1 \quad 8 \quad 1 \quad 1 \quad 1 \quad 1)_{-1}$$

Consequently, the appropriate serial matrix is given by

$$\begin{pmatrix} * & * & 1 & 8 & 1 & 1 & 1 & 1 \\ * & * & 1 & 1 & 9 & 1 & 9 & * \\ * & * & 1 & 8 & 1 & 1 & * & * \\ * & 1 & 8 & 1 & 1 & 1 & 1 & * \\ * & 1 & 1 & 9 & 1 & 9 & * & * \\ 1 & 8 & 1 & 1 & 1 & 1 & * & * \end{pmatrix}_{-1,5}$$

Taking extended maximums down the columns gives the same result as equation (1).

In the previous example the basis for the kernel was finite, consisting of only six terms. As the next example illustrates, however, finiteness is not a necessary characteristic of a basis.

Example 7.15:

For any extended digital signal f, define $[\Omega(f)](i)$ to be the average of the two values $f(i)$ and $f(i + 1)$, where, for fractional averages, we round up. Clearly, Ω is an increasing τ-mapping. Moreover, the kernel of Ω consists of all signals for which the unquantized average value of $f(0)$ and $f(1)$ is greater than -1. Thus, the signals

$$h_1 = (2 \quad -3 \quad -8)_0$$
$$h_2 = (7 \quad -5 \quad 9 \quad 2)_{-1}$$

and

$$h_3 = (2 \quad 1 \quad +\infty \quad -99)_{-2}$$

lie in Ker $[\Omega]$, whereas the signals

$$k_1 = (-2 \quad 0)_0$$
$$k_2 = (8 \quad 99)_{-1}$$

and

$$k_3 = (2 \quad 3 \quad 4)_1$$

do not lie in Ker $[\Omega]$.

Signals of the form

$$f^n = (n \quad -n - 1)_0$$

where $n \in Z$, form a basis for the kernel. Typical basis signals are

$$(0 \quad -1)_0$$
$$(1 \quad -2)_0$$
$$(-1 \quad 0)_0$$

and

$$(98 \quad -99)_0$$

Plainly, even though the basis is infinite, its structure is relatively simple.

Let us employ Theorem 7.4 to compute $\Omega(g)$ for the signal

$$g = (1 \quad 2 \quad -5)_{-1}$$

We obtain

$$\Omega(g) = \text{EXTSUP}[\text{ERODE}(g, f^n)]$$
$$= \text{EXTMAX}[\text{ERODE}(g, f^{-1}), \text{ERODE}(g, f^3)]$$
$$= \text{EXTMAX}[(2 \quad -5)_{-1}, (-2 \quad -1)_{-1}]$$
$$= (2 \quad -1)_{-1}$$

Consequently, only two elements from the basis are required, which of course we know after having found the appropriate basis elements. In fact, for the mapping Ω, we can determine the maximum (finite) number of terms required for any given input signal. (See Section 8.4.)

7.4 ALGEBRAIC OPENINGS OF SIGNALS

In this section we consider the signal-to-signal version of Theorem 5.10, the Matheron representation theorem for τ-openings. We shall again be considering mappings from $\mathcal{S}^*$, the collection of extended signals, to $\mathcal{S}^*$. As in the case of the τ-mapping representation given in Theorem 7.1, the theory is somewhat abstract; however, once we pass to digital applications, the consequences are most concrete in that they result in straightforward machine-implementable algorithms. Moreover, the ultimate result, Theorem 7.5, has profound implications for the practical application of morphological filtering. (Those interested only in application should read the definitions carefully, skim the intervening theory, and go directly to the statement of Theorem 7.5.)

A mapping $\Psi: \mathcal{S}^* \to \mathcal{S}^*$ is said to be *antiextensive* if $\Psi(f) \leqslant f$ for all $f \in \mathcal{S}^*$ and *extensive* if $\Psi(f) \geqslant f$ for all $f \in \mathcal{S}^*$. Ψ is called *idempotent* if $\Psi(\Psi(f)) = \Psi(f)$. An *algebraic opening* (*closing*) on the class of extended signals is a signal-to-signal mapping that is antiextensive (extensive), increasing, and idempotent. An algebraic opening (closing) is called a τ-*opening* (τ-*closing*) if it is translation invariant. By this definition, the gray-scale opening (closing) is a τ-opening (τ-closing).

The *invariant class* of an algebraic opening (closing) Ψ is the collection of all extended signals for which $\Psi(f) = f$. We denote the invariant class by Inv $[\Psi]$. Since Ψ is idempotent, $\Psi(f) \in$ Inv $[\Psi]$ for all $f \in \mathcal{S}^*$. As in the case of constant images, the invariant class plays a key role in the study of algebraic openings. From a practical standpoint, an algebraic opening is a special type of morphological filter. If $f \in$ Inv $[\Psi]$, then the filter Ψ has no effect on f. Since Ψ is idempotent, once filtered by Ψ, a signal is no longer affected by Ψ. This is precisely the situation with the opening $O(\cdot, g)$.

Suppose Ψ is an algebraic opening. A subcollection $\mathcal{B}_0$ of signals in Inv $[\Psi]$ is called a *base* for Inv $[\Psi]$ if every signal in Inv $[\Psi]$ can be represented as an extended supremum of translations of elements in $\mathcal{B}_0$. In other words, every f in Inv $[\Psi]$ can be expressed using EXTSUP on some collection of inputs $g_x + y$, where all the signals g come from $\mathcal{B}_0$. As in the case of the base for a τ-opening on Euclidean constant images, the notion is somewhat abstract. As in that case also, in practice the finding of a base is very often quite straightforward.

In the constant-image setting, the elements in the invariant class of an opening $O(\cdot, B)$ were said to be B-open. Similarly, in the gray-scale setting, if $O(f, g) = f$, we say that the signal f is g-open. Consequently, the invariant class of the gray-scale opening $O(\cdot, g)$ consists of all g-open signals.

We shall now give gray-scale versions of properties M-23 and M-25. The properties will be given in the fully digital setting so that we can employ the umbra homomorphism for OPEN, which was proved in Section 6.5. Note that GF-13 states that composing an opening by a g-open signal with an opening by g, in either order, gives the same filtered output as simply opening by the g-open signal.

We begin by showing that g-openness induces $\mathbf{U}[g]$-openness.

Property GF-12. If f is g-open, then $\mathbf{U}[f]$ is $\mathbf{U}[g]$-open.

Proof. Applying the umbra homomorphism for the opening,

$$\text{OPEN}(\mathbf{U}[f], \mathbf{U}[g]) = \mathbf{U}[\text{OPEN}(f, g)] = \mathbf{U}[f]$$

Property GF-13. If f is g-open, then for any signal h,

(i) $\text{OPEN}[\text{OPEN}(h, f), g] = \text{OPEN}(h, f)$

(ii) $\text{OPEN}[\text{OPEN}(h, g), f] = \text{OPEN}(h, f)$

Proof. For (i),

$$\begin{aligned}
\text{OPEN}(h, f) &= \mathbf{S}[\text{OPEN}(\mathbf{U}[h], \mathbf{U}[f])] \\
&= \mathbf{S}[\text{OPEN}(\text{OPEN}(\mathbf{U}[h], \mathbf{U}[f]), \mathbf{U}[g])] \\
&= \mathbf{S}[\text{OPEN}(\mathbf{U}[\text{OPEN}(h, f)], \mathbf{U}[g])] \\
&= \mathbf{S}[\mathbf{U}[\text{OPEN}(\text{OPEN}(h, f), g)]] \\
&= \text{OPEN}(\text{OPEN}(h, f), g)
\end{aligned}$$

where M-23 (i) is applied to obtain the second equality. The proof of (ii) is similar: apply M-23 (ii) with the roles of f and g interchanged.

Property GF-14. If f is g-open, then for any signal h,

$$\text{OPEN}(h, f) \ll \text{OPEN}(h, g)$$

Proof. By GF-12, $\mathbf{U}[f]$ is $\mathbf{U}[g]$-open. Applying M-25, a constant-image opening by $\mathbf{U}[f]$ must be a subimage of an opening by $\mathbf{U}[g]$. Thus,

$$\begin{aligned}
\text{OPEN}(h, f) &= \mathbf{S}[\text{OPEN}(\mathbf{U}[h], \mathbf{U}[f])] \\
&\ll \mathbf{S}[\text{OPEN}(\mathbf{U}[h], \mathbf{U}[g])] \\
&= \text{OPEN}(h, f)
\end{aligned}$$

Example 7.16:

Let

$$f = (0 \quad 1 \quad 1)_0$$
$$g = (0 \quad 0)_0$$

and

$$h = (0\ 1\ 4\ 9\ 8\ 0\ *\ 1\ 1)_0$$

Then

$$\text{OPEN}(h, f) = (0\ 1\ 4\ 5\ 5\ 0)_0$$
$$\leqslant (0\ 1\ 4\ 8\ 8\ 0\ *\ 1\ 1)_0$$
$$= \text{OPEN}(h, g)$$

We shall now consider general gray-scale τ-openings for Euclidean (and therefore, as a special case, digital) signals. However, before proceeding to the theory, some comments regarding geometry are in order. Referring to Figure 6.19, we can see how the sharp edges of the original signal f have been smoothed by "rolling the ball" g underneath the graph of f. Just as Theorem 1.2 characterized the rolling ball notion for the constant-image opening, Theorem 6.2 provides a rigorous specification for the gray-scale opening. In terms of Figure 6.19, the curvature of the structuring element g has resulted in a corresponding smoothing, and for this reason, $O(\cdot, g)$ might be called a low-pass filter. Of course, should the curvature of g be greater, as is the case in Figure 6.20, the smoothing process is finer and the filtered signal tracks more closely the original.

As has been mentioned in several different contexts, as a filter, the opening behaves well with respect to certain fundamental geometric characteristics. Moreover, in essence, there is little difference between the gray-scale and two-valued settings. In the case of Theorem 5.10, constant-image filters that are antiextensive, increasing, idempotent, and translation invariant are completely characterized in terms of the basic set-theoretic opening operation. In Theorem 7.5, gray-scale filters that possess the same four properties (with respect to gray-scale signals) are completely characterized in terms of the basic gray-scale opening.

As was discussed following Theorem 5.10, that theorem gives a practical prescription for the construction of a class of two-valued morphological filters. In the case of Theorem 7.5, similar remarks apply, except that whereas in the two-valued case finer τ-openings result from forming unions of elementary constant-image openings, in the gray-scale case, finer τ-openings are constructed through the application of EXTSUP to elementary gray-scale openings.

The original Matheron representation theorems for two-valued increasing τ-mappings and τ-openings provide a cohesive theoretical framework for constant-image morphological filtering; Theorems 7.1 and 7.5 do the same for gray-scale morphological filtering. Since this framework is geometric in character, it is perceptually relevant, especially in the case of image processing. Moreover, it provides a clear demarcation of the region of application where morphological filtering can be of benefit. Specifically, in the gray-scale setting, as was the case in the two-valued setting, the basis for the kernel and the invariant class determine the action of an increasing τ-mapping and a τ-opening, respectively.

Property GF-15. $S[O(U[A], U[B])] = O(S[A], S[B])$

Proof. We use GF-9 and GF-10 together with the definition of the opening in terms of dilation and erosion:

$$S[O(U[A], U[B])] = S[\mathcal{D}(\mathcal{E}(U[A], U[B]), U[B])]$$
$$= \mathcal{D}(S[\mathcal{E}(U[A], U[B])], S[B])$$
$$= \mathcal{D}(\mathcal{E}(S[A], S[B]), S[B])$$
$$= O(S[A], S[B])$$

Property GF-16. Suppose Φ is a τ-opening on subsets of R^2. Then $\tilde{\Phi} = \mathbf{S} \circ \Phi \circ \mathbf{U}$ is a τ-opening on signals.

Proof. Since Φ is antiextensive,

$$\tilde{\Phi}(f) = S[\Phi(U[f])] \leqslant S[U[f]] = f$$

Hence, $\tilde{\Phi}$ is also antiextensive. By GF-4 and GF-5, $\tilde{\Phi}$ is an increasing τ-mapping. Thus, we need only show that $\tilde{\Phi}$ is idempotent. Now, since $\mathbf{S}$ followed by $\mathbf{U}$ yields an extensive mapping, the idempotence of Φ implies that

$$\tilde{\Phi}(\tilde{\Phi}(f)) = S[\Phi(U[S[\Phi(U[f])]])]$$
$$\geqslant S[\Phi(\Phi(U[f]))]$$
$$= S[\Phi(U[f])]$$
$$= \tilde{\Phi}(f)$$

On the other hand, Φ is antiextensive. Hence,

$$\tilde{\Phi}(\tilde{\Phi}(f)) = S[\Phi(U[S[\Phi(U[f])]])]$$
$$\leqslant S[\Phi(U[S[U[f]]])]$$
$$= S[\Phi(U[f])] \qquad (\mathbf{U} \circ \mathbf{S} \circ \mathbf{U} = \mathbf{U})$$
$$= \tilde{\Phi}(f)$$

Before proceeding to Theorem 7.5, we require a property concerning the opening of an umbra. Specifically, if A is an umbra and B is any set, then

$$O(A, U[B]) = O(A, B)$$

It is easy to see that $U[B]$ is B-open. Hence by M-25,

$$O(A, U[B]) \subset O(A, B)$$

Moreover, suppose the point (x, y) lies in $O(A, B)$. Then there exists a point (u, v) such that

$$(x, y) \in B + (u, v) \subset A$$

Therefore

$$(x, y) \in U[B + (u, v)] \subset U[A] = A$$

By the translation invariance of the umbra transform,

$$(x, y) \in U[B] + (u, v) \subset A$$

Sec. 7.4   Algebraic Openings of Signals

which means that $(x, y) \in O(A, \mathbf{U}[B])$, and hence,

$$O(A, B) \subset O(A, \mathbf{U}[B])$$

Theorem 7.5. A mapping $\Psi: \mathscr{S}^* \to \mathscr{S}^*$ is a τ-opening if and only if there exists a class of signals $\mathscr{B}$ such that

$$\Psi(f) = \text{EXTSUP}_{g \in \mathscr{B}} O(f, g)$$

Moreover, $\mathscr{B}$ is a base for Inv $[\Psi]$.

Proof. Suppose Ψ is a τ-opening. Let

$$h = \text{EXTSUP}\{g_\mathbf{z}: \mathbf{z} \in R^2, g \in \text{Inv}[\Psi], g_\mathbf{z} \leqslant f\}$$

Since Ψ is idempotent, $\Psi(f) \in \text{Inv}[\Psi]$. Since Ψ is antiextensive,

$$\Psi(f) = \Psi(f)_{(0,0)} \leqslant f$$

Hence, $\Psi(f)$ is a member of the collection on which EXTSUP is applied to obtain h. Thus, $\Psi(f) \leqslant h$.

To show the reverse ordering, let $x \in D_h$. Then there exist points $\mathbf{z}_j \in R^2$ and signals $g^j \in \text{Inv}[\Psi]$ such that

$$g^j_{\mathbf{z}_j} \leqslant f \qquad (1)$$

and

$$\lim_{j \to \infty} g^j_{\mathbf{z}_j}(x) = h(x)$$

from below. By the translation invariance and idempotence of Ψ,

$$\Psi(g^j_{\mathbf{z}_j}) = \Psi(g^j)_{\mathbf{z}_j} = g^j_{\mathbf{z}_j}$$

and therefore

$$\lim_{j \to \infty} [\Psi(g^j_{\mathbf{z}_j})](x) = h(x)$$

By (1) and the increasing monotonicty of Ψ,

$$\Psi(g^j_{\mathbf{z}_j}) \leqslant \Psi(f)$$

and hence, $h(x) \leq [\Psi(f)](x)$, and we conclude that $h \leqslant \Psi(f)$.

Employing Theorem 6.2, we can rewrite the defining relation for h as

$$h = \text{EXTSUP}_{g \in \text{Inv}[\Psi]}\{g_\mathbf{z}: \mathbf{z} \in R^2 \text{ and } g_\mathbf{z} \leqslant f\}$$
$$= \text{EXTSUP}_{g \in \text{Inv}[\Psi]} O(f, g)$$

If we can show that Inv$[\Psi]$ is a base for itself, then it will have been shown that $\Psi(f)$ is of the desired form. Let $\mathscr{H}$ denote the collection of all signals that can be expressed as extended suprema of translations of Ψ-invariant signals. Since it is obvious that Inv$[\Psi] \subset \mathscr{H}$, we need only show that $\mathscr{H} \subset \text{Inv}[\Psi]$. Now, if $h \in \mathscr{H}$, then

$$h = \text{EXTSUP}[g^k_{\mathbf{z}_k}]$$

where $g^k \in \text{Inv}[\Psi]$ and $\mathbf{z}_k \in R^2$. Since Ψ is increasing, for each k,

$$\Psi(h) \geqslant \Psi(g^k_{\mathbf{z}_k}) = \Psi(g^k)_{\mathbf{z}_k} = g^k_{\mathbf{z}_k}$$

Hence, $h \leqslant \Psi(h)$. Since Ψ is antiextensive, $h = \Psi(h)$, and we have $h \in \text{Inv}[\Psi]$.

As for the converse,

$$\Psi(f) = \text{EXTSUP}_{g\in\mathcal{B}} O(f, g)$$
$$= \text{EXTSUP}_{g\in\mathcal{B}} O(S[U[f]], S[U[g]])$$
$$= \text{EXTSUP}_{g\in\mathcal{B}} S[O(U[f], U[g])] \qquad \text{(GF-15)}$$
$$= S\left[\bigcup_{g\in\mathcal{B}} O(U[f], U[g])\right] \qquad \text{(GF-8)}$$

According to Theorem 5.10,

$$\Phi = \bigcup_{g\in\mathcal{B}} O(\cdot, U[g])$$

is a τ-opening on subsets of R^2, and GF-16 states that

$$\Psi = S \circ \Phi \circ U$$

is a τ-opening on signals.

Example 7.17:

Let

$$g = (0 \ \ 1 \ \ 2 \ \ 1 \ \ 0)_0$$
$$h = (0 \ \ 0 \ \ 0)_0$$

and define the mapping Φ by

$$\Phi(f) = \text{EXTMAX}[\text{OPEN}(f, g), \text{OPEN}(f, h)]$$

Then, according to Theorem 7.5, Φ is a τ-opening with base $\{g, h\}$. Suppose

$$k = (6 \ \ 7 \ \ 8 \ \ 9 \ \ 8 \ \ 8 \ \ 5 \ \ 8 \ \ 8 \ \ 8 \ \ 5 \ \ 5)_0$$

Then

$$\text{OPEN}(k, g) = (6 \ \ 7 \ \ 8 \ \ 9 \ \ 8 \ \ 7 \ \ 5 \ \ 6 \ \ 7 \ \ 6 \ \ 5 \ \ 4)_0$$
$$\text{OPEN}(k, h) = (6 \ \ 7 \ \ 8 \ \ 8 \ \ 8 \ \ 8 \ \ 5 \ \ 8 \ \ 8 \ \ 8 \ \ 5 \ \ 5)_0$$

and $\Phi(k) = k$, which means that $k \in \text{Inv}[\Phi]$.

According to Theorem 7.5, k should equal an extended maximum of (morphological) translations of g and h. In fact,

$$k = \text{EXTMAX}[g_0 + 6, g_1 + 7, h_4 + 8, h_5 + 5, h_8 + 8, h_{10} + 5]$$

7.5 GRAY-SCALE MORPHOLOGICAL FILTERS FOR IMAGES

Put simply, the entire theory of the present chapter applies to images with no alteration except for the use of bound matrices instead of bound vectors. The notion of kernel, basis, and invariant class, properties GF-1 though GF-16, Theorems 7.1 through 7.5,

Sec. 7.5 Gray-Scale Morphological Filters for Images

and all the proofs go over without change. Because of the greater scale of the calculations, image manipulations are much more cumbersome, and for that reason we have not treated them in this chapter. However, in Chapter 8 we shall examine the representation of some morphological filters defined on images.

EXERCISES

7.1. Let
$$f(t) = \begin{cases} t^2, & \text{for } 0 \le t < 2 \\ +\infty, & \text{for } 2 \le t < 4 \\ -2, & \text{for } 4 \le t \le 6 \end{cases}$$

and
$$g(t) = \begin{cases} 1, & \text{for } 0 \le t < 3 \\ +\infty, & \text{for } 3 \le t \le 5 \end{cases}$$

Find:
(a) EXTMAX(f, g) (c) U$[f]$
(b) MIN(f, g) (d) $f_{-3}(t) + 5$

7.2. Let
$$f_1 = (1 \quad 2 \quad 3 \quad 1 \quad +\infty \quad +\infty \quad 2 \quad 1 \quad -2)_0$$
$$f_2 = (1 \quad 3 \quad 2 \quad * \quad * \quad 1 \quad 1 \quad 2 \quad 2 \quad 9)_1$$
$$f_3 = (+\infty \quad +\infty \quad +\infty \quad 1 \quad 1 \quad 1 \quad * \quad * \quad 2)_{-1}$$

Find:
(a) TRAN$(f_1; 4)$ (f) MIN(f_1, f_2)
(b) OFF$(f_1; -3)$ (g) MIN(f_1, f_3)
(c) EXTMAX(f_1, f_2) (h) MIN(f_2, f_3)
(d) EXTMAX(f_1, f_3) (i) U$[f_3]$
(e) EXTMAX(f_2, f_3) (j) SUB(f_3)

7.3. Let
$$f = \begin{pmatrix} 2 & 1 & * & 1 \\ +\infty & 0 & -2 & +\infty \end{pmatrix}_{0,1}$$

and
$$g = \begin{pmatrix} * & +\infty & +\infty \\ 1 & -2 & 1 \\ +\infty & 4 & * \end{pmatrix}_{0,1}$$

Find EXTMAX(f, g) and MIN(f, g).

7.4. Let f_1, f_2, and f_3 be the signals given in Exercise 7.2, and let
$$g_1 = (1 \quad 2 \quad 1)_0$$
$$g_2 = (1 \quad +\infty \quad 1)_{-1}$$
$$g_3 = (1 \quad * \quad -1)_{-2}$$

For $i, j = 1, 2, 3$, find:
(a) ERODE(f_i, g_j)
(b) DILATE(f_i, g_j)
(c) OPEN(f_i, g_j)
(d) CLOSE(f_i, g_j)
(e) CLOSE($f_i, -g_j$)
(f) OPEN($f_i, g_j^\wedge$)
(g) ERODE[DILATE($f_i, g_j^\wedge$), $g_j^\wedge$]
(h) DILATE[ERODE(f_i, g_j), g_j]

7.5. For any extended signal f, define $\Omega(f)$ by: $[\Omega(f)](i)$ is the second-largest value among the gray values $f(i)$, $f(i + 1)$, $f(i + 2)$, $f(i + 3)$. Make certain that Ω is defined consistently with the order relation $\leqslant$. (See Example 7.13.) Ω is an increasing τ-mapping. Find $\Omega(f_1)$, $\Omega(f_2)$, and $\Omega(f_3)$ for the signals given in Exercise 7.2.

7.6. Consider the constant image

$$T = \begin{pmatrix} 1 & * & * & 1 & 1 \\ * & * & * & 1 & * \\ * & 1 & * & 1 & * \\ 1 & 1 & * & * & * \end{pmatrix}_{0,1}$$

Find $\hat{\Omega}(T)$. Leave your answer in umbra matrix format.

7.7. Find the kernel of the mapping Ω of Exercise 7.5.

7.8. For any signal f, let $[\Sigma(f)](i)$ be given by the weighted average

$$[\Sigma(f)](i) = \frac{1}{4}f(i - 1) + \frac{1}{2}f(i) + \frac{1}{4}f(i + 1)$$

(Note that $+\infty + *$ should be defined to be $*$.) Find Ker $[\Sigma]$, and evaluate $\Sigma(f_i)$ for each of the signals f_i given in Exercise 7.2.

7.9. Find the bases for the kernels of the mappings Ω and Σ of Exercises 7.5 and 7.8.

7.10. Apply Theorem 7.4 to the signals f_i of Exercise 7.2, employing the τ-mappings Ω and Σ of Exercises 7.5 and 7.8. In each case, find the minimal number of terms needed in the expansion. Note that the situation is different for Ω and Σ, since the basis for Ω is finite whereas the basis for Σ is infinite.

7.11. What happens to the basis for Ω of Exercise 7.5 if, instead of defining Ω at star-valued pixels in a manner consistent with the order relation $\leqslant$, we define $[\Omega(f)](i)$ to be $*$ if any of the relevant gray values is a star?

7.12. Let

$$f = \begin{pmatrix} 1 & 1 & 1 & 3 & 2 \\ 2 & 2 & +\infty & 3 & 2 \\ 1 & 0 & 2 & -1 & 0 \\ 1 & * & 1 & 0 & -1 \\ 1 & 1 & 0 & 0 & 0 \end{pmatrix}_{0,3}$$

and

$$g = \begin{pmatrix} 0 & 0 & 0 \\ 0 & 9 & 0 \\ * & 0 & 0 \end{pmatrix}_{-1,1}$$

Find ERODE(f, g), DILATE(f, g), OPEN(f, g), and CLOSE($f, -g$).

8

Morphological Filters: Applications

8.1 CLASSICAL FILTERING

The principal motivation for filtering is the selective removal and attenuation of specific characteristics of signals or images with the intent of making the information within the signal or image more apparent. Of the filters presented herein, several can be duplicated perfectly utilizing morphological theory, the actual construction being specified by Theorem 7.1, the version of the Matheron Representation Theorem for signals. In other instances, morphological filters can be employed to approximate nonmorphological filters. Since construction of a morphological filter is based on a geometric fitting procedure, the filter can be readily tuned for specific noise removal or attenuation through the customization of those structuring elements which generate it.

In general terms, a filter is an operator $\mathcal{T}$ that maps input signals belonging to a given class into a unique output signal belonging to some other designated class. The mapping, of course, is usually designed with some purpose in mind, and it is often realized as a collection of physical elements. For instance, in an electrical network the elements might include resistors, capacitors, and inductors, while in a mechanical system they might include springs, pulleys, and gears. In any case, the elements comprising the system cause the system to behave in some desired manner. The behavior in question then defines the operator $\mathcal{T}$. A filter $\mathcal{T}$, with inputs $f_1, f_2, \ldots,$

f_n and output g, is denoted by

$$g = \mathcal{T}(f_1, f_2, \ldots, f_n)$$

and can be represented by a block diagram as in Figure 8.1.

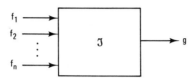

Figure 8.1 N-ary filter

The inputs and the output of a filter will be signals, and hence functions of the single variable t. In image processing systems they are functions of two variables, x and y. Because the characterizations of most filter-theoretic notions are independent of the number of variables, we shall restrict our attention to signals; the generalizations to images are essentially trivial.

Given a filter $\mathcal{T}$, the number of inputs is called the *arity* of the filter. For $n = 1$, the filter is called *unary*, for $n = 2$, *binary*, for $n = 3$, *ternary*, and so on. The output g is often called the *response* (of $\mathcal{T}$ to the inputs $f_1, f_2, \ldots, f_n$).

Example 8.1:

Consider the *delay-type* unary filter $g = \mathcal{T}(f)$ defined, for any signal f, by

$$g(t) = f(t - 1)$$

Figure 8.2 illustrates the delay operation on

$$f(t) = \begin{cases} 0, & \text{for } t < 0 \\ t, & \text{for } t \geq 0 \end{cases}$$

The output is

$$g(t) = \begin{cases} 0, & \text{for } t < 1 \\ t - 1, & \text{for } t \geq 1 \end{cases}$$

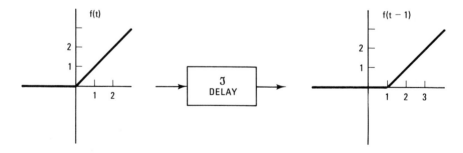

Figure 8.2 Delay-type filter

Example 8.2:

A *square-law* device is a unary filter characterized by the rule $\mathcal{T}(f) = f^2$. Figure 8.3 illustrates the operation of this device on the input $f(t)$ given in Example 8.1

Sec. 8.1 Classical Filtering

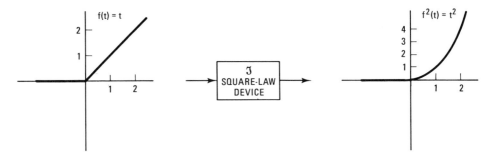

Figure 8.3 Square-law device

Numerous square-law devices arise if we consider inputs from other function classes besides $R \to R$. Many of these perform in a manner similar to $\mathcal{T}$ in Example 8.2. For instance, signals of the form

$$f: [0, 10] \longrightarrow [0, \infty)$$

give rise to a filter similar to $\mathcal{T}$.

Another square-law device operates on fully digital signals, i.e., those mapping $Z \to Z$. Such a filter is called a *fully digital* filter. Figure 8.4 illustrates the application of this system to the signal

$$f(t) = \begin{cases} -1, & \text{for } t < 0 \\ t, & \text{for } t \geq 0 \end{cases}$$

If a filter involves signals with uncountable domains and countable codomains, it is called a *sample-data* filter. Filters involving signals possessing uncountable domains and uncountable codomains are called *Euclidean*, *continuous*, or *analog* filters.

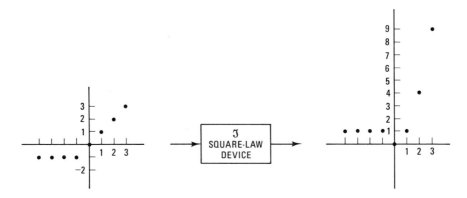

Figure 8.4 Digital square-law device

Example 8.3:

Let $\mathcal{T}$ be the maximum-waveform binary filter in analog form; i.e.,

$$[\mathcal{T}(f, g)](t) = \max [f(t), g(t)]$$

where

$$f, g: R \longrightarrow R$$

The application of $\mathcal{T}$ to the signal $f(t)$ of Example 8.1 and $g(t) = \sin t$ is given in Figure 8.5.

A filter is *additive* if, whenever

$$g_1 = \mathcal{T}(f_{1,1}, f_{1,2}, \ldots, f_{1,n})$$

and

$$g_2 = \mathcal{T}(f_{2,1}, f_{2,2}, \ldots, f_{2,n})$$

then

$$g_1 + g_2 = \mathcal{T}(f_{1,1} + f_{2,1}, \ldots, f_{1,n} + f_{2,n})$$

where $f_{1,i} + f_{2,i}$ belongs to the relevant input class for $i = 1, 2, \ldots, n$. A filter is *homogeneous* if, whenever

$$g = \mathcal{T}(f_1, f_2, \ldots, f_n)$$

then

$$cg = \mathcal{T}(cf_1, cf_2, \ldots, cf_n)$$

where c is any constant such that $cf_1, cf_2, \ldots, cf_n$ belong to the relevant input class. A filter is said to be *linear* if it is both additive and homogeneous.

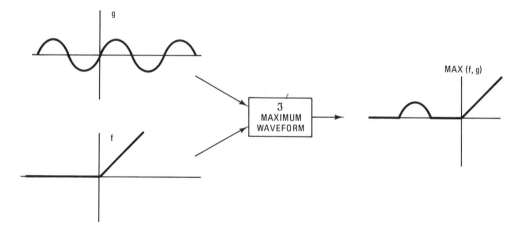

Figure 8.5 Maximum waveform filter

Example 8.4:

The delay-type filter described in Example 8.1 is linear, whereas the square-law device given in Example 8.2 and the maximum waveform filter of Example 8.3 are not linear.

Example 8.5:

Consider the circuit comprised of the resistor and inductor illustrated in Figure 8.6. It is

Sec. 8.1 Classical Filtering

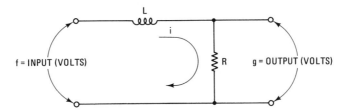

Figure 8.6 R-L circuit

known that the input voltage equals the sum of the voltages dissipated in the resistor and the inductor; that is,

$$f = L\frac{di}{dt} + iR, \, t \geq 0$$

The output voltage is given by $g = iR$; therefore, we obtain the differential equation

$$\frac{L}{R}\dot{g} + g = f$$

or

$$\dot{g} + \frac{R}{L}g = \frac{R}{L}f$$

Circuits containing energy storing elements such as inductors and capacitors are memory-type systems, which give rise to initial conditions. If we assume that $g(0) = 0$, the solution to the differential equation is given by the standard convolution integral

$$g(t) = \frac{R}{L}\int_0^t e^{(-R/L)(t-x)}f(x)\,dx, \, t \geq 0$$

This expression can be employed as the rule

$$g = \mathcal{T}(f)$$

defining the filter $\mathcal{T}$. Since the convolutional integration operation is both additive and homogeneous, $\mathcal{T}$ is linear. It has input functions belonging to the family of piecewise continuous functions defined on $[0, \infty)$. For instance, if

$$f(t) = \begin{cases} 1, & \text{for } 0 \leq T_1 \leq t < T_2 \\ 0, & \text{elsewhere in } [0, \infty) \end{cases}$$

then

$$g(t) = \begin{cases} 0, & \text{if } 0 < t < T_1 \\ 1 - e^{(-R/L)(t-T_1)}, & \text{if } T_1 \leq t < T_2 \\ e^{(-R/L)t}(e^{(R/L)T_2} - e^{(R/L)T_1}), & \text{if } T_2 \leq t \end{cases}$$

In particular, if $T_1 = 0$ and $T_2 \to \infty$, then $f(t) = 1$ and $g(t) = 1 - e^{(-R/L)t}$ on $[0, \infty)$. Figure 8.7 illustrates the use of this filter, employing unit inputs for various values of R/L.

For those familiar with Fourier and Laplace transforms, the transfer function for the R-L circuit is

$$\tilde{H}(s) = \frac{R/L}{s + R/L}$$

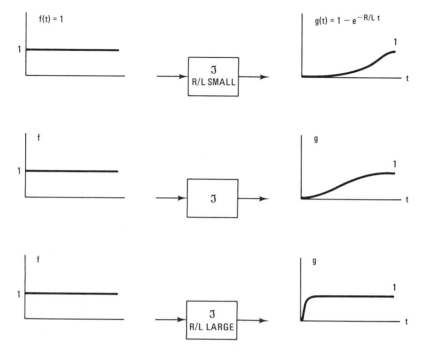

Figure 8.7 Convolutional filter

Letting $s = j\omega$, we obtain the frequency representation

$$H(\omega) = \tilde{H}(j\omega) = \frac{R/L}{j\omega + R/L}$$

and the corresponding transfer amplitude spectrum is

$$|H(\omega)| = \frac{R/L}{\sqrt{\omega^2 + R^2/L^2}} = \frac{1}{\sqrt{\left(\frac{\omega}{R/L}\right)^2 + 1}}$$

For $|\omega| = R/L$, $|H(\omega)| = 1/\sqrt{2}$; for $|\omega| > R/L$, $|H(\omega)|$ is small; and for $-R/L < \omega < R/L$, $|H(\omega)| \cong 1$. Figure 8.8 gives illustrations for $|H(\omega)|$ for various values of R/L. When $R/L \to \infty$, $|H(\omega)|$ approximates unity, and, at least intuitively, the output signal should almost equal the input signal. Compare this figure with Figure 8.7, where it is clearly seen that the output $g(t) = 1 - e^{-R/L t}$, $t > 0$, approximates $f(t) = 1$, $t \geq 0$, quite closely (outside a small neighborhood of the origin) when R/L is large.

For those familiar with Fourier transforms, some concepts discussed in Example 8.5 can be idealized. For instance, a system for which $H(\omega) = 1$ is said to be an *all-pass* system. In such a system, if we let F and G be the Fourier transforms of f and g, respectively, then $G(\omega) = F(\omega)$, and all frequency values of the input remain unchanged. An *ideal low-pass* filter has a transfer amplitude spectrum for which $|H(\omega)| = 1$ for $-\omega_c < \omega < \omega_c$, and $H(\omega) = 0$ outside this range. Low frequencies are passed unaltered through such a filter, and high frequencies ($|\omega| \geq \omega_c$) are zeroed

Sec. 8.1 Classical Filtering   261

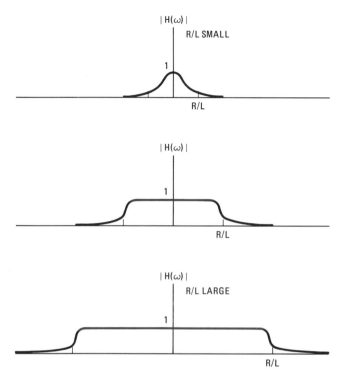

Figure 8.8 Transfer amplitude spectra

out. Here, ω_c is called the *cutoff frequency*. In Example 8.5, R/L acted somewhat like the cutoff frequency and the filter given there is low pass, but not ideally so.

A *bandpass filter* has a transfer amplitude spectrum such that $|H(\omega)| = 1$ for $\omega_a < |\omega| < \omega_b$, $0 < \omega_a < \omega_b$, and $H(\omega) = 0$ outside this frequency range. A *highpass filter* has a transfer amplitude spectrum such that $|H(\omega)| = 1$ for $|\omega| > \omega_c$ and $H(\omega) = 0$ elsewhere. Figure 8.9 illustrates some of the idealizations just mentioned.

It is important to note that all of the preceding ideal filters are idempotent. However, they are not realizable using solutions of linear differential equations. This is crucial, since the property of idempotence is highly desirable. An advantage of morphological filters constructed from openings is that they, too, are idempotent. In fact, the import of Theorem 7.5 is that all filters that are antiextensive, translation invariant, increasing, and idempotent must be constructed morphologically through the use of elementary openings.

Numerous discrete linear filters arise from approximations of analog linear filters. Many of these are described by difference equations. A popular method for obtaining such equations is to replace the derivatives in the general linear equation by difference quotients. The difference quotients can be backward, forward, or central. Other methods, such as the impulse invariance method and the bilinear transform technique, also result in difference equation approximations to specific differential equations.

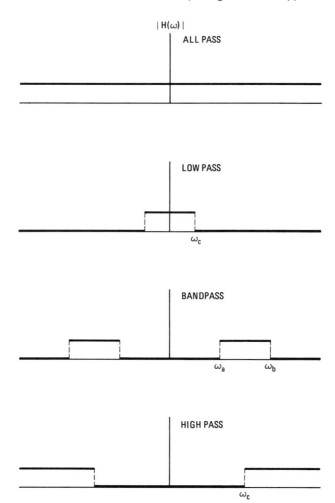

Figure 8.9 Ideal amplitude spectra

Example 8.6:

A simple illustration of the difference equation can be given using a forward difference approximation to the derivative. Consider, for instance, the linear filter discussed in Example 8.5, where, it was seen that the input f had a response g given by

$$\dot{g} + \frac{R}{L}g = \frac{R}{L}f, \ t \geq 0$$

For small $\Delta t > 0$, it makes sense to replace the derivative $\dot{g}(t)$ by the approximating difference quotient

$$\frac{g(t + \Delta t) - g(t)}{\Delta t}$$

We assume that Δt, which is called the *sampling interval*, satisfies the restraint

Sec. 8.1 Classical Filtering

$(R/L) \Delta t < 1$. After simplification, we obtain the functional equation

$$g(t + \Delta t) = g(t)\left[1 - \frac{R}{L}\Delta t\right] + \frac{R}{L}\Delta t \cdot f(t)$$

for $t \geq 0$.

Functional equations are very difficult to work with and solve. For instance, $h(t) = h(t + \pi)$ has an infinite number of independent solutions—$\sin 2t$ and $\cos 4t$, to name just two. For this reason, and because of digital computer implementations, we shall replace the functional equation by a difference equation. We restrict the functions involved in the equation to $t = 0, \Delta t, 2\Delta t, \ldots$, and let $t_k = k\Delta t$, denoting $f(k\Delta t)$ by f_k and $g(k\Delta t)$ by g_k to obtain the difference equation

$$g_{k+1} = g_k\left[1 - \frac{R}{L}\Delta t\right] + \frac{R}{L}\Delta t \cdot f_k$$

with initial condition

$$g_0 = g(0) = 0$$

The solution to this difference equation specifies the discrete linear filter. Moreover, we have

$$g_{n+1} = \left(1 - \frac{R}{L}\Delta t\right)^{n+1} \frac{R}{L}\Delta t \sum_{j=0}^{n} \frac{f_j}{(1 - (R/L)\Delta t)^{j+1}}$$

For instance, if $f_k = 1$, then $g_n = 1 - (1 - R/L\ \Delta t)^n$. (See Figure 8.10.) If, now, $\Delta t \to 0$ and $n \to \infty$ in such a manner that the product $n\Delta t$ is fixed (i.e., let $n\Delta t = x$), then we obtain the same solution in the limit as that given in the continuous case, viz.,

$$g(x) = 1 - e^{(-R/L)x}, \quad \text{for } x \geq 0$$

(See Figure 8.11.)

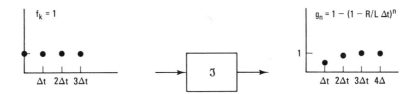

Figure 8.10 Discrete linear filter

Results similar to those obtained in Example 8.6 can be given using backward difference, impulse response, bilinear transform, and other techniques.

A filter $\mathcal{T}$ defined by $g = \mathcal{F}(f_1, f_2, \ldots, f_n)$ is said to be *time invariant* if

$$g_\lambda = \mathcal{T}[(f_1)_\lambda, (f_2)_\lambda, \ldots, (f_n)_\lambda]$$

for any inputs f_i, $i = 1, \ldots, n$, where, as usual, $(f_i)_\lambda(t) = f_i(t - \lambda)$. This means that the system rule $\mathcal{T}$ commutes with translation relative to the variable t—i.e., it does not matter if the input is translated and then the response is obtained, or if the response to the original input is first found and the result then translated. In either case,

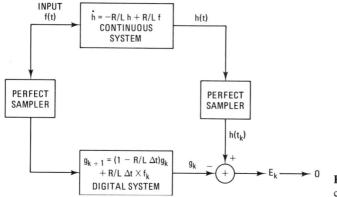

Figure 8.11 Analog system and corresponding digital system

the same output results. A system that is not time invariant is said to be time varying. Figure 8.12 illustrates the input–output relation for a unary time-invariant filter.

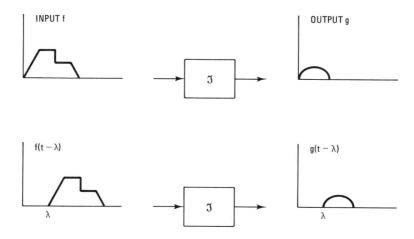

Figure 8.12 Time invariant filter

Example 8.7:

The delay-type unary system specified in Example 8.1 is time invariant. So, too, are the square-law device given in Example 8.2 and the binary maximum-waveform system of Example 8.3.

Digital linear time-invariant filters are usually classified as being *finite impulse response* (FIR) or *infinite impulse response* (IIR). These classifications correspond to which coefficients (a_i or b_i) are present in the digital linear time-invariant system with output g_k satisfying the difference equation

$$a_n g_{k-n} + a_{n-1} g_{k-(n-1)} + \cdots + a_1 g_{k-1} + a_0 g_k = b_m f_{k-m} + \cdots + b_1 f_{k-1} + b_0 f_k$$

where we employ the difference equation notation $g_k = g(k\Delta t)$. Since we can always insert $a_i = 0$ or $b_i = 0$, no generality is lost by assuming that $a_0 = 1$ and $n = m = N$.

Thus, we can write

$$g_k = \sum_{i=0}^{N} b_i f_{k-i} - \sum_{i=1}^{N} a_i g_{k-i}$$

If all the a_i's are zero, then the system is said to be an *FIR filter*, otherwise, it is said to be an *IIR filter*.

8.2 MORPHOLOGICAL FILTERING OPERATIONS

Morphological filters can be applied in a manner similar to the way other image and signal filters are applied. They can be employed to enhance certain features of an input signal or image and to attenuate other features. Since they work directly upon the geometric characteristics of the input, they are fairly easy to use. There is no need for abstract transform techniques such as Fourier, Mellin, Laplace, and Hilbert transforms, nor is there a requirement for transform-world concepts such as frequency and effective bandwidth. As with any filtering technique, in order to apply a morphological procedure one must know what type of information in the signal is of significance and what characteristics are of no relevance. However, in the morphological approach, it is the geometric nature of the signal or image that is of consequence. Thus, by means of a judicious choice of a structuring element based on *a priori* information, the output that results when the observed image or signal is acted upon by the structuring element will contain more features that are considered beneficial and fewer that are undesirable than will the input. Accordingly, in image processing, where geometric content is of the essence, morphological filters have a prima facie advantage over filters that operate in the frequency domain. It is the purpose of this section to intuitively describe the action of some basic morphological filters through the use of schematics.

Consider, then, the RL-circuit given in Example 8.5, the purpose of which is to attenuate high-frequency noise. In Figure 8.7 a unit step signal was employed as input, and the corresponding outputs were given for various values of R/L. A morphological approach is taken in Figure 8.13, where the unit step is filtered by $O(\cdot, g)$, where g is any one of three semicircular structuring elements. The outputs should be compared with the outputs of Figure 8.7.

Given the goal of reducing high-frequency noise, a possible approach utilizing morphological elements is illustrated in the block diagram of Figure 8.14. In Figures 8.15 through 8.18, the same input signal is probed by four rectangular structuring elements of various heights and widths in order to demonstrate the effect of the methodology of Figure 8.14. Notice how erosion and dilation are used in conjunction. As mentioned previously, erosion smooths from below while dilation smooths from above. A similar procedure can be implemented by using the opening and the closing.

As in the two-valued situation illustrated in Figures 1.27 and 1.28, smoothing can be accomplished by iterating the opening and the closing. Consider the morphological filters $C[O(f, g), -g]$ and $O[C(f, -g), g]$. By definition,

$$C[O(f, g), -g] = -O[-O(f, g), g]$$

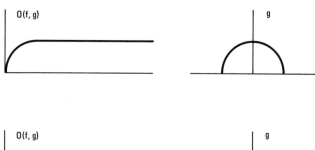

Figure 8.13 Opening by semicircle

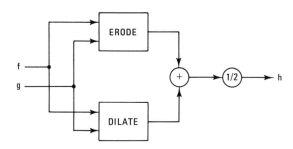

Figure 8.14 Dilation-erosion hybrid filter

and

$$O[C(f, -g), g] = O[-O(-f, g), g]$$

Either of these can be applied directly, or a hybrid filter can be utilized which averages the two outputs. Figure 8.19 shows a block diagram specifying such a hybrid filter, while Figure 8.20 depicts the action of $O[C(f, -g), g]$ and Figure 8.21 does the same for $C[O(f, g), -g]$.

The next example considers the hybrid case in the nonextended digital setting.

Example 8.8:

Let $f = (1\ \ 1\ \ 1.2\ \ 1\ \ 1\ \ .7\ \ 1)_0$ be a signal that is to be filtered using the average of, first, the opening followed by the closing, and then, the closing followed by the opening. Let $g = (0\ \ 0\ \ 0)_0$. Then

$$\text{OPEN}(f, g) = (1\ \ 1\ \ 1\ \ 1\ \ 1\ \ .7\ \ .7)_0$$

Sec. 8.2   Morphological Filtering Operations

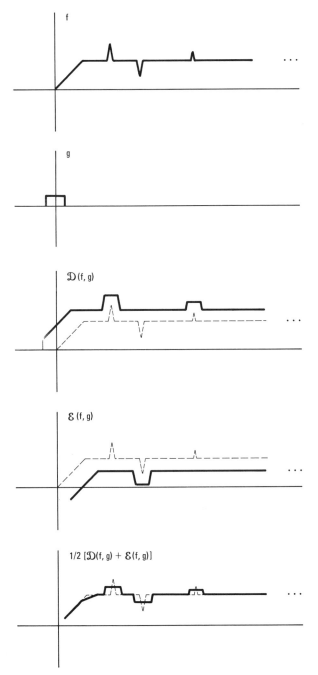

Figure 8.15 Dilation-erosion hybrid filter illustration I

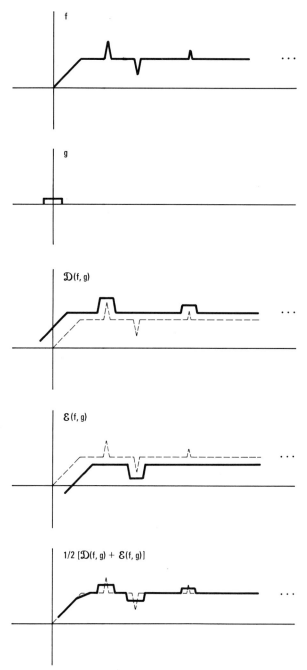

Figure 8.16 Dilation-erosion hybrid filter illustration II

Sec. 8.2 Morphological Filtering Operations **269**

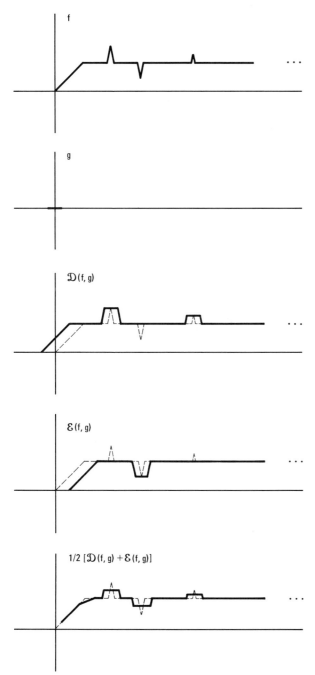

Figure 8.17 Dilation-erosion hybrid filter illustration III

270 Morphological Filters: Applications Chap. 8

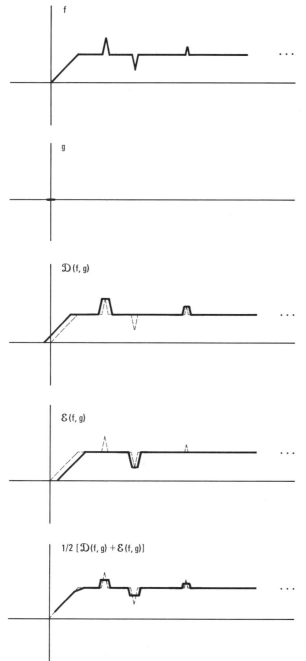

Figure 8.18 Dilation-erosion hybrid filter illustration IV

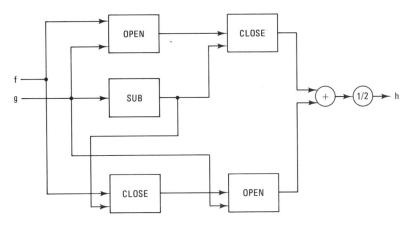

Figure 8.19 Opening-closing hybrid filter

and
$$\text{CLOSE}(\text{OPEN}(f, g), -g) = (1 \quad 1 \quad 1 \quad 1 \quad 1 \quad 1 \quad 1)_0$$

Also,
$$\text{CLOSE}(f, -g) = (1.2 \quad 1.2 \quad 1.2 \quad 1 \quad 1 \quad 1 \quad 1)_0$$

and
$$\text{OPEN}(\text{CLOSE}(f, -g), g) = (1.2 \quad 1.2 \quad 1.2 \quad 1 \quad 1 \quad 1 \quad 1)_0$$

Hence, the output signal defined by the diagram in Figure 8.19 is
$$h = (1.1 \quad 1.1 \quad 1.1 \quad 1 \quad 1 \quad 1 \quad 1)_0$$

Although the morphological opening is often perceived to be a low-pass filtering operation, this need not be true in general. In fact, filtering depends on the shape of the structuring element, as the next example shows.

Example 8.9:

Let the observed signal be $f(t) = \sin(\pi t)$, and let the structuring signal be

$$g(t) = \begin{cases} 1, & \text{if } 0 < t < 1 \\ \text{undefined}, & \text{elsewhere} \end{cases}$$

Then $O(f, g)$ acts like a rectifier-type filter in that it zeros out the positive part of f but allows the rest of f to pass through unaltered. We have

$$O(f, g)(t) = \begin{cases} f(t), & \text{if } f(t) \leq 0 \\ 0, & \text{otherwise} \end{cases}$$

Note that whereas the Fourier series of f contains only one harmonic, the Fourier series of $O(f, g)$ contains an infinite number of harmonics. (See Figure 8.22.)

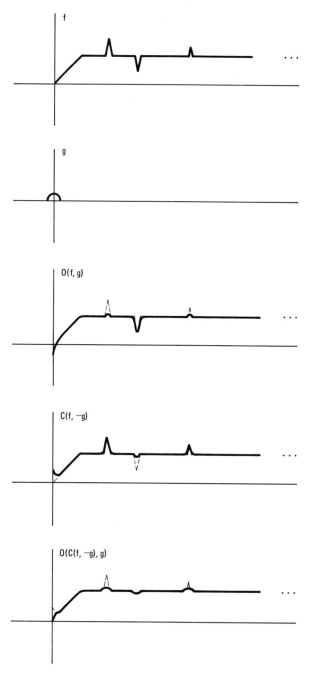

Figure 8.20 Filtering by opening following closing

Sec. 8.2 Morphological Filtering Operations

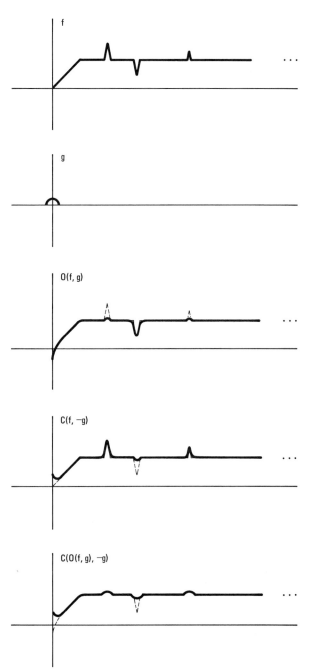

Figure 8.21 Filtering by closing following opening

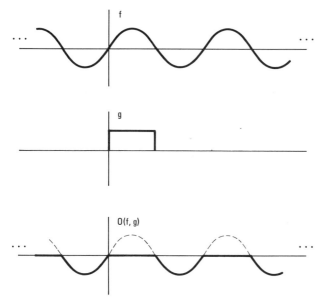

Figure 8.22 Opening as a rectifier-type filter

8.3 ORDER-STATISTIC FILTERS

In this section we consider a method of filtering that is based upon the numerical ordering within certain collections of gray values of a signal (or image). At first we shall assume that the signal is defined and real valued at all points of Z so that we need not concern ourselves with boundary conditions.

Given such a signal together with an input integer N, consider, at each point i, a collection of N points comprising some neighborhood of i, where it is assumed that all neighborhoods are of the same form. Let

$$f(i)^{\langle 0 \rangle}, f(i)^{\langle 1 \rangle}, \ldots, f(i)^{\langle N-1 \rangle}$$

denote the ordering, from lowest to highest, of the N values of f in the neighborhood. Then the mth *order statistic of f at i* is defined to be $f(i)^{\langle m \rangle}$.

Example 8.10:

Let $f(i) = i^2$, let $N = 3$, let the neighborhood over which the ordering is to be taken be $\{i, i+1, i+2\}$, and let $m = 1$. Then we need to order the values $f(i), f(i+1)$, and $f(i+2)$. Having done this, we obtain the output of the filter, call it L, to be $[L(f)](i) = (i+1)^2$ for $i \neq -1$, and $[L(f)](-1) = 1$. Written out,

$$f = \cdots \quad 25 \quad 16 \quad 9 \quad 4 \quad 1 \quad \textcircled{0} \quad 1 \quad 4 \quad 9 \quad 16 \quad 25 \quad \cdots$$

$$L(f) = \cdots \quad 16 \quad 9 \quad 4 \quad 1 \quad 1 \quad \textcircled{1} \quad 4 \quad 9 \quad 16 \quad 25 \quad 36 \quad \cdots$$

For instance, if $i = -3$, then $f(i) = 9, f(i+1) = 4$, and $f(i+2) = 1$. Consequently $f(i)^{\langle 0 \rangle} = 1, f(i)^{\langle 1 \rangle} = 4$, and $f(i)^{\langle 2 \rangle} = 9$, resulting in $[L(f)](i) = 4$.

In the preceding example, we took the middle value in the ordering. Consequently, L is known as a *median filter*. In general, if N is odd, then the order-statistic filter of order $(N-1)/2$ is a median filter.

Sec. 8.3 Order-Statistic Filters

There are numerous sorting methods for computing order statistics—for instance, tree sorts, binary sorts, shell sorts, and quick sorts. However, the mth order statistic can be computed directly by using maxima and minima. To see this, consider N real numbers $a_1, a_2, \ldots, a_N$. The mth largest of these a_k is given by

$$a^{\langle m \rangle} = \bigwedge_{k=1}^{\binom{N}{m+1}} [a_{j_{k_1}} \vee a_{j_{k_2}} \vee \cdots \vee a_{j_{k_{(m+1)}}}]$$

where the j_{k_i} are distinct integers chosen from among $1, 2, \ldots, N$, $\binom{N}{m+1}$ denotes the number of combinations of N objects taken $m + 1$ at a time, and the minimum is taken over all distinct subsets $\{j_{k_1}, j_{k_2}, \ldots, j_{k_{(m+1)}}\}$ containing $m + 1$ integers from $\{1, 2, \ldots, N\}$. When N is odd, the median of the N numbers, denoted by $a^{\langle (N-1)/2 \rangle}$, is found by first finding the maxima for all possible sets of $(N + 1)/2$ of the a_i, and then, among these maxima, choosing the minimum. Thus, for $N = 3$, we have

$$a^{\langle 1 \rangle} = (a_1 \vee a_2) \wedge (a_1 \vee a_3) \wedge (a_2 \vee a_3)$$

which is illustrated in Figure 8.23. A distinct, but equivalent, formulation is illustrated in Figure 8.24.

In Example 7.13, we considered the increasing τ-mapping Y which took the second highest value among $f(i - 1)$, $f(i)$, $f(i + 1)$, and $f(i + 2)$. In terms of the current concerns, this means that Y is an order statistic with $N = 4$ and $m = 2$. In the example cited, we addressed the problem of boundary conditions since we assumed, as in practice we must, that the domain was finite. We saw that it was beneficial to treat $*$ as if it were $-\infty$ in the ordering: doing so gave us a means of

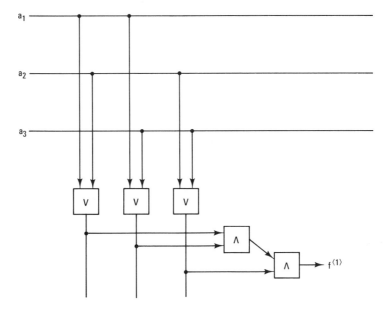

Figure 8.23 Median filter

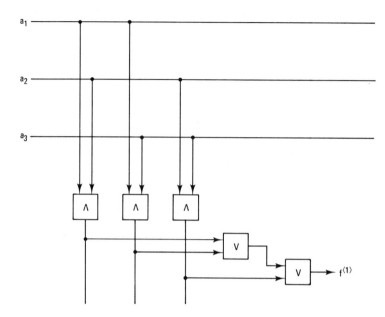

Figure 8.24 Median filter—alternate formulation

generating a basis. In Example 7.12, we saw that if we defined the output $[Y(f)](i)$ to be $*$ whenever one of the values of f in the collection $\{f(i-1), f(i), f(i+1), f(i+2)\}$ was $*$, then no basis would exist for the kernel. Consequently, we shall treat $*$ as $-\infty$ when considering order-statistic filters. This convention is quite practical since, numerically, it only affects values on the boundary, and it allows a uniform representation in terms of Theorem 7.4 whenever the codomain consists of integers, an assumption that holds in what is to follow.

The foregoing remarks can be made rigorous: if f is a fully digital extended signal, and $\mathcal{T}(f) = f^{\langle m \rangle}$ is the mth order-statistic filter for a given neighborhood, then $\mathcal{T}$ is an increasing τ-mapping and there exists a basis Bas $[\mathcal{T}]$. The full implications of this fundamental proposition concerning filtering will be seen in the next section.

We conclude this section with a consideration of some interesting points regarding the interplay between the median and the opening and closing. Suppose g is a structuring element with domain D_g consisting of p consecutive integers, and suppose that we consider order statistics of f over neighborhoods of i that are symmetric at i and that contain $N = 2p - 1$ points. Then

$$f^{\langle 0 \rangle} \leqslant \text{OPEN}(f, g) \leqslant \text{CLOSE}(f, -g) \leqslant f^{\langle N-1 \rangle}$$

where we assume that all operations take place in the extended context.

Example 8.11:

Let

$$f = (0\ 1\ 2\ 3\ 2\ 1\ 0)_0$$

Sec. 8.3 Order-Statistic Filters

and
$$g = (0 \quad 1 \quad 0)_0$$

Then
$$-f = (0 \quad -1 \quad -2 \quad -3 \quad -2 \quad -1 \quad 0)_0^s$$
$$\text{OPEN}(f, g) = (0 \quad 1 \quad 2 \quad 3 \quad 2 \quad 1 \quad 0)_0$$
$$\text{OPEN}(-f, g) = (0 \quad -1 \quad -2 \quad -3 \quad -2 \quad -1 \quad 0)_0^s$$
$$\text{CLOSE}(f, -g) = -\text{OPEN}(-f, g)$$
$$= (0 \quad 1 \quad 2 \quad 3 \quad 2 \quad 1 \quad 0)_0$$
$$f^{(0)} = (0 \quad 1 \quad 0)_2$$

and
$$f^{(4)} = (0 \quad 1 \quad 2 \quad 3 \quad 3 \quad 3 \quad 3 \quad 3 \quad 2 \quad 1 \quad 0)_{-2}$$

Then, clearly,
$$f^{(0)} \leqslant \text{OPEN}(f, g) \leqslant \text{CLOSE}(f, -g) \leqslant f^{(4)}$$

In Example 8.8 we considered a hybrid filter in which we averaged an opening and a closing. In doing so, we neither assumed a fully digital signal nor worked in the extended context. In the next example we construct a similar filter, except that we take the median of the opening, the closing, and the signal itself. Moreover, we work in the fully digital context and we do pay attention to the extended definitions. For notational clarity, O and C are used instead of OPEN and CLOSE.

Example 8.12:

Let
$$f = (1 \quad 1 \quad 1 \quad 2 \quad 1 \quad 1 \quad 4 \quad 4 \quad 4)_0$$

and consider the hybrid filter
$$h = \text{MEDIAN}[O[C(f, -g), g], f, C[O(f, g), -g]]$$

where $g = (1 \quad 1)_0$. (See Figure 8.25.) Then
$$O(f, g) = (1 \quad 1 \quad 1 \quad 1 \quad 1 \quad 1 \quad 4 \quad 4 \quad 4)_0$$
$$-f = (-1 \quad -1 \quad -1 \quad -2 \quad -1 \quad -1 \quad -4 \quad -4 \quad -4)_0^s$$
$$O(-f, g) = (-1 \quad -1 \quad -1 \quad -2 \quad -1 \quad -1 \quad -4 \quad -4 \quad -4)_0^s$$
$$C(f, -g) = -O(-f, g) = (1 \quad 1 \quad 1 \quad 2 \quad 1 \quad 1 \quad 4 \quad 4 \quad 4)_0$$
$$-O(f, g) = (-1 \quad -1 \quad -1 \quad -1 \quad -1 \quad -1 \quad -4 \quad -4 \quad -4)_0^s$$
$$O(-O(f, g), g) = (-1 \quad -1 \quad -1 \quad -1 \quad -1 \quad -1 \quad -4 \quad -4 \quad -4)_0^s$$
$$C(O(f, g), -g) = -O(-O(f, g), g)$$
$$= (1 \quad 1 \quad 1 \quad 1 \quad 1 \quad 1 \quad 4 \quad 4 \quad 4)_0$$

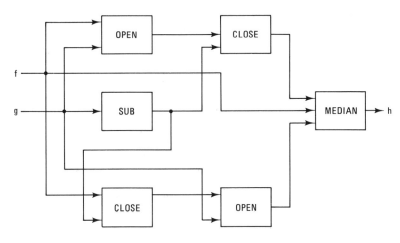

Figure 8.25 Hybrid median filter

and
$$O(C(f, -g), g) = (1\ 1\ 1\ 1\ 1\ 1\ 4\ 4\ 4)_0$$

The output h is found by employing the median:
$$h = (1\ 1\ 1\ 1\ 1\ 1\ 4\ 4\ 4)_0$$

8.4 APPLICATIONS OF THE BASIS REPRESENTATION

In this section we consider the expansions of numerous filters in terms of erosions over the basis for the kernel. We shall be applying Theorem 7.4, and we assume throughout that we are working with extended signals and images, and that we are in the fully digital setting.

To begin with, consider the advance-type filter defined by
$$[\mathcal{T}(f)](i) = f(i + 1)$$

In Section 8.1, we saw that $\mathcal{T}$ is time invariant, and since $\mathcal{T}$ is also offset invariant, i.e., $\mathcal{T}(f + j) = \mathcal{T}(f) + j$, it is an increasing τ-mapping. Moreover, its kernel is given by the set of signals f for which $f(1) \geq 0$, where $+\infty$ is allowed since we are in the extended situation. Finally, the singleton set $\{(0)_1\}$ is a basis for the kernel, and consequently, according to Theorem 7.4,
$$\mathcal{T}(f) = \text{ERODE}(f, (0)_1)$$

Example 8.13:

Suppose
$$f(i) = \begin{cases} i^2, & \text{if } i = 0, 1, 2, \ldots \\ *, & \text{if } i = -1, -2, \ldots \end{cases}$$

Then $[\mathcal{T}(f)](i) = (i + 1)^2$ for $i = -1, 0, 1, 2, \ldots$, and it is easy to see that
$$[\text{ERODE}(f, (\circledast\ \ 0))](i) = (i + 1)^2$$

for $i = -1, 0, 1, 2, \ldots$.

Sec. 8.4 Applications of the Basis Representation

A two-time-unit advance $[\mathcal{T}(f)](i) = f(i + 2)$ can obviously be obtained by eroding twice in succession utilizing the structuring element $(0)_1$. However, a direct implementation of Theorem 7.4 yields the representation

$$\mathcal{T}(f) = \text{ERODE}(f, (0)_2)$$

Similarly, an n-unit advance is given by

$$[\mathcal{T}(f)](i) = f(i + n) = [\text{ERODE}(f, (0)_n)](i), n \geq 0$$

and the n-unit delay is given by

$$[\mathcal{T}(f)](i) = f(i - n) = [\text{ERODE}(f, (0)_{-n})](i), n > 0$$

Now consider the translation operation for (extended) images. It, too, is an increasing τ-mapping, and we have

$$\text{TRAN}(f; i, j) = \text{ERODE}(f, (0)_{-i, -j})$$

The next example compares the left- and right-hand sides of this equation.

Example 8.14:

Let

$$f = \begin{pmatrix} * & * & * & * \\ 2 & 4 & * & * \\ 3 & 1 & * & * \end{pmatrix}_{0,0}$$

Then

$$\text{TRAN}(f; 2, 1) = \begin{pmatrix} * & * & 2 & 4 \\ * & * & 3 & 1 \\ * & * & * & * \end{pmatrix}_{0,0}$$

Erosion of f by

$$(0)_{-2, -1} = \begin{pmatrix} * & * & \circledast \\ 0 & * & * \end{pmatrix}$$

gives the same result.

In Section 8.1 we saw that the square-law device is time invariant; however, it is not increasing, i.e., $f \ll g$ does not imply that $f^2 \ll g^2$. By comparsion, the cube-law device

$$[\mathcal{T}(f)](i) = [f(i)]^3$$

is both time invariant and increasing. However, it cannot be represented by erosions utilizing the Matheron representation theorem: since it is not offset invariant, it is not a τ-mapping. That is, for the cube-law filter, we do not in general have

$$\mathcal{T}(f + j) = (f + j)^3 = f^3 + j = \mathcal{T}(f) + j$$

Now consider the median filter $\mathcal{F}$ applied to digital images. For a given pixel (i, j), we define $[\mathcal{F}(f)](i, j)$ to be the median value, in the extended sense, of (i, j) taken together with its four strong neighbors. $\mathcal{F}$ is an increasing τ-mapping and its kernel consists of all images that have gray values defined at three or more pixels in the strong neighborhood about the origin, and for which at least three of the defined gray values are nonnegative. A basis for the kernel is given by the following ten images:

$$g_1 = \begin{pmatrix} * & 0 & * \\ 0 & \circledast & 0 \\ * & * & * \end{pmatrix} \quad g_2 = \begin{pmatrix} * & 0 & * \\ 0 & \circledast & * \\ * & 0 & * \end{pmatrix}$$

$$g_3 = \begin{pmatrix} * & 0 & * \\ 0 & \textcircled{0} & * \\ * & * & * \end{pmatrix} \quad g_4 = \begin{pmatrix} * & * & * \\ 0 & \circledast & 0 \\ * & 0 & * \end{pmatrix}$$

$$g_5 = \begin{pmatrix} * & * & * \\ 0 & \textcircled{0} & 0 \\ * & * & * \end{pmatrix} \quad g_6 = \begin{pmatrix} * & * & * \\ 0 & \textcircled{0} & * \\ * & 0 & * \end{pmatrix}$$

$$g_7 = \begin{pmatrix} * & 0 & * \\ * & \circledast & 0 \\ * & 0 & * \end{pmatrix} \quad g_8 = \begin{pmatrix} * & 0 & * \\ * & \textcircled{0} & 0 \\ * & * & * \end{pmatrix}$$

$$g_9 = \begin{pmatrix} * & 0 & * \\ * & \textcircled{0} & * \\ * & 0 & * \end{pmatrix} \quad g_{10} = \begin{pmatrix} * & * & * \\ * & \textcircled{0} & 0 \\ * & 0 & * \end{pmatrix}$$

According to Theorem 7.4,

$$\mathcal{F}(f) = \text{EXTMAX}_{k=1,2,\ldots,10}[\text{ERODE}(f, g_k)]$$

If the median filter were used where the weak neighbors are also employed in the smoothing process, then a basis for the kernel would also exist; however, the kernel would consist of all images with gray values defined in at least five pixels in the three by three square neighborhood about the origin and for which at least five of the defined gray values are nonnegative. The basis would be comprised of 126 three by three bound matrices of the form

$$\begin{pmatrix} a_{11} & a_{12} & a_{13} \\ a_{21} & a_{22} & a_{23} \\ a_{31} & a_{32} & a_{33} \end{pmatrix}_{-1,1}$$

for which exactly five of the a_{ij} equal 0 and the rest of the a_{ij} are stars. In general, if the median filter is computed employing a neighborhood consisting of n pixels, n odd, then the basis is comprised of

$$\binom{n}{(n+1)/2}$$

images, each of which is zero-valued at $(n+1)/2$ of the neighborhood pixels, and star valued elsewhere.

Example 8.15:

Let

$$f = \begin{pmatrix} * & * & 3 & 4 & 5 & * & * \\ * & * & 3 & 3 & 3 & * & 4 \\ * & * & 3 & 3 & 4 & * & * \\ * & * & 4 & 4 & 8 & * & * \\ * & * & 5 & 4 & 3 & * & * \\ * & * & 3 & 4 & 4 & * & * \end{pmatrix}_{0,0}$$

To find the median filter $\mathcal{F}$, using the strong neighbors, $\text{ERODE}(f, g_k)$ must be computed for $k = 1, 2, \ldots, 10$, and EXTMAX must be applied. For $(i, j) = (2, -3)$, we obtain

$$[\text{ERODE}(f, g_1)](2, -3) = *$$
$$[\text{ERODE}(f, g_2)](2, -3) = *$$
$$[\text{ERODE}(f, g_3)](2, -3) = *$$
$$[\text{ERODE}(f, g_4)](2, -3) = *$$
$$[\text{ERODE}(f, g_5)](2, -3) = *$$
$$[\text{ERODE}(f, g_6)](2, -3) = *$$
$$[\text{ERODE}(f, g_7)](2, -3) = 3$$
$$[\text{ERODE}(f, g_8)](2, -3) = 3$$
$$[\text{ERODE}(f, g_9)](2, -3) = 3$$
$$[\text{ERODE}(f, g_{10})](2, -3) = 4$$

Applying EXTMAX gives

$$[\mathcal{F}(f)](2, -3) = 4$$

The full output is given by

$$\mathcal{F}(f) = \begin{pmatrix} * & * & 3 & 3 & 3 & * & * \\ * & * & 3 & 3 & 3 & * & * \\ * & * & 3 & 3 & 3 & * & * \\ * & * & 4 & 4 & 4 & * & * \\ * & * & 4 & 4 & 4 & * & * \\ * & * & 3 & 4 & 3 & * & * \end{pmatrix}_{0,0}$$

From a practical and intuitive point of view, the median filter preserved edge-type information while smoothing out the original image. It also removed possible pinpoint-type noise at pixels $(6, -1)$ and $(4, -3)$. Moreover, whereas a moving-average filter might have been overly sensitive to the gray value 8 at pixel $(4, -3)$, $\mathcal{F}$ was not.

In Example 7.15 we considered the filter Ω, where $[\Omega(f)](i)$ was defined to be average of the two input values $f(i)$ and $f(i + 1)$, with the convention that fractional values are to be rounded up. It was seen there that signals of the form

$$f^n = (n \quad -n - 1)_0, n \in \mathbb{Z}$$

form a basis for the kernel. Of interest is the fact that in the expansion of Theorem 7.4, we can always be certain that, given a finite input bound vector of length m, no more than $m - 1$ erosions will be necessary. In particular, it can be shown that for the signal

$$g = (a_1 \quad a_2 \quad \cdots \quad a_m)_p$$

with all a_i real,

$$\Omega(g) = \text{EXTSUP}_{n \in S}[\text{ERODE}(g, f^n)]$$

where

$$S = \left\{ -\left\lceil \frac{a_{i+1} - a_i}{2} \right\rceil : i = 1, 2, \ldots, m - 1 \right\}$$

and where $\lceil x \rceil$ denotes the smallest integer greater than or equal to x.

Example 8.16:

If

$$g = (1 \quad 2 \quad -5 \quad 8 \quad 7 \quad -2 \quad 5 \quad 4)_0$$

then the indices to be used in the erosion of g by

$$f^n = (n \quad -n - 1)_0$$

are specified by the relation

$$n \in \{-1, 3, -7, 0, 4, -4\} = S$$

Each erosion appears as a row in the serial-type bound matrix

$$\begin{pmatrix} 2 & -5 & -4 & 7 & -2 & -1 & 4 \\ -2 & -1 & -8 & 5 & 2 & -5 & 2 \\ -4 & -11 & 2 & 1 & -8 & -1 & -2 \\ 1 & -4 & -5 & 8 & -1 & -2 & 5 \\ -3 & -2 & -9 & 4 & 3 & -6 & 1 \\ -1 & -8 & -1 & 4 & -5 & 2 & 1 \end{pmatrix}_{0,5}$$

and the solution is found by determining the maximum value in each column. We obtain

$$\Omega(g) = (2 \quad -1 \quad 2 \quad 8 \quad 3 \quad 2 \quad 5)_0$$

Note in the previous example that values of the output $\Omega(f)$ appear entry for entry along the main diagonal of the serial matrix, with the exception of the final entry. This results from the fact that the values of $-\lceil (a_{i+1} - a_i)/2 \rceil$ were placed in the set S in order for $i = 1, 2, \ldots, m - 2$, whereas the value 0 that is obtained for $i = m - 1$ is not placed again in S, it already being there. In point of fact, had we treated S as an array instead of a set, it would have taken the form

$$S = [-1, 3, -7, 0, 4, -4, 0]$$

and the fourth row of the serial matrix would have appeared again at the bottom. Then the value $[\Omega(g)](6) = 5$ would have occurred in the lower right position of the serial matrix. In sum, one does not actually have to proceed through the full erosions to find

Sec. 8.4 Applications of the Basis Representation

$\Omega(g)$. Indeed, for

$$g = (a_1 \quad a_2 \quad \cdots \quad a_m)_p$$

we have

$$[\Omega(g)](p - 1 + i) = [\text{ERODE}(g, f^{n(i)})](p - 1 + i)$$

where

$$n(i) = -\left\lceil \frac{a_{i+1} - a_i}{2} \right\rceil, \quad i = 1, 2, \ldots, m - 1$$

To prove the validity of the preceding remarks, we need only show that when the aforementioned $n(i)$ are employed, the appropriate quantized average is obtained. At index i, using $n(i)$, the erosion is defined to be the largest value of j such that the following simultaneous inequalities hold:

$$n(i) + j \leq a_i$$
$$-n(i) - 1 + j \leq a_{i+1}$$

Solving these inequalities is equivalent to finding the largest value of j such that

$$j \leq a_i + \left\lceil \frac{a_{i+1} - a_i}{2} \right\rceil$$

and

$$j \leq a_{i+1} - \left\lceil \frac{a_{i+1} - a_i}{2} \right\rceil + 1$$

Combinations of the four possibilities, $a_{i+1} - a_i$ even, odd, positive, and nonpositive, are set out in the following table:

	POSITIVE	NONPOSITIVE
EVEN	$j \leq \frac{a_{i+1} + a_i}{2}$	$j \leq \frac{a_{i+1} + a_i}{2}$
EVEN	$j \leq \frac{a_{i+1} + a_i}{2} + 1$	$j \leq \frac{a_{i+1} + a_i}{2} + 1$
ODD	$j \leq \frac{a_{i+1} + a_i}{2} + \frac{1}{2}$	$j \leq \frac{a_{i+1} + a_i}{2} + \frac{1}{2}$
ODD	$j \leq \frac{a_{i+1} + a_i}{2} + \frac{1}{2}$	$j \leq \frac{a_{i+1} + a_i}{2} + \frac{1}{2}$

The unique solution is given by

$$j = \left\lceil \frac{a_{i+1} + a_i}{2} \right\rceil$$

thereby proving that erosion by $f^{n(i)}$ gives the desired result.

Various FIR filters discussed in Section 8.1 can be implemented without change using morphological techniques. However, before doing so, we must ensure that the output is quantized so that Theorem 7.4 can be applied. While all FIR filters (in one dimension) are time invariant, the subclass of low-pass filters defined by

$$\mathcal{F}(f) = \left[\sum_{i=-N}^{N} a_i f_i \right]$$

where the a_i are nonnegative real numbers such that

$$\sum_{i=-N}^{N} a_i = 1$$

is in fact a set of increasing τ-mappings. Letting

$$f_i = \text{TRAN}(f; i)$$

we find that the kernel consists of all signals g for which

$$a_N g_N(0) + \cdots + a_0 g_0(0) + a_{-1} g_{-1}(0) + \cdots + a_{-N} g_{-N}(0) > -1$$

An example of a basis associated with this type of filtering scheme can be found in Example 7.15.

A similar FIR-type mapping for images is given by

$$\mathcal{F}(f) = \left[\sum_{i=-N}^{N} \sum_{j=-M}^{M} a_{ij} f_{ij} \right]$$

with $a_{ij} \geq 0$,

$$\sum_{i=-N}^{N} \sum_{j=-M}^{M} a_{ij} = 1$$

and $f_{i,j} = \text{TRAN}(f; i, j)$. The kernel is comprised of all images g such that

$$\sum_{i=-N}^{N} \sum_{j=-N}^{N} a_{ij} g_{ij}(0, 0) > -1$$

A special case is the moving-average filter

$$\mathcal{F}(f)(i, j) = \left[\frac{1}{3} f(i, j) + \frac{1}{3} f_{1,0}(i, j) + \frac{1}{3} f_{0,-1}(i, j) \right]$$

The kernel for this filter consists of all images g such that

$$g(0, 0) + g_{1,0}(0, 0) + g_{0,-1}(0, 0) > -3$$

or

$$g(0, 0) + g_{1,0}(0, 0) + g_{0,-1}(0, 0) \geq -2$$

The basis for the kernel is the set of all bound matrices of the form

$$h^{n,m} = \begin{pmatrix} * & n \\ m & -(n + m + 2) \end{pmatrix}_{-1, 1}, \quad n, m \text{ integers}$$

Although the preceding basis is infinite in cardinality, in practical applications only a finite subset of it need be employed in Theorem 7.4.

Sec. 8.4   Applications of the Basis Representation   **285**

Example 8.17:

Consider the image

$$f = \begin{pmatrix} 0 & 3 & 6 \\ 3 & 3 & 4 \\ 2 & 1 & 0 \end{pmatrix}_{0,0}$$

and suppose we wish to find the three-gray-value moving-average filter $\mathcal{F}$ discussed above. In theory, the expansion of Theorem 7.4 is infinite, but as usual, only a small number of basis elements are actually required. Indeed, let

$$h^0 = \begin{pmatrix} * & 0 \\ 0 & -2 \end{pmatrix}_{-1,1}$$

$$h^1 = \begin{pmatrix} * & 1 \\ -2 & -1 \end{pmatrix}_{-1,1}$$

$$h^2 = \begin{pmatrix} * & 2 \\ -2 & -2 \end{pmatrix}_{-1,1}$$

Proceeding with Theorem 7.4, we get

$$\text{ERODE}(f, h^0) = \begin{pmatrix} 3 & 3 \\ 2 & 1 \end{pmatrix}_{1,-1}$$

$$\text{ERODE}(f, h^1) = \begin{pmatrix} 2 & 5 \\ 2 & 1 \end{pmatrix}_{1,-1}$$

$$\text{ERODE}(f, h^2) = \begin{pmatrix} 1 & 4 \\ 1 & 2 \end{pmatrix}_{1,-1}$$

and EXTMAX yields

$$\mathcal{F}(f) = \begin{pmatrix} 3 & 5 \\ 2 & 2 \end{pmatrix}_{1,-1}$$

In a manner analogous to the preceding remarks, only a finite number of—and in fact, relatively few—erosions need be employed in determining the output of a weighted-average-type FIR filter.

Consider, now, the Wilcoxon filter $\mathcal{W}$ on extended signals defined by

$$[\mathcal{W}(f)](i) = \text{MEDIAN}\left(\left\lceil \frac{f_1(i) + f_{-1}(i)}{2} \right\rceil, \left\lceil \frac{f(i) + f_1(i)}{2} \right\rceil, \left\lceil \frac{f(i) + f_{-1}(i)}{2} \right\rceil \right)$$

where f_1 is the translate of f one unit to its right and f_{-1} is the translate of f one unit to its left. This filter, a combination of the median filter discussed previously and the two-point-average filter given in Example 7.15, is an increasing τ-mapping. Its kernel consists of all signals g such that the sum of any two pairs of distinct values among the three values $g(0)$, $g(-1)$, and $g(1)$ must be greater than -2. (Cf. Example 7.15.) The basis for the kernel is the set S of all bound vectors

$$S = \{(n \quad -n-1 \quad -n-1)_{-1}, (-m-1 \quad m \quad -m-1)_{-1},$$
$$(-k-1 \quad -k-1 \quad k)_{-1} : n, m, k \in Z\}$$

and, by Theorem 7.4,

$$\mathcal{W}(f) = \text{EXTSUP}[\text{ERODE}_{g \in S}(f, g)]$$

As with the previous filters, not all the basis elements need be used. Moreover, the structuring elements that are needed can be found by utilizing an expression of the form

$$-\left\lceil \frac{a_i - a_j}{2} \right\rceil$$

to find n, m, and k, where i and j are distinct integers that differ by at most 2 and

$$f = (a_1, a_2, \ldots, a_N)_p$$

Example 8.18:

For

$$f = (3 \quad 7 \quad 4 \quad 1)_0$$

we obtain

$$\mathcal{W}(f) = (5 \quad 4)_1$$

On the other hand,

$$\mathcal{W}(f) = \text{EXTMAX}[\text{ERODE}(f, g^1), \text{ERODE}(f, g^2)]$$

where

$$g^1 = (-2 \quad 1 \quad -2)_{-1}$$

and

$$g^2 = (3 \quad -4 \quad -4)_{-1}$$

since

$$\text{ERODE}(f, g^1) = (5 \quad 3)_1$$

and

$$\text{ERODE}(f, g^2) = (0 \quad 4)_1$$

so that EXTMAX provides the same result for $\mathcal{W}$ as before.

The *smoothing median filter* is defined by

$$\mathcal{S}(f)(i) = \text{MEDIAN}\left(\left\lceil \frac{f_{-2}(i) + f_{-1}(i)}{2} \right\rceil, f(i), \left\lceil \frac{f_2(i) + f_1(i)}{2} \right\rceil\right)$$

where f_j is the translate of f j units to its right.

Clearly, $\mathcal{S}$ is an increasing τ-mapping whose kernel consists of all signals g that satisfy certain properties at -2, -1, 0, 1, and 2—specifically, either $g(0) \geq 0$ and $g(-1) + g(-2) > -2$, or $g(0) \geq 0$ and $g(1) + g(2) > -2$, or $g(-1) + g(-2) > -2$ and $g(1) + g(2) > -2$. The basis for the kernel is the set S

of all signals given by

$$S = \{(* \quad * \quad 0 \quad n \quad -n-1)_{-2}, (m \quad -m-1 \quad 0 \quad * \quad *)_{-2},$$
$$(p \quad -p-1 \quad * \quad q \quad -q-1)_{-2} : m, n, p, q \in Z\}$$

Thus,

$$\mathcal{S}(f) = \text{EXTSUP}_{g \in S}[\text{ERODE}(f, g)]$$

Again, only a finite number of signals in S need be applied for a fixed bound vector f.

Example 8.19:

Applying the smoothing median filter $\mathcal{S}$ to the bound vector

$$f = (-3 \quad 7 \quad 20 \quad -10 \quad 1 \quad 8)_0$$

gives

$$\mathcal{S}(f) = (-3 \quad 5 \quad 2 \quad 5 \quad 1 \quad -4)_0$$

If we erode f in turn by the basis elements

$$g^1 = (0 \quad 0 \quad -1)_0$$
$$g^2 = (0 \quad 14 \quad -15)_0$$
$$g^3 = (-6 \quad 5 \quad 0)_{-2}$$
$$g^4 = (-7 \quad 6 \quad * \quad -4 \quad 3)_{-2}$$
$$g^5 = (14 \quad -15 \quad 0)_{-2}$$

then, in accordance with Theorem 7.4, $\mathcal{S}(f)$ is given by taking extended maximums down the columns of the serial matrix

$$\begin{pmatrix} -3 & -9 & -10 & -10 & * & * \\ -7 & 5 & -24 & -13 & * & * \\ * & * & 2 & -10 & -15 & -4 \\ * & * & -6 & 5 & * & * \\ * & * & -17 & -10 & 1 & -24 \end{pmatrix}_{0, 4}$$

EXERCISES

8.1. Employ the methodologies of Figures 8.23 and 8.24 to compute the order-statistic filter $f^{(2)}$ for $N = 4$, the neighborhood defined by the mask

$$M = \begin{pmatrix} 1 & 1 \\ \boxed{1} & 1 \end{pmatrix}$$

and the extended image

$$f = \begin{pmatrix} 1 & 2 & 1 & +\infty \\ 1 & 3 & -7 & 1 \\ 1 & 2 & 1 & 0 \\ * & 1 & 1 & 9 \end{pmatrix}_{2, 8}$$

8.2. Find the basis for the mapping $\Phi: f \to f^{(2)}$ considered in Exercise 8.1.

8.3. Repeat Example 8.15; however, this time employ the image

$$f = \begin{pmatrix} 9 & -1 & * & * & * & * & * & * & * & * \\ * & * & 0 & 0 & 0 & 0 & 4 & 4 & 4 & 4 \\ * & * & 0 & 0 & 0 & 0 & 4 & 4 & 4 & 4 \\ * & * & 0 & 0 & 0 & 0 & 4 & 4 & 4 & 4 \\ * & * & * & * & * & * & * & * & * & 8 \\ * & * & * & * & * & 1 & * & * & * & 9 \end{pmatrix}_{0,5}$$

Find $\mathcal{F}(f)$, first by inspection and then by employing Theorem 7.4.

8.4. Apply the methodology following Example 8.15 in the text to the quantized weighted filter $\mathcal{F}$ defined by

$$[\mathcal{F}(f)](i) = \frac{3}{4}f(i) + \frac{1}{4}f(i+1)$$

Employ the results to find $\mathcal{F}(g)$, where

$$g = (2\ 2\ 4\ 0\ 0\ 0)_0$$

8.5. Repeat Exercise 8.4, except this time use the unweighted filter $\mathcal{M}$ defined by

$$[\mathcal{M}(f)](i) = \frac{1}{3}[f(i-1) + f(i) + f(i+1)]$$

Appendix

Given two images A and B in R^2, the *Minkowski addition of A by B* is defined to be

$$A \oplus B = \bigcup_{x \in B} A + x$$

Proposition A.1

$$A \oplus B = \bigcup_{\substack{x \in A \\ y \in B}} \{x + y\}$$

Proof.

$$\begin{aligned}
A \oplus B &= \bigcup_{y \in B} (A + y) \\
&= \bigcup_{y \in B} \left(\left[\bigcup_{x \in A} \{x\} \right] + y \right) \\
&= \bigcup_{y \in B} \left(\bigcup_{x \in A} \{x + y\} \right) \\
&= \bigcup_{\substack{x \in A \\ y \in B}} \{x + y\}
\end{aligned}$$

Proposition A.2

$\quad$ (i) $A \oplus B = B \oplus A \quad$ (Commutativity)

$\quad$ (ii) $(A \oplus B) \oplus C = A \oplus (B \oplus C) \quad$ (Associativity)

Proof. By Proposition A.1,

$$A \oplus B = \bigcup_{\substack{x \in A \\ y \in B}} \{x + y\} = \bigcup_{\substack{y \in B \\ x \in A}} \{y + x\} = B \oplus A$$

As for associativity,

$$(A \oplus B) \oplus C = \left(\bigcup_{\substack{x \in A \\ y \in B}} \{x + y\} \right) \oplus C$$

$$= \bigcup_{\substack{x \in A \\ y \in B \\ z \in C}} \{(x + y) + z\}$$

$$= \bigcup_{\substack{x \in A \\ y \in B \\ z \in C}} \{x + (y + z)\}$$

$$= A \oplus \left(\bigcup_{\substack{y \in B \\ z \in C}} \{y + z\} \right)$$

$$= A \oplus (B \oplus C)$$

Proposition A.3

$$A \oplus [B + x] = [A \oplus B] + x$$

Proof. This is a special case of the associative law, Proposition A.2(ii). Specifically,

$$A \oplus [B + x] = A \oplus [B \oplus \{x\}]$$
$$= [A \oplus B] \oplus \{x\}$$
$$= [A \oplus B] + x$$

The second fundamental morphological operation is *Minkowski subtraction*, defined to be

$$A \ominus B = \bigcap_{x \in B} A + x$$

Proposition A.4

$$A \oplus B = [A^c \ominus B]^c \quad \text{(Duality)}$$

Proof. Let x be any point in R^2. Then $x \in [A^c \ominus B]^c$ if and only if $x \notin [A^c \ominus B]$ if and only if there exists $b \in B$ such that $x \notin A^c + b$ if and only if there exists $b \in B$ such that $x \in A + b$ if and only if $x \in A \oplus B$.

Proposition A.5

$$A \ominus B = [A^c \oplus B]^c \qquad \text{(Duality)}$$

Proof. Applying Proposition A.4 and the fact that successive complementations produce the original set, we have

$$A^c \oplus B = [(A^c)^c \ominus B]^c = [A \ominus B]^c$$

Complementing both sides yields Proposition A.5.

Proposition A.6

(i) $A \ominus [B + x] = [A \ominus B] + x$

(ii) $[A + x] \ominus B = [A \ominus B] + x$

Proof. We apply duality together with translational invariance for Minkowski addition:

$$A^c \oplus [B + x] = [A^c \oplus B] + x$$

Taking complements, we obtain (i); that is,

$$A \ominus [B + x] = ([A^c \oplus B] + x)^c$$
$$= [A^c \oplus B]^c + x = [A \ominus B] + x$$

where we have used the fact that complementation can be interchanged with translation. As for (ii), apply complementation to

$$[A^c + x] \oplus B = [A^c \oplus B] + x$$

The result follows in a similar manner to the proof of part (i).

Proposition A.7

$$A \ominus B = \{x: -B + x \subset A\}$$

Proof.

$$\{x: -B + x \subset A\} = \bigcap_{y \in B} \{x: -y + x \in A\}$$
$$= \bigcap_{y \in B} \{x: x \in A + y\}$$
$$= \bigcap_{y \in B} A + y = A \ominus B$$

Since $-(-B) = B$, the preceding proposition could also be written as

$$A \ominus (-B) = \{x: B + x \subset A\}$$

We define the operation of *erosion* of A by B as

$$\mathcal{E}(A, B) = \{x: B + x \subset A\}$$

Thus, Proposition A.7 may be reformulated as Proposition A.8.

Proposition A.8

$$\mathcal{E}(A, B) = A \ominus (-B)$$

Corresponding to the terminology of erosion, there is a term commonly employed with regard to Minkowski addition. Instead of $A \oplus B$, it is common in image processing to say the *dilation* of A by B, written $\mathcal{D}(A, B)$. Then $\mathcal{D}(A, B) = A \oplus B$, while $\mathcal{E}(A, B) = A \ominus (-B)$.

Proposition A.9

$$A \oplus B = \{x: [(-B) + x] \cap A \neq \emptyset\}$$

Proof.

$\{x: [(-B) + x] \cap A \neq \emptyset\}$

$= \{x: \text{there exists an } a \in A \text{ such that } a \in [(-B) + x]\}$

$= \{x: \text{there exists an } a \in A \text{ such that } a - x \in (-B)\}$

$= \{x: \text{there exists an } a \in A \text{ such that } x \in B + a\}$

$= \left\{x: x \in \bigcup_{a \in A} B + a\right\} = B \oplus A = A \oplus B$

Theorem A.1 collects some of the earlier propositions and puts them in terms of dilation and erosion.

Theorem A.1

(i) $\mathcal{D}(A, B) = \mathcal{D}(B, A)$

(ii) $\mathcal{D}(A, B) = [\mathcal{E}(A^c, -B)]^c$

(iii) $\mathcal{E}(A, B) = [\mathcal{D}(A^c, -B)]^c$

(iv) $\mathcal{D}(A, B + x) = \mathcal{D}(A, B) + x$

(v) $\mathcal{E}(A, B + x) = \mathcal{E}(A, B) - x$

An operation $P(A)$ on sets is said to be *increasing* if $A \subset B$ implies $P(A) \subset P(B)$. $P(A)$ is said to be *decreasing* if $A \subset B$ implies $P(A) \supset P(B)$.

Proposition A.10. Suppose B is fixed and $A_1 \subset A_2$. Then

(i) $\mathcal{D}(A_1, B) \subset \mathcal{D}(A_2, B)$

(ii) $\mathcal{E}(A_1, B) \subset \mathcal{E}(A_2, B)$

Proof. (i) follows at once from the definition of Minkowski addition since each of $\mathcal{D}(A_1, B)$ and $\mathcal{D}(A_2, B)$ is the union over the elements of B and in each case $A_1 + x \subset A_2 + x$. The second statement follows in like manner since each of $\mathcal{E}(A_1, B)$ and $\mathcal{E}(A_2, B)$ is an intersection over the elements of B.

Proposition A.11. Suppose A is fixed and $B_1 \subset B_2$. Then $\mathcal{E}(A, B_1) \supset \mathcal{E}(A, B_2)$.

Proof. Using the definition of erosion directly, $B_2 + x \subset A$ implies $B_1 + x \subset A$. Hence, $x \in \mathcal{E}(A, B_2)$ implies $x \in \mathcal{E}(A, B_1)$, and the proposition follows.

Proposition A.12. Let t be a real number, $t \neq 0$. Then

$$t\left[\left(\frac{1}{t}\right)A \oplus B\right] = A \oplus (tB)$$

Proof.

$$z \in t\left[\left(\frac{1}{t}\right)A \oplus B\right]$$

if and only if there exists $a \in A$ and $b \in B$ such that

$$z = t\left[\left(\frac{1}{t}\right)a + b\right] = a + tb$$

which means precisely that $z \in A \oplus (tB)$.

Proposition A.13. Let t be a real number, $t \neq 0$. Then

$$t\left[\left(\frac{1}{t}\right)A \ominus B\right] = A \ominus (tB)$$

Proof.

$$z \in t\left[\left(\frac{1}{t}\right)A \ominus B\right]$$

if and only if

$$\left(\frac{1}{t}\right)z \in \left(\frac{1}{t}\right)A \ominus B.$$

But the latter is true if and only if

$$-B + \left(\frac{1}{t}\right)z \subset \left(\frac{1}{t}\right)A$$

which is equivalent to asserting that for any point b in B, there is a corresponding point a_b in A such that

$$-b + \left(\frac{1}{t}\right)z = \left(\frac{1}{t}\right)a_b$$

or, in other words,
$$-tb + z = a_b$$
But this means precisely that $-tB + z \subset A$, which in turn means that $z \in A \ominus tB$.

Proposition A.14

(i) $A \oplus (B \cup C) = (A \oplus B) \cup (A \oplus C)$

(ii) $(B \cup C) \oplus A = (B \oplus A) \cup (C \oplus A)$

(Distributivity of Minkowski addition over union)

Proof.

$$A \oplus (B \cup C) = \bigcup_{z \in A} (B \cup C) + z = \bigcup_{z \in A} (B + z) \cup (C + z)$$
$$= [\bigcup_{z \in A} B + z] \cup [\bigcup_{z \in A} C + z]$$
$$= (A \oplus B) \cup (A \oplus C)$$

Hence, (i) is proven. Part (ii) follows at once from the commutativity of Minkowski addition.

Proposition A.15

(i) $A \oplus (B \cap C) \subset (A \oplus B) \cap (A \oplus C)$

(ii) $(B \cap C) \oplus A \subset (B \oplus A) \cap (C \oplus A)$

Proof.

$$A \oplus (B \cap C) = \bigcup_{z \in A} (B \cap C) + z$$
$$= \bigcup_{z \in A} (B + z) \cap (C + z)$$
$$\subset \left[\bigcup_{z \in A} B + z\right] \cap \left[\bigcup_{z \in A} C + z\right]$$
$$= (A \oplus B) \cap (A \oplus C)$$

Thus, (i) is proven. Part (ii) follows by the commutativity of Minkowski addition.

Proposition A.16

$$A \ominus (B \cup C) = (A \ominus B) \cap (A \ominus C)$$

Proof. Applying duality, the previous result, and De Morgan's Law, we have
$$A \ominus (B \cup C) = [A^c \oplus (B \cup C)]^c$$

Appendix

$$= [(A^c \oplus B) \cup (A^c \oplus C)]^c$$
$$= (A^c \oplus B)^c \cap (A^c \oplus C)^c$$
$$= (A \ominus B) \cap (A \ominus C)$$

Proposition A.17
$$(B \cap C) \ominus A = (B \ominus A) \cap (C \ominus A)$$

Proof.

$$(B \cap C) \ominus A = [(B \cap C)^c \oplus A]^c = [(B^c \cup C^c) \oplus A]^c$$
$$= [(B^c \oplus A) \cup (C^c \oplus A)]^c$$
$$= (B^c \oplus A)^c \cap (C^c \oplus A)^c$$
$$= (B \ominus A) \cap (C \ominus A)$$

Proposition A.18
$$(A \ominus B) \ominus C = A \ominus (B \oplus C)$$

Proof.

$$(A \ominus B) \ominus C = \bigcap_{z \in C} (A \ominus B) + z$$
$$= \bigcap_{z \in C} \left[\left(\bigcap_{y \in B} A + y \right) + z \right]$$
$$= \bigcap_{z \in C} \left[\bigcap_{y \in B} A + (z + y) \right]$$
$$= \bigcap_{x \in B \oplus C} A + x = A \ominus (B \oplus C)$$

The *opening* and *closing* of one image by another are respectively defined as
$$O(A, B) = [A \ominus (-B)] \oplus B$$
and
$$C(A, B) = [A \oplus (-B)] \ominus B$$
It follows immediately from the definitions of dilation and erosion that the preceding definitions can be written as
$$O(A, B) = \mathcal{D}[\mathcal{E}(A, B), B]$$
and
$$C(A, B) = \mathcal{E}[\mathcal{D}(A, -B), -B]$$

Proposition A.19

$$C(A, B)^c = O(A^c, B) \quad \text{(Duality)}$$

Proof. Applying the duality criteria for Minkowski subtraction and Minkowski addition, we obtain

$$C(A, B)^c = [(A \oplus (-B)) \ominus B]^c$$
$$= [A \oplus (-B)]^c \oplus B$$
$$= [A^c \ominus (-B)] \oplus B = O(A^c, B)$$

Proposition A.20

$$O(A, B)^c = C(A^c, B) \quad \text{(Duality)}$$

Proof.

$$O(A, B)^c = O[(A^c)^c, B]^c$$
$$= [C(A^c, B)^c]^c \quad \text{(by Proposition A.19)}$$
$$= C(A^c, B)$$

Proposition A.21. $z \in O(A, B)$ if and only if $[(-B) + z] \cap [A \ominus (-B)] \neq \emptyset$.

Proof. $z \in O(A, B)$ if and only if $z \in \cup_{y \in B} [A \ominus (-B)] + y$ if and only if there exists $b \in B$ with $z \in [A \ominus (-B)] + b$ if and only if there exists $b \in B$ and $w \in [A \ominus (-B)]$ with $z = w + b$, or, equivalently, with $w = -b + z$. But this means precisely that $w \in [(-B) + z] \cap [A \ominus (-B)]$, which means that the intersection in the proposition is nonempty.

Theorem A.2

$$O(A, B) = \cup\{B + y : B + y \subset A\}$$

Proof. By Proposition A.21, $z \in O(A, B)$ if and only if there exists a point y such that $y \in [(-B) + z] \cap [A \ominus (-B)]$, that is, if and only if there is a point y such that $z \in B + y$ and $B + y \subset A$. But this last statement means precisely that z is in the union specified in the statement of the theorem.

Proposition A.22

$$C(A, B) = \cap\{(B + y)^c : B + y \subset A^c\}$$

Proof. We apply duality to Theorem 3.2, yielding

$$C(A, B)^c = O(A^c, B) = \cup\{B + y : B + y \subset A^c\}$$

Application of De Morgan's Law gives Proposition A.22.

Proposition A.23. $z \in C(A, B)$ if and only if $(B + y) \cap A \neq \emptyset$ for any translate $B + y$ containing z.

Proof. By Proposition A.22, $z \in C(A, B)$ if and only if $B + y \subset A^c$ implies $z \in (B + y)^c$, which is itself true, by contraposition, if and only if $z \in B + y$ implies $(B + y) \cap A \neq \emptyset$.

Theorem A.3. The opening satisfies

(i) $O(A, B) \subset A$ (Antiextensive)

(ii) $A_1 \subset A_2$ implies $O(A_1, B) \subset O(A_2, B)$ (Increasing)

(iii) $O[O(A, B), B] = O(A, B)$ (Idempotent)

Proof.

(i) $z \in O(A, B)$ implies that there exists a translate $B + y$ such that $z \in B + y \subset A$. Hence, $O(A, B) \subset A$.

(ii) Once again applying Theorem A.2, $B + y \subset A_1$ implies $B + y \subset A_2$.

(iii) By (i), $O[O(A, B), B] \subset O(A, B)$. For the converse,

$$O[O(A, B), B] = \{[((A \ominus (-B)) \oplus B] \ominus (-B)\} \oplus B$$
$$= C[A \ominus (-B), (-B)] \oplus B$$
$$\supset [A \ominus (-B)] \oplus B = O(A, B)$$

where the containment follows from the fact that the closing of a given set by another contains the original set. (This will be proven independently in Theorem A.4, part i.)

Theorem A.4. The closing satisfies

(i) $C(A, B) \supset A$ (Extensive)

(ii) $A_1 \subset A_2$ implies $C(A_1, B) \subset C(A_2, B)$ (Increasing)

(iii) $C[C(A, B), B] = C(A, B)$ (Idempotent)

Proof.

(i) $C(A, B)^c = O(A^c, B) \subset A^c$ by duality and the antiextensivity of the opening. Taking complements reverses the inclusion sign, and (i) follows.

(ii) $C(A_1, B)^c = O(A_1^c, B) \supset O(A_2^c, B) = C(A_2, B)^c$ by duality. Taking complements gives (ii).

(iii) $C[C(A, B), B] = O[C(A, B)^c, B]^c$ (by duality)

$$= O[O(A^c, B), B]^c$$
$$= O(A^c, B)^c \quad \text{(Theorem A.3, part (iii))}$$
$$= C(A, B)$$

Set A is said to be *open with respect to set B* if $O(A, B) = A$. Similarly, set A is said to be *closed with respect to set B* if $C(A, B) = A$.

Proposition A.24. A is open with respect to B if and only if A^c is closed with respect to B.

Proof. A is open with respect to B if and only if $O(A, B) = A$. Taking complements and applying duality gives logical equivalence with $A^c = C(A^c, B)$, which is precisely the criterion for A^c being closed with respect to B.

Henceforth we shall say that A is *B-open* if A is open with respect to B and that A is *B-closed* if A is closed with respect to B.

Proposition A.25. A is B-open if and only if there exists a set E such that $A = E \oplus B$.

Proof. If A is B-open, then
$$A = O(A, B) = [A \ominus (-B)] \oplus B$$
Simply letting $E = A \ominus (-B)$ yields the forward implication. As for the converse, suppose that $A = E \oplus B$. Then
$$O(A, B) = O(E \oplus B, B) = [(E \oplus B) \ominus (-B)] \oplus B$$
$$= C(E, -B) \oplus B \supset E \oplus B = A$$
But by antiextensivity,
$$O(A, B) = O(E \oplus B, B) \subset E \oplus B = A$$
Consequently, $O(A, B) = A$, and A is B-open.

Proposition A.26. A is B-closed if and only if there exists a set E such that $A = E \ominus B$.

Proof. If A is B-closed, then
$$A = C(A, B) = [A \oplus (-B)] \ominus B$$
Simply take $E = A \oplus (-B)$. Conversely, suppose $A = E \ominus B$. Then, by duality, $A = [E^c \oplus B]^c$, and taking complements gives $A^c = E^c \oplus B$. By Proposition A.25, it follows that A^c is B-open, and hence, by Proposition A.24, A is B-closed.

Proposition A.27. Suppose A is B-open. Then, for any image F,
$$O(F, A) \subset O(F, B) \subset F \subset C(F, B) \subset C(F, A)$$

Proof. The principal part of the proof is to show the first inclusion. To that end, let $z \in O(F, A)$. Then, by the characterization of opening given in Theorem A.2, there exists some point y such that
$$z \in A + y \subset F$$
However, since A is B-open, there exists an E such that $A = E \oplus B$. Consequently,
$$z \in (E \oplus B) + y \subset F$$

The first inclusion means that there exist points b' in B and e' in E such that $z = e' + b' + y$. The second inclusion means that $[B + (e + y)] \subset F$ for any e in E. Since $e' \in E$, this applies to e'. Hence, taken together, the two inclusions yield the corresponding relations

$$z = (e' + y) + b'$$

and

$$B + (e' + y) \subset F$$

Together, these imply that

$$z \in B + (e' + y) \subset F$$

which in turn implies that $z \in O(F, B)$. Hence, $O(F, A) \subset O(F, B)$.

The next two inclusions follow respectively from the antiextensivity and extensivity of the opening and the closing. We obtain the last inclusion by duality; in particular,

$$C(F, B) = O(F^c, B)^c \subset O(F^c, A)^c = C(F, A)$$

The inclusion follows from the first part of the proof applied to F^c, where one must keep in mind that complementing both sides of a containment relation yields an inclusion relation.

Proposition A.28. If A is B-open, then for any image F,

(i) $O[O(F, B), A] = O(F, A)$

(ii) $O[O(F, A), B] = O(F, A)$

Proof.

(i) By antiextensivity of the opening, $O(F, B) \subset F$. But the opening is increasing; hence,

$$O[O(F, B) A] \subset O(F, A)$$

Moreover, by Proposition A.27, since A is B-open, $O(F, A) \subset O(F, B)$. Consequently, by idempotence,

$$O(F, A) = O[O(F, A), A] \subset O[O(F, B), A]$$

and (i) is proven.

(ii) Suppose $z \in O(F, A)$. Then there exists y such that $z \in A + y \subset F$. But A is B-open. Hence,

$$z \in O(A, B) + y \subset F$$

Consequently, there exists some point w such that

$$z \in (B + w) + y \subset A + y \subset F$$

Since $O(F, A)$ consists of the union of all translates of A which are in F, this last

relation implies that
$$z \in B + (w + y) \subset O(F, A)$$
Therefore, $z \in O[O(F, A), B]$, and it follows that
$$O(F, A) \subset O[O(F, A), B]$$
The reverse inclusion is immediate by the antiextensivity of the opening, and (ii) follows thereupon.

Proposition A.29. If A is B-open, then for any image F,

(i) $C[C(F, B), A] = C(F, A)$

(ii) $C[C(F, A), B] = C(F, A)$

Proof. From part (i) of Proposition A.28, with F^c in place of F, we obtain, by duality,
$$C[O(F^c, B)^c, A] = C(F, A)$$
But, again by duality, $O(F^c, B)^c = C(F, B)$, and (i) follows. The second part follows similarly; in particular, by duality and part (ii) of Proposition A.28, we have
$$C[O(F^c, A)^c, B] = C(F, A)$$
The desired result follows immediately by another application of duality.

Proposition A.30. If A and B are convex, then so are $A \oplus B$, $A \ominus B$, $O(A, B)$, and $C(A, B)$.

Proof. Let $z, w \in A \oplus B$, $r + s = 1$, and $r, s \geq 0$. Since z and w are elements of the Minkowski addition, $z = a + b$ and $w = a' + b'$, where $a, a' \in A$ and $b, b' \in B$. Hence, due to the convexity of A and B,
$$rz + sw = r(a + b) + s(a' + b')$$
$$= (ra + sa') + (rb + sb') \in A \oplus B$$
and $A \oplus B$ is convex. Since the intersection of convex sets is again a convex set, and since translation is a special case of Minkowski addition,
$$A \ominus B = \bigcap_{b \in B} A + b$$
is convex. The convexity of the opening and closing is immediate since each is an iteration of a Minkowski sum and a Minkowski subtraction.

It should be noted that the proof of Proposition A.30 actually shows that $A \ominus B$ is convex whenever A alone is convex. Before proceeding, we mention the fact that for any $t > 0$, tA is convex whenever A is convex.

In general, $(r + s)A \subset rA \oplus sA$ since $x \in (r + s)A$ means that there exists an

$a \in A$ such that
$$x = (r + s)a = ra + sa \in rA \oplus sA$$
What is of interest is that actual equality is achieved if A is convex.

Proposition A.31. If A is convex, then $(r + s)A = rA \oplus sA$ for $r, s \geq 0$.

Proof. Let $x = ra + sa' \in rA \oplus sA$. Since A is convex,
$$\left(\frac{r}{r+s}\right)a + \left(\frac{s}{r+s}\right)a' \in A$$
Scalar multiplication by $r + s$ then shows that $x \in (r + s)A$, so that $rA \oplus sA \subset (r + s)A$. It follows immediately that $(r + s)A = rA \oplus sA$.

Now suppose $r \geq s > 0$; then, according to the preceding result, if A is convex,
$$rA = [s + (r - s)]A = sA \oplus (r - s)A$$
Consequently, according to Proposition A.25, rA is sA-open. Hence, we have the following proposition, which is fundamental to the method of granulometries.

Proposition A.32. If $r \geq s > 0$ and B is convex, then for any set A, $O(A, rB) \subset O(A, sB)$.

Proof. By the remarks preceding Proposition A.32, rB is sB-open. The result follows at once from Proposition A.27.

Let X denote the class of all two-dimensional constant Euclidean images (subsets of R^2). An image-to-image mapping $\Psi: X \to X$ is said to be *compatible with translation* if the application of Ψ can be interchanged with translation, i.e., if
$$\Psi(A + x) = \Psi(A) + x$$
for all images A and points x. According to Propositions A.3 and A.6, Minkowski addition and subtraction are both compatible with translation. Any such mapping will be called a *τ-mapping*.

Given a τ-mapping Ψ, we define the *kernel* of Ψ by
$$\text{Ker}[\Psi] = \{A: \overline{0} \in \Psi(A)\}$$

Proposition A.33. $z \in \Psi(A)$ if and only if $A \in \text{Ker}[\Psi] + z$.

Proof. $z \in \Psi(A)$ if and only if $\overline{0} \in \Psi(A) - z$ if and only if $\overline{0} \in \Psi(A - z)$ if and only if $A - z \in \text{Ker}[\Psi]$ if and only if $A \in \text{Ker}[\Psi] + z$.

Proposition A.34. Suppose image B is fixed and Ψ is defined in terms of Minkowski addition by $\Psi(A) = A \oplus B$. Then
$$\text{Ker}[\Psi] = \{A: A \cap (-B) \neq \varnothing\}$$

Proof. As mentioned before, Ψ is a τ-mapping. Now, $A \in \text{Ker}\,[\Psi]$ if and only if $\bar{0} \in \Psi(A)$, which means precisely that $\bar{0} \in A \oplus B$. But the latter holds if and only if there exists an $a \in A$ and a $b \in B$ such that $\bar{0} = a + b$, which means that $a = -b$.

Proposition A.35. Suppose image B is fixed and Ψ is defined in terms of Minkowski subtraction by $\Psi(A) = A \ominus B$. Then Ψ is a τ-mapping, and

$$\text{Ker}\,[\Psi] = \{A: A \supset -B\}$$

Proof. The fact that Ψ is τ-mapping has already been noted. Moreover, $A \in \text{Ker}\,[\Psi]$ if and only if $\bar{0} \in \Psi(A)$, which means, by Proposition A.7, that

$$-B = -B + \bar{0} \subset A$$

Proposition A.36. Suppose B is fixed and Ψ is defined in terms of the opening by $\Psi(A) = O(A, B)$. Then Ψ is a τ-mapping, and the kernel of Ψ is given by

$$\text{Ker}\,[\Psi] = \bigcup_{z \in B} \{A: B - z \subset A\}$$

Proof. That the opening $O(\cdot, B)$ is a τ-mapping follows from its definition in terms of the two primitive operators and the fact that they are both τ-mappings; indeed,

$$O(A, B) + x = [(A \ominus (-B)) \oplus B] + x$$
$$= [(A \ominus (-B)) + x] \oplus B$$
$$= [(A + x) \ominus (-B)] \oplus B$$
$$= O(A + x, B)$$

now, $\bar{0} \in \Psi(A)$ if and only if (Theorem A.2) there exists a w such that

$$\bar{0} \in B + w \subset A$$

But the latter means precisely that $w \in (-B)$ and $B + w \subset A$, i.e., that there exists a $z \in B$ with $B - z \subset A$.

Proposition A.37. Suppose B is fixed and Ψ is defined in terms of the closing by $\Psi(A) = C(A, B)$. Then Ψ is a τ-mapping, and

$$\text{Ker}\,[\Psi] = \bigcap_{z \in B} \{A: A \cap (B - z) \neq \varnothing\}$$

Proof. That is the closing $C(\cdot, B)$ is a τ-mapping can be shown in a manner similar to that employed in Proposition A.36 to show that $O(\cdot, B)$ is a τ-mapping. By Proposition A.23, $\bar{0} \in C(A, B)$ if and only if $(B + y) \cap A \neq \varnothing$ for any translate $B + y$ containing $\bar{0}$. However, the latter condition means that $-y \in B$. Hence, $\bar{0} \in C(A, B)$ is equivalent to $(B - z) \cap A \neq \varnothing$ for any $z \in B$. But this is precisely what is to be proved.

Proposition A.38. Let Q_0 denote the collection of all images containing the origin, and let Ψ be a τ-mapping. Then

(i) Ψ is extensive if and only if $Q_0 \subset \text{Ker}\,[\Psi]$

(ii) Ψ is antiextensive if and only if $Q_0 \supset \text{Ker}\,[\Psi]$

Proof. Suppose there exists some image A such that A is not properly contained in $\Psi(A)$. Then there exists a point z in $A - \Psi(A)$, where "$-$" denotes set difference. Since Ψ is a τ-mapping,

$$\overline{0} \in [A - \Psi(A)] - z = (A - z) - (\Psi(A) - z)$$
$$= (A - z) - \Psi(A - z)$$

Hence, $A - z$ contains $\overline{0}$, but $\Psi(A - z)$ does not. In other words, $A - z \in Q_0$ but $A - z \notin \text{Ker}\,[\Psi]$; i.e., the kernel of Ψ does not contain Q_0. The reverse implication in (i) follows by contradiction. Suppose $Q_0 \not\subset \text{Ker}\,[\Psi]$. Then there exists an image A such that $\overline{0} \in A$ but $\overline{0} \notin \Psi(A)$. It is immediate that A is not contained in $\Psi(A)$; hence, Ψ is not extensive. The second part of the proposition may be demonstrated by a similar argument.

Proposition A.39. Let Ψ_1 and Ψ_2 be τ-mappings. Then $\text{Ker}\,[\Psi_1] \subset \text{Ker}\,[\Psi_2]$ if and only if $\Psi_1(A) \subset \Psi_2(A)$ for any image A.

Proof. Suppose there exists some image A in the kernel of Ψ_1 which is not in the kernel of Ψ_2. Then $\overline{0}$ is in $\Psi_1(A)$ but not $\Psi_2(A)$, and hence, $\Psi_1(A) \not\subset \Psi_2(A)$. Conversely, suppose there exists an element z such that

$$z \in \Psi_1(A) - \Psi_2(A).$$

Then, since Ψ is a τ-mapping,

$$\overline{0} \in [\Psi_1(A) - \Psi_2(A)] - z = \Psi_1(A - z) - \Psi_2(A - z)$$

In other words, $A - z \notin \text{Ker}\,[\Psi_2]$, but $A - z \in \text{Ker}\,[\Psi_1]$.

An immediate consequence of the preceding proposition is that two τ-mappings with the same kernel must be identical. The next proposition states, in terms of Minkowski subtraction, the first part of the Matheron Representation Theorem for increasing τ-mappings.

Proposition A.40. Suppose Ψ is an increasing τ-mapping. Then for any image A,

$$\Psi(A) = \bigcup_{B \in \text{Ker}[\Psi]} A \ominus (-B)$$

Proof. Suppose $B \in \text{Ker}\,[\Psi]$. If A contains B, then $A \in \text{Ker}\,[\Psi]$ since $\overline{0} \in \Psi(B) \subset \Psi(A)$. In other words, if B is in the kernel of Ψ, then $\{A: A \supset B\}$ is a subclass of $\text{Ker}\,[\Psi]$. Hence,

$$\bigcup_{B \in \text{Ker}[\Psi]} \{A: A \supset B\} \subset \text{Ker}\,[\Psi]$$

The reverse inclusion is trivial since $B \in \{A: A \supset B\}$. Consequently,

$$\bigcup_{B \in \operatorname{Ker}[\Psi]} \{A: A \supset B\} = \operatorname{Ker}[\Psi]$$

But, by Proposition A.35, the collection to the left of the equal sign is the kernel of the mapping Φ, where $\Phi(A)$ is defined as

$$\bigcup_{B \in \operatorname{Ker}[\Psi]} A \ominus (-B)$$

Therefore, by the remark following Proposition A.39, Proposition A.40 is proved.

In order to obtain a dual result for the preceding theorem, we introduce the notion of a *dual mapping* Ψ^*. For any τ-mapping Ψ, Ψ^* is defined by

$$\Psi^*(A) = [\Psi(A^c)]^c$$

Proposition A.41. If Ψ is an increasing τ-mapping, then so is its dual Ψ^*. Moreover, the kernel of the dual is given by

$$\operatorname{Ker}[\Psi^*] = \{A: A^c \notin \operatorname{Ker}[\Psi]\}$$

Proof. To see that the dual is increasing, suppose $A \subset B$. Then $B^c \subset A^c$. Since Ψ is increasing, $\Psi(B^c) \subset \Psi(A^c)$. Taking complements yields

$$\Psi^*(A) = [\Psi(A^c)]^c \subset [\Psi(B^c)]^c = \Psi^*(B)$$

Moreover, the dual is a τ-mapping since, if Ψ is a τ-mapping, then for any point x,

$$\Psi(A^c) + x = \Psi(A^c + x)$$

which implies, by taking complements, that

$$\Psi^*(A + x) = [\Psi((A + x)^c)]^c = [\Psi(A^c) + x]^c$$
$$= [\Psi(A^c)]^c + x = \Psi^*(A) + x$$

Finally, the relation involving the kernel of the dual follows from

$$\operatorname{Ker}[\Psi^*] = \{A: \overline{0} \in \Psi^*(A)\}$$
$$= \{A: \overline{0} \notin \Psi(A^c)\}$$
$$= \{A: A^c \notin \operatorname{Ker}[\Psi]\}$$

The dual of Proposition A.40 can now be obtained. It gives a characterization of an increasing τ-mapping in terms of Minkowski addition.

Proposition A.42. Suppose Ψ is an increasing τ-mapping. Then for any image A,

$$\Psi(A) = \bigcap_{B \in \operatorname{Ker}[\Psi^*]} A \oplus (-B)$$

Proof. By Proposition A.40 applied to Ψ^* and A^c,

$$\Psi^*(A^c) = \bigcup_{B \in \mathrm{Ker}[\Psi^*]} A^c \ominus (-B)$$

Taking complements, applying duality, and recognizing that the dual of the dual is the original mapping, we have

$$\Psi(A) = \bigcap_{B \in \mathrm{Ker}[\Psi^*]} [A^c \ominus (-B)]^c$$

$$= \bigcap_{B \in \mathrm{Ker}[\Psi^*]} A \oplus (-B)$$

Theorem A.5. (Matheron Representation Theorem). Suppose Ψ is an increasing τ-mapping. Then for any image A,

$$\Psi(A) = \bigcup_{B \in \mathrm{Ker}[\Psi]} \mathcal{E}(A, B) = \bigcap_{B \in \mathrm{Ker}[\Psi^*]} \mathcal{D}(A, -B)$$

In general, if Q and Q' are collections of sets, then a mapping $\Psi: Q \to Q'$ is called an *algebraic opening* if the following conditions are satisfied:

(i) Ψ is antiextensive, i.e., $\Psi(A) \subset A$.

(ii) Ψ is increasing, i.e., $A \subset B$ implies $\Psi(A) \subset \Psi(B)$.

(iii) Ψ is idempotent, i.e., $\Psi[\Psi(A)] = \Psi(A)$.

On the other hand, a mapping Ψ is called an *algebraic closing* if Ψ is extensive ($\Psi(A) \supset A$), increasing and idempotent. According to Theorems A.3 and A.4, the opening $O(\cdot, B)$ and the closing $C(\cdot, B)$, are an algebraic opening and an algebraic closing, respectively. In each case the domain and range of the mappings are X, the collection of all sets in R^2. In the sequel, we shall assume that $Q = Q' = X$.

Proposition A.43. Ψ^*, the dual of Ψ, is an algebraic opening if and only if Ψ is an algebraic closing.

Proof. The three aforementioned conditions need to be checked. (i) Ψ^* is antiextensive if and only if $\Psi^*(A) \subset A$ for any A if and only if $\Psi^*(A^c) \subset A^c$ for any A if and only if $\Psi(A) = [\Psi^*(A^c)]^c \supset A$ (by taking complements). But the latter means that Ψ is extensive. (ii) Let $A \subset B$. Then $A^c \supset B^c$. Now, $\Psi^*(A) \subset \Psi^*(B)$ if and only if $[\Psi^*(A)]^c \supset [\Psi^*(B)]^c$ if and only if $\Psi(A^c) \supset \Psi(B^c)$ (by the definition of the dual). (iii) is simply an exercise in complementation and will be left to the reader.

Note that since the dual of the dual is the original mapping, Proposition A.43 holds with the roles of Ψ and Ψ^* reversed.

If Ψ is an algebraic opening (closing) on X, then it is called a τ-*opening* (τ-*closing*) if it is compatible with translation. Thus, $O(\cdot, B)$ is a τ-opening and $C(\cdot, B)$ is a τ-closing. Matheron's theorem states that all τ-openings and τ-closings can be represented in terms of these elementary openings and closings, respectively.

Now suppose Ψ is either an algebraic opening or an algebraic closing. Then the class of *invariant* sets under Ψ is the class Inv $[\Psi]$ of all images A such that $\Psi(A) = A$. That is, Inv $[\Psi]$ consists precisely of those images which are unaffected by Ψ. For the opening $O(\cdot, B)$, the invariant images are those which are B-open; for the closing $C(\cdot, B)$, the invariant images are those which are B-closed.

Proposition A.44. An algebraic opening is a τ-opening if and only if Inv $[\Psi]$ is closed under translation. (Inv $[\Psi]$ is closed under translation means that $A \in$ Inv $[\Psi]$ if and only if $A + x \in$ Inv $[\Psi]$ for any $x \in R^2$.)

Proof. Suppose Ψ is a τ-opening. Then $\Psi(A) = A$ if and only if $\Psi(A) + x = A + x$ if and only if $\Psi(A + x) = A + x$ if and only if $A + x \in$ Inv $[\Psi]$. The converse is just as easy.

A similar proposition to Proposition A.44 holds for τ-closings.

Now suppose that Ψ is an algebraic opening. A class $\mathcal{B}$ of images is said to be a *base* for Inv $[\Psi]$ if Inv $[\Psi]$ is the class generated by $\mathcal{B}$ under translations and infinite unions.

Note that every invariant set of a τ-opening has a base: Inv $[\Psi]$ is a base for itself. Proposition A.44 guarantees that Inv $[\Psi]$ is closed under translations. But it is also closed under unions. Indeed, if $\{A_k\} \subset$ Inv $[\Psi]$, $\Psi(A_k) = A_k$ for all k. Moreover,

$$\Psi\left[\bigcup_k A_k\right] \supset \Psi(A_k) = A_k$$

and thus,

$$\bigcup_k A_k \subset \Psi\left[\bigcup_k A_k\right]$$

The reverse inclusion follows from antiextensivity, and consequently the union is an element of Inv $[\Psi]$.

Theorem A.6 (Matheron). An image-to-image mapping Ψ is a τ-opening if and only if there exists a class of sets $\mathcal{B}$ such that

$$\Psi(A) = \cup \{O(A, B): B \in \mathcal{B}\}$$

Moreover, $\mathcal{B}$ is a base for Inv $[\Psi]$.

Proof. We first show that if Ψ is given by the above union, then Ψ is a τ-opening.

(i) Ψ is antiextensive:

$$\Psi(A) = \cup \{O(A, B): B \in \mathcal{B}\} \subset \cup \{A: B \in \mathcal{B}\} = A$$

(ii) Ψ is increasing: $E \subset F$ implies that
$$\Psi(E) = \cup \{O(E, B): B \in \mathcal{B}\} \subset \cup \{O(F, B): B \in \mathcal{B}\} = \Psi(F)$$

(iii) Ψ is idempotent: Since Ψ is antiextensive, $\Psi[\Psi(A)] \subset \Psi(A)$. As for the reverse inclusion, if $x \in \Psi(A)$, then there exists $B_0 \in \mathcal{B}$ and $y \in R^2$ such that
$$x \in B_0 + y \subset O(A, B_0) \subset \cup \{O(A, B): B \in \mathcal{B}\} \subset A$$

and this implies that
$$x \in \cup \{O[\cup \{O(A, B): B \in \mathcal{B}\}, B]: B \in \mathcal{B}\} = \Psi[\Psi(A)]$$

(iv) Ψ is translation compatible:
$$\Psi(A) + x = [\cup \{O(A, B): B \in \mathcal{B}\}] + x$$
$$= \cup \{O(A, B) + x: B \in \mathcal{B}\}$$
$$= \cup \{O(A + x, B): B \in \mathcal{B}\} = \Psi(A + x)$$

To see that $\mathcal{B}$ is a base for Inv $[\Psi]$, suppose $A \in$ Inv $[\Psi]$ and $x \in A$. Since $A = \cup \{O(A, B): B \in \mathcal{B}\}$, there exists z_x and $B_x \in \mathcal{B}$ such that $x \in B_x + z_x \subset A$. Consequently,
$$A = \cup \{B_x + z_x: x \in A\}$$

and hence, Inv $[\Psi]$ is generated by $\mathcal{B}$.

We must now show that any τ-opening is of the form specified in the statement of the theorem. To that end, let Ψ be a τ-opening. Since Ψ is idempotent, $\Psi(A) \in$ Inv $[\Psi]$. Let
$$Q = \cup \{B + x: x \in R^2, B \in \text{Inv}[\Psi], B + x \subset A\}$$

Since $\Psi[A] + \overline{0} = \Psi[A] \subset A$, $\Psi[A] \subset Q$. Now suppose $z \in Q$. Then there exists $B' \in$ Inv $[\Psi]$ and $x' \in R^2$ such that $z \in B' + x' \subset A$. Since Ψ is increasing and compatible with translation,
$$\Psi(A) \supset \Psi(B' + x') = \Psi(B') + x' = B' + x'$$

where the last equality follows from the fact that $B' \in$ Inv $[\Psi]$. Therefore, $z \in \Psi(A)$ and $Q \subset \Psi(A)$. Hence, by Theorem A.2,
$$\Psi(A) = Q = \bigcup_{B \in \text{Inv}[\Psi]} [\cup \{B + x: x \in R^2, B + x \subset A\}]$$
$$= \bigcup_{B \in \text{Inv}[\Psi]} O(A, B)$$

which is precisely what is to be proved since Inv $[\Psi]$ is a base for itself.

Proposition A.45. If Ψ is a τ-opening, then the dual of Ψ has the representation
$$\Psi^*(A) = \cap \{C(A, B): B \in \mathcal{B}\}$$

where $\mathcal{B}$ is a base for Inv $[\Psi]$.

Proof. Applying Theorem A.6, together with duality, we have

$$\Psi^*(A) = [\Psi(A^c)]^c = [\cup \{O(A^c, B): B \in \mathcal{B}\}]^c$$
$$= \cap \{O(A^c, B)^c: B \in \mathcal{B}\}$$
$$= \cap \{C(A, B): B \in \mathcal{B}\}$$

A *granulometry on X*, the collection of two-dimensional Euclidean images, is a family of mappings $\Psi_t: X \to X$, $t > 0$, such that

(i) $\Psi_t(A) \subset A$ for any $t > 0$ (Ψ_t is antiextensive)

(ii) $A \subset B$ implies that $\Psi_t(A) \subset \Psi_t(B)$ (Ψ_t is increasing)

(iii) $\Psi_t \circ \Psi_{t'} = \Psi_{t'} \circ \Psi_t = \Psi_{\max(t,t')}$ for all $t, t' > 0$

where "$\circ$" denotes function composition. It should be noted at once that a mapping Ψ_t within a granulometry $\{\Psi_t\}$ satisfies the first two requirements of an algebraic opening. For the special case $t = 0$, we define $\Psi_0(A) = A$.

Proposition A.46. Suppose $\{\Psi_t\}$ is a granulometry. If $r \geq s$, then $\Psi_r(A) \subset \Psi_s(A)$.

Proof. By axiom (i) of a granulometry, $\Psi_r(A) \subset A$. Therefore, applying (iii) and (ii) in that order,

$$\Psi_r(A) = \Psi_s[\Psi_r(A)] \subset \Psi_s(A)$$

Proposition A.47. $\{\Psi_t\}$ is a granulometry if and only if

(i') Ψ_t is an algebraic opening for all $t > 0$.

(ii') If $r \geq s > 0$, then Inv $[\Psi_r] \subset$ Inv $[\Psi_s]$.

Proof. First assume that $\{\Psi_t\}$ is a granulometry. Then the granulometry axioms (i) and (ii) hold, and all we need do to show that Ψ_t is an algebraic opening is show idempotence. But this follows from axiom (iii):

$$\Psi_t[\Psi_t(A)] = \Psi_{\max(t,t)}(A) = \Psi_t(A)$$

For the proof of (ii'), suppose $A \in$ Inv $[\Psi_r]$ and $r \geq s$. Then

$$\Psi_s(A) = \Psi_s[\Psi_r(A)] = \Psi_{\max(s,r)}(A) = \Psi_r(A) = A$$

Hence, $A \in$ Inv $[\Psi_s]$.

For the proof of the converse of the proposition, suppose that Ψ_t is an opening. Then, immediately, the first two requirements of a granulometry hold. To prove requirement (iii), suppose $r \geq s > 0$. By idempotence and (ii'),

$$\Psi_r(A) \in \text{Inv } [\Psi_r] \subset \text{Inv } [\Psi_s]$$

Therefore, $\Psi_s[\Psi_r(A)] = \Psi_r(A)$. Consequently,

$$\Psi_r(A) = \Psi_r(\Psi_r(A)) = \Psi_r(\Psi_s(\Psi_r(A))) \subset \Psi_r(\Psi_s(A)) \subset \Psi_r(A)$$

where the two inclusions hold because Ψ_t is both antiextensive and increasing for any $t > 0$. Accordingly, it follows that

$$\Psi_r(A) \subset \Psi_r[\Psi_s(A)] \subset \Psi_r(A)$$

and hence,

$$\Psi_r[\Psi_s(A)] = \Psi_r(A) = \Psi_{\max(s,r)}(A)$$

However, in almost exactly the same way,

$$\Psi_r(A) = \Psi_r(\Psi_r(A)) = \Psi_s(\Psi_r(\Psi_r(A))) \subset \Psi_s(\Psi_r(A)) \subset \Psi_r(A)$$

where the last inclusion follows directly from the antiextensivity of Ψ_s. But then, it follows that

$$\Psi_s[\Psi_r(A)] = \Psi_r(A) = \Psi_{\max(r,s)}(A)$$

and requirement (iii) of a granulometry is satisfied.

If the following two axioms are added to those for a general granulometry, we obtain a *Euclidean granulometry*:

(iv) For any $t > 0$, Ψ_t is compatible with translation.

(v) For any $t > 0$ and image A, $\Psi_t(A) = t\Psi_1\left(\dfrac{1}{t}A\right)$.

Note that condition (iv), together with Proposition A.47, asserts that Ψ_t is a τ-opening. According to Proposition A.44, Ψ_t is a τ-opening if and only if Inv $[\Psi_t]$ is closed under translation.

Proposition A.48. Let $\{\Psi_t\}$ be a granulometry for which axiom (iv) holds. Then axiom (v) is equivalent to Inv $[\Psi_t] = t \cdot$ Inv $[\Psi_1]$, which means that $A \in$ Inv $[\Psi_t]$ if and only if $(1/t)A \in$ Inv $[\Psi_1]$.

Proof. Assuming (v) to hold, $A \in$ Inv $[\Psi_t]$ if and only if $\Psi_t(A) = A$, which means precisely that

$$\Psi_1\left(\frac{1}{t}A\right) = \frac{1}{t}\Psi_t(A) = \frac{1}{t}A$$

i.e., $(1/t)A \in$ Inv $[\Psi_1]$. For the converse, since (iv) holds, Ψ_t is a τ-opening. Hence, the mapping $\Phi_t(A) = t\Psi_1((1/t)A)$ is also a τ-opening; indeed, that Φ is antiextensive and increasing follows at once from the corresponding properties of Ψ_t. As for idempotence,

$$\Phi_t[\Phi_t(A)] = t\Psi_1\left[\frac{1}{t}\left(t\Psi_1\left(\frac{1}{t}A\right)\right)\right]$$

$$= t\Psi_1\left[\Psi_1\left(\frac{1}{t}A\right)\right] = t\Psi_1\left(\frac{1}{t}A\right) = \Phi_t(A)$$

Moreover, compatibility with translation follows from

$$\Phi_t(A + x) = t\Psi_1\left[\frac{1}{t}(A + x)\right] = t\left[\Psi_1\left(\frac{1}{t}A + \frac{x}{t}\right)\right]$$

$$= t\left[\Psi_1\left(\frac{1}{t}A\right) + \frac{x}{t}\right]$$

$$= t\Psi_1\left(\frac{1}{t}A\right) + x = \Phi_t(A) + x$$

Now, $A \in \text{Inv}[\Phi_t]$ if and only if $(1/t)A \in \text{Inv}[\Psi_1]$, which, by the hypothesis of the proposition, means precisely that $A \in \text{Inv}[\Psi_t]$. Hence, Φ_t and Ψ_t are τ-openings having the same invariant sets. But, it is an immediate consequence of Theorem A.6 that τ-openings with identical invariant sets are themselves identical.

Proposition A.49. Let $\mathcal{B}$ be a collection of subsets of R^2. Then there exists a Euclidean granulometry $\{\Psi_t\}$ such that $\mathcal{B} = \text{Inv}[\Psi_1]$ if and only if $\mathcal{B}$ satisfies the closure conditions

(i) If $A_i \in \mathcal{B}$ for $i \in I$, then $\cup_i A_i \in \mathcal{B}$.
(ii) If $A \in \mathcal{B}$, then $A + x \in \mathcal{B}$ for all $x \in R^2$.
(iii) If $A \in \mathcal{B}$, then $tA \in \mathcal{B}$ for all $t \geq 1$.

Proof. Suppose there exists a Euclidean granulometry such that $\mathcal{B} = \text{Inv}[\Psi_1]$. To prove (i), consider a collection of sets $A_j \in \text{Inv}[\Psi_1]$. Since Ψ_1 is an algebraic opening, it is antiextensive. Hence,

$$\Psi_1\left(\bigcup_j A_j\right) \subset \bigcup_j A_j$$

On the other hand, Ψ_1 is increasing, so that

$$\bigcup_k A_k = \bigcup_k \Psi_1(A_k) \subset \bigcup_k \left[\Psi_1\left(\bigcup_j A_j\right)\right] = \Psi_1\left(\bigcup_j A_j\right)$$

the last equality following from the redundancy of the outer union. Therefore, $\cup_j A_j \in \text{Inv}[\Psi_1]$, and $\text{Inv}[\Psi_1]$ is closed under arbitrary unions.

Next, note that property (ii), closure under translations, holds by Proposition A.44. To prove property (iii), suppose that $A \in \text{Inv}[\Psi_1]$ and $t \geq 1$. Then, by axiom (v), $tA \in \text{Inv}[\Psi_t]$. Therefore, by axiom (iii),

$$\Psi_1(tA) = \Psi_1[\Psi_t(tA)] = \Psi_{\max(1,t)}(tA) = \Psi_t(tA) = tA$$

and $tA \in \text{Inv}[\Psi_1]$. We now proceed to the converse of the proposition. According to Theorem A.6, Ψ_t defined by

$$\Psi_t(A) = \cup \{O(A, B): B \in t\mathcal{B}\}$$

is a τ-opening with base $t\mathcal{B}$. Since $\mathcal{B}$ is closed under unions and translations, Inv $[\Psi_t] = t\mathcal{B}$ and hence $\mathcal{B} = $ Inv $[\Psi_1]$. In order to show that $\{\Psi_t\}$ is a Euclidean granulometry, it remains to show that (ii') of Proposition A.47 and axiom (v) hold. For (ii'), suppose $r \geq s > 0$ and $A \in$ Inv $[\Psi_r]$. Then $A = rB$ for some $B \in \mathcal{B}$. Since $r/s \geq 1$, by the hypothesis of condition (iii) above, $(r/s)B \in \mathcal{B}$, which implies that $rB = sB'$ for some $B' \in \mathcal{B}$, which in turn implies that $A \in s\mathcal{B} = $ Inv $[\Psi_s]$. Consequently, Inv $[\Psi_r] \subset$ Inv $[\Psi_s]$, and (ii') holds. Finally, axiom (v) holds by Proposition A.48 since, by construction, Inv $[\Psi_t] = t$ Inv $[\Psi_1]$.

The previous result gives a complete characterization of classes of images which can serve as invariant classes for some sieving function Ψ_1 from a Euclidean granulometry $\{\Psi_t\}$. Taken together with Proposition A.48, which states that the invariant classes of a Euclidean granulometry are determined by the invariant class for Ψ_1, Proposition A.49 characterizes the invariant classes of Euclidean granulometries. In fact, much more has been shown within the proof, viz., that $\{\Psi_t\}$ is a Euclidean granulometry if and only if there exists some collection $\mathcal{B}$ of sets such that

$$\Psi_t(A) = \cup \{O(A, B): B \in t\mathcal{B}\}$$
$$= \cup \{O(A, tB): B \in \mathcal{B}\}$$

and in such a case, Inv $[\Psi_1] = \mathcal{B}$. In other words, Euclidean granulometries themselves have been characterized in terms of the invariant sets of Ψ_1.

Suppose, now, that $\mathcal{B}$ is some collection of sets closed under unions, translations, and scalar multiplications by $t \geq 1$. By Proposition A.49, $\mathcal{B}$ is the invariant set for Ψ_1 of some Euclidean granulometry. A collection of sets $\mathcal{B}_0$ is called a *generator* of $\mathcal{B}$ if the class closed under arbitrary union, translation, and scalar multiplication by $t \geq 1$ which is generated by $\mathcal{B}_0$ is the class $\mathcal{B}$. If $\{\Psi_t\}$ is the Euclidean granulometry with Inv $[\Psi_1] = \mathcal{B}$, then we also call $\mathcal{B}_0$ a generator of $\{\Psi_t\}$. Using these concepts, we arrive at Matheron's Representation Theorem for Euclidean granulometries.

Theorem A.7 (Matheron). A family of image-to-image mappings $\{\Psi_t\}$, $t > 0$, is a Euclidean granulometry if and only if there exists a class of images $\mathcal{B}_0$ such that

$$\Psi_t(A) = \bigcup_{B \in \mathcal{B}_0} \bigcup_{r \geq t} O(A, rB)$$

Moreover, $\mathcal{B}_0$ is a generator of $\{\Psi_t\}$ and Inv $[\Psi_t] = t$ Inv $[\Psi_1]$ for all $t > 0$.

Proof. Let $\mathcal{B}$ be the class generated by $\mathcal{B}_0$. Applying Proposition A.49, let $\{\Psi_t\}$ be the Euclidean granulometry with Inv $[\Psi_1] = \mathcal{B}$. We need to show that $\Psi_t(A)$ is given by the double union in the statement of the theorem, i.e., that

$$\cup \{O(A, tB): B \in \mathcal{B}\} = \cup \{O(A, t'B'): t' \geq t, B' \in \mathcal{B}_0\}$$

Accordingly, let z be an element of the left-hand union. Then there exists $\overline{B} \in \mathcal{B}$ and $x \in R^2$ such that $z \in t\overline{B} + x \subset A$. Since $\mathcal{B}$ is generated by $\mathcal{B}_0$, there exist some index sets I, J, and K such that

$$\bar{B} = \bigcup_{\substack{i \in I \\ j \in J \\ k \in K}} t_i(B_j + x_k)$$

where $t_i \geq 1$, $B_j \in \mathcal{B}_0$, and $x_k \in R^2$. Consequently, for some i, j, and k,

$$z \in t \cdot t_i(B_j + x_k) + x = (t \cdot t_i)B_j + [(t \cdot t_i)x_k + x] \subset A$$

Since $t \cdot t_i \geq t$, it follows that z is an element of the right-hand union above.

Now suppose z is an element of the right-hand union. Then there exists a $t' \geq t$, $B' \in \mathcal{B}_0$, and x such that $z \in t'B' + x \subset A$. But this can be rewritten as

$$z \in t \cdot \left(\frac{t'}{t}\right)B' + x \subset A$$

Now, since $t'/t \geq 1$, $(t'/t)B' \in \mathcal{B}$ and hence z is an element of the left-hand union. Therefore, the two unions are equal, and one part of the theorem is demonstrated.

For the converse, suppose $\{\Psi_t\}$ is a Euclidean granulometry. Then, by the remarks preceding the theorem,

$$\Psi_t(A) = \cup \{O(A, tB) : B \in \text{Inv}[\Psi_1]\}$$

Certainly, Inv $[\Psi_1]$ is a generator of itself, and hence the representation of $\Psi_t(A)$ as a double union given in the statement of the theorem holds by simply applying the already completed part of the theorem, to Ψ_t.

Actually, more has been shown in the proof than has been stated in the theorem, viz., if $\{\Psi_t\}$ is a Euclidean granulometry and $\mathcal{B}_0$ is *any* generator of Inv $[\Psi_1]$, then

$$\Psi_t(A) = \bigcup_{B \in \mathcal{B}_0} \bigcup_{r \geq t} O(A, rB)$$

Thus, not only does the proof of the theorem guarantee the existence of some generating set, which trivially can be considered to be Inv $[\Psi_1]$ itself, but any generator will also do.

REFERENCES

BEUCHER, S. "Random Processes Simulations on the Texture Analyser," *Lecture Notes in Biomathematics,* No. 23 (1977): Springer-Verlag.

BEUCHER, S., and LANTUEJOUL, CH. "Use of Watersheds in Contour Detection," *Int. Workshop on Image Processing,* CCETT. Rennes, France (1979).

BLASCHKE, W. V. "Vorlesungen über Integral Geometrie," Teubner, Leipzig (1936).

BOOKSTEIN, F. L., ed. "The Measurement of Biological Shape and Shape Change," *Lecture Notes in Biomathematics,* No. 24, Springer-Verlag, Berlin, Heidelber, New York (1978).

CALABI, L. "A Study of the Skeleton of Plane Figures," *Parke Math. Lab, Inc.* (1965).

COLEMAN, R. "An Introduction to Mathematical Stereology," *Memoirs n° 3,* Department of theoretical statistics, Univ. of Aarhus (1979).

CROFTON, M. W. "On the Theory of Local Probability, Applied to Straight Lines Drawn at Random in a Plane, the Method Used Being Also Extended to the Proof of Certain New Theorems in the Integral Calculus," *Philos. Trans. Roy. Soc., London,* 158 (1868): 181–189.

CRUZ-ORIVE, L. M. "Particle Size-Shape Distribution: The General Spheroid Problem II, Stochastic Model and Practice Guide," *J. of Micro.,* 112, part 2 (1978): 153–168.

DOUGHERTY, E. R., and GIARDINA, C. R. *Matrix-Structured Image Processing.* Englewood Cliffs, N.J: Prentice-Hall, Inc., 1987.

DOUGHERTY, E. R., and GIARDINA, C. R. "Error Bounds for Morphologically Derived Feature Measurements," *SIAM Journal on Applied Mathematics.* In Press.

DOUGHERTY, E. R., and GIARDINA, C. R. "Binary Euclidean Images and Convergence," *1986 Conference on Intelligent Systems and Machines,* Published in Conference Proceedings, Oakland Univ., Rochester, Michigan.

DOUGHERTY, E. R., and GIARDINA, C. R. "Sampling Criteria for Euclidean Images," *39th Annual Conference, SPSE,* Minnesota (May 1986).

DOUGHERTY, E. R., and GIARDINA, C. R. "A Digital Version of Matheron's Theorem for Increasing τ-mappings in Terms of a Basis for the Kernel," *IEEE Computer Vision and Pattern Recognition,* Miami (June 1986).

DOUGHERTY, E. R., and GIARDINA, C. R. "A Structurally Induced Image Algebra," *Computers and Mathematics,* Stanford (1986).

DOUGHERTY, E. R., and GIARDINA, C. R. "An Investigation of the Relationship Between Morphological and Linear Filters," *Electronic Imaging 86,* Boston (1986).

DOUGHERTY, E. R., and GIARDINA, C. R. "Image Algebra, Final Phase I Report," U. S. Air Force, Contract F08635-84-C-0296, SOW Task Reference 5.0, Kearfott Division, The Singer Company, Wayne, NJ (1986).

DOUGHERTY, E. R., and GIARDINA, C. R. *"Image Processing: Continuous to Discrete, Volume I,* Englewood Cliffs: Prentice-Hall, Inc., 1987.

DUFF, M. J. B., and WATSON, D. M. "The Cellular Logic Array Image Processor," *Dept. of Physics and Astronomy,* Univ. College, London (1974).

ELLAS, H., and WEIBEL, E. R. *Quantitative Methods in Morphology,* Springer-Verlag, Berlin (1967).

FEDERER, H. *Geometric Measure Theory.* Berlin: Springer-Verlag, 1969.

FLOOK, A. G. "The Use of Dialtion Logic on the Quantimet to Achieve Fractal Dimension Characterisation of Textured and Structured Profiles," *Powder Technology,* 21 (1978): 295–298.

GOETCHARIAN, V. "From Binary to Grey Level Tone Image Processing by Using Fuzzy Logic Concepts," *Pattern Recognition,* Vol. 12 (1980): 7–15.

GOETCHERIAN, V. "Parallel Image Processes and Real-Time Texture Analysis," Thesis Doctor of Philosophy, Univ. College, London (1980).

GOLAY, M. J. E. "Hexagonal Parallel Pattern Transformation," *IEEE Trans. Comput.,* C-18 (1969): 733–740.

HADWIGER, H. *Vorslesungen Über Inhalt, Oberfläche and Isoperimetrie.* Berlin: Springer-Verlag, 1957.

HARALICK, R. M., STERNBERG, S. R., and ZHUANG, X. "Grayscale Morphology," *IEEE Computer Vision and Pattern Recognition,* Miami (June 1986).

HARALICK, R. M., STERNBERG, S. R., and ZHUANG, X. "Image Analysis Using Mathematical Morphology: Part I," Machine Vision International, Ann Arbor (1985).

HARALICK, R. M., and ZHUANG, X. "Morphological Structuring Element Decomposition," Machine Vision International, Ann Arbor (1985).

KENDALL, M. G. G., and MORAN, P. A. P. *Geometrical Probability.* London: Griffin, 1963.

KENDALL, D. G. "Foundations of a Theory of Random Sets," in *Stochastic Geometry,* E. F. Harding and D. G. Kendall, eds. New York: John Wiley & Sons, 1974.

KIMMEL, M. J., JAFFE, R. S., MANDERVILLE, J. R., and LAVIN, M. A., "MITE: Morphic Image Transform Engine, An Architecture for Reconfigurable Pipelines of Neighborhood Processors," *IEEE Computer Society Workshop on Computer Architecture For Pattern Analysis and Image Database Management,* Miami Beach (1985).

KLEIN, J. C., and SERRA, J. "The Texture Analyser," *J. of Micr.,* 95, part 2 (April 1973): 349–356.

LAY, STEVEN. *Convex Sets and Their Applications.* New York: John Wiley & Sons, 1982.

MANDEELBROT, B. B. *Fractals: Form, Chance, Dimension.* San Francisco & London: W. H. Freeman & Co., 1977.

MARAGOS, P. A., and SHAFER, R. W. " A Unification of Linear, Median, Order-Statistics and Morphological Filters Under Mathematical Morphology," *IEEE International Conference on Acoustic, Speech, and Signal Processing* (1985).

MATHERON, G. "Eléments pour une Théorie des Milieux Poreux," Masson, Paris (1967).

MATHERON, G., "Estimer et Choisir," *Centre de Morphologie Mathematique de Fontainebleau,* Fontainebleau, France (1978).

MATHERON, G., "Filters and Lattices," *Centre de Geostatistique et de Morphologie Mathematique,* Fontainebleau, France (1983).

MATHERON, G., "Random Sets Theory and Its Applications to Stereology," *J. of Micr.,* 95, part 1 (Feb 1972): 15–23.

MATHERON, G. *Random Sets and Integral Geometry.* New York: John Wiley & Sons, 1975.

MATHERON, G., "Remarques Sur Les Filtres-Partitions," *Centre de Geostatistique et de Morphologie Mathematique,* Fontainebleau, France (1985).

MATHERON, G. "La Formule de Crofton pour les Sections Épaisses," *J. Appl. Prob.,* 13, (1976): 707–713.

MCCUBBERY, D. L., and LOUGHEED, R. M. "Morphological Image Analysis Using a Raster Pipeline Processor," *IEEE Computer Society Workshop on Computer Architecture For Pattern Recognition and Image Database Management,* Miami Beach (1985).

MILES, R. E., and SERRA, J., eds. "Geometrical Probabilities and Biological Structures," *Lecture Notes in Biomathematics,* No. 23 (1978): Springer-Verlag.

References

MINKOWSKI, H. "Volumen and Oberfläche," *Math. Ann.,* Vol. 57 (1903): 447–495.

MORAN, P. A. P. "The Probabilistic Basis of Stereology," *Suppl. Adv. Appl. Prob.* 69–91.

POTTER, J. L. "Image Processing on the Massively Parallel Processor," *Computer, Vol. 16, No. 1,* (1983): 62–67.

SANTALO, L. A. *Introduction to Integral Geometry.* Paris: Hermann, 1953.

SANTALO, L. A. *Integral Geometry and Geometric Probability.* Reading, MA: Addison-Wesley, 1976.

SERRA, J. *Image Analysis and Mathematical Morphology.* New York: Academic Press, 1983.

SERRA, J. "Semi-groupes de Filtres Morphologiques," *Centre de Geostatistique et de Morphologie Mathematique,* Fontainebleau, France (1984).

STERNBERG, S. R. "Biomedical Image Processing," *Computer, Vol. 16, No. 1* (1983): 22–34.

STERNBERG, S. R. "Parallel Architectures for Image Processing," *Proceedings IEEE COMPSAC,* Chicago (1979).

STERNBERG, S. R. "Pipeline Architectures For Image Processing," *Multicomputers and Image Processing,* (K. Preston and L. Uhr, Eds.), New York: Academic Press, 1982

UNDERWOOD, E. E. *Quantitative Stereology.* Reading, MA: Addison-Wesley, 1970.

VALENTINE, F. A. *Convex Sets.* New York: McGraw-Hill Book Co. 1964.

WATSON, G. "Mathematical Morphology," Tech. Report No. 21, Dept. of Stat., Princeton Univ., New Jersey (1973)

WEIBEL, E. R. "Practical Methods for Biological Morphometry, Vol. 1: Stereological Methods," New York: Academic Press, 1980.

Index

A

Activated, 48
Additive, 84, 258
Algebraic closing, 146
Algebraic opening, 146, 247
All-pass filter, 260
Amplitude spectrum, 260
Analog filter, 257
Antidistributivity, 17
Antiextensive, 151, 247
Antiextensivity, 23, 171
Arity, 256
Associativity, 9, 50, 198
Augmented bound matrix, 63, 196
Augmented morphological basis, 49

B

Bandpass filter, 261
Base, 247
Basis for the kernel, 142, 242
B-closed, 25
Binary image, 48
Black-and-white image, 48
B-open, 25
BOUND, 113
Boundary, 29, 113, 116
Bounded set, 29
Bound matrix, 36
Bound row vector, 181

C

C-additivity, 84
Cantor Intersection Theorem, 87
Cauchy Projection Theorem, 90
Center of the structuring element, 161
Circuit, 258
CLOSE, 67, 222, 231
Closed set, 29

Closed umbra, 175, 215
Closing, 20, 171, 212, 247
Closure, 29
Codomain, 36
Commutativity, 8, 50, 198
COMP, 44, 49
Compact set, 29
Complement, 44
Complementary bound matrix, 50
Complementary image, 10, 49
Connected, 112
Constant image, 48
Continuous from above, 87, 243
Continuous image functional, 86
Convex hull, 28
Convex ring, 92
Convex set, 27
Covariance, 95
CREATE, 46
Creation operator, 46
Cutoff frequency, 261

D

Deactivated, 48
Decreasing operator, 13
Delay-type filter, 278
Delay-type unary filter, 256
De Morgan's Laws, 50
Difference equation, 263
Digital convex hull, 118
Digital linear granulometric size distribution, 103
Digital linear structuring element, 68
Digitizable, 74
Digitization problem, 73
Digitize, 72
DILATE, 57, 217, 230
Dilation, 8, 161, 209
Directly connected, 115
Direct neighbor, 112
Discrete linear filter, 261
Distance from a point to a set, 85
Distributivity, 14
Domain, 36, 175

DOMAIN, 45
Duality, 10, 21, 129

E

Equivalent umbra matrices, 193
ERODE, 58, 219, 230
Erosion, 7, 165, 209
Euclidean filter, 257
Euclidean granulometry, 152
Expansive, 31
Extended Euclidean signal, 225
Extended image, 231
Extended maximum operator, 41
Extended signal, 231
Extensive operator, 24, 146, 247
External boundary, 113
External edge, 117
EXTMAX, 41
EXTSUP, 209, 226

F

Feature measurements, 79
Feature parameter, 81
Filling, 114
Finite impulse response (FIR) filter, 264, 265
Fitting, 6, 21, 220
Fully digital filter, 257
Fully digital image, 216
Fully digitized, 191
Functional equations, 263
Fuzzy measure, 92

G

Generates, 147
Generator, 153
Geometric probability, 106
Granulometry, 97, 150
Graph, 174
Gray value, 36

Index

H

Hadwiger Theorem, 91
Hausdorff metric, 74, 85
High-pass filter, 261
HIT, 125
Hit-and-miss operator, 125
Homogeneity, 83
Homogeneous, 258
Homogeneous of degree k, 83
Horizontal convexity, 115
Horizontal digital covariance, 105
Horizontal digital linear structuring element, 68
Horizontal hull, 140
Horizontal section, 218
HULL, 118
Hybrid filter, 266

I

Ideal low-pass filter, 260
Idempotence, 23, 24, 50, 171, 175
Idempotent, 247
Image functional, 81
Increasing, 11, 85, 133, 151, 233
Increasing monotonicity, 23, 24, 171, 177
Indirect neighbor, 112
INF, 209, 226
Infinite impulse response (IIR) filter, 264, 265
Infinitely divisible, 32
Infinite spike, 230
Initial pixel, 112
Interior of a set, 29
Internal boundary, 114
Invariant class, 147, 247

K

Kernel, 135, 240

L

Lambda-bounded, 145
Lebesgue measure, 72
Linear digital granulometry, 103
Linear filter, 258

M

Matheron Representation Theorem, 136
Medial axis, 120
Median filter, 274
MIN, 42
Minimal bound matrix, 40
Minimal bound vector, 182
Minimal complementary bound matrix, 51
Minimal origin-containing bound matrix, 63
Minimal umbra matrix, 193
Minimum operator, 42
Minkowski addition, 4, 161, 165, 209
Minkowski functionals, 92
Minkowski subtraction, 5, 166, 209
MOREDGE, 117
Morphological algebra, 2, 61
Morphological basis, 20, 44
Morphological filters, 2, 134, 233
Morphology, 1

N

Neighbor, 112
Nested, 31, 87
NINETY, 43

O

OFF, 182, 226, 230
Offset, 182, 209
OPEN, 66, 220, 230

Opening, 20, 168, 209
Order statistic, 274

P

Path, 112
Perimeter, 83
Pixel, 36
Polytope, 29
Proj, 82
Projection, 82
Pseudoconvex hull, 140
Pseudoextensive, 171

Q

Quantized image, 216

R

RANGE, 45
Rectifier, 271
Reflection through the origin, 167
Region, 112
Regular umbra, 175, 212
R-L circuit, 259
Rotated image, 6
Rotationally invariant, 83
Rotation operator, 43

S

Sample, 72
Sample-data, 257
Sampled image, 216
Sampling interval, 262
Sampling problem, 73
Sampling rate, 72
Scalar multiplication, 9

Serial matrix, 218, 245
Serial matrix for dilation, 190
Serial matrix for erosion, 188
Sieving, 150
Signal ordering, 160
Signature, 81
Simple path, 112
Size distribution, 97
Skeleton, 120
Smoothing, 20, 265
Smoothing median filter, 286
Square-law device, 256
Square neighbor mask, 117
Stereological, 90
Strong neighbor, 112
Strong neighbor mask, 117
Structural basis, 47
Structural operators, 46
Structuring element, 1, 7
SUB, 182, 216
Support, 30
Surface, 175, 212
Surface operator, 194

T

Tau-closing, 146, 247
Tau-mapping, 134, 233
Tau-opening, 146, 247
Terminal pixel, 112
THICK, 129
THIN, 125
Thinning, 118
Time invariant, 263
Time varying, 264
T-oversize, 150
TRAN, 42, 230
Transfer function, 259
Translation, 3, 42, 160, 209
Translationally invariant image functional, 82
Translation invariance, 171, 177
Translation invariant, 134, 233

U

Umbra, 154
Umbra homomorphism theorems, 192
Umbra matrix, 192, 222
Umbra transform, 174, 227

V

Vertical convexity, 115
Vertical digital covariance, 105
Vertical digital linear structuring element, 68

W

Weakly grid convex, 116
Weak neighbor, 112
Wilcoxon filter, 285